U.S. Climate Action Report

2010

*Fifth National Communication
of the United States of America
Under the United Nations Framework
Convention on Climate Change*

You may electronically download or order
a hard copy of this document from the
following U.S. Department of State Web site:
http://www.state.gov/g/oes/rls/rpts/car/index.htm

This document may be cited as follows:
United States Department of State.
U.S. Climate Action Report 2010.
Washington: Global Publishing Services, June 2010.

3 Greenhouse Gas Inventory 22

4 Policies and Measures 39

5 Projected Greenhouse Gas Emissions 76

6 Vulnerability Assessment, Climate Change Impacts, and Adaptation Measures 86

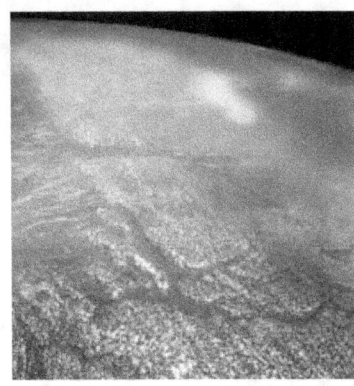

Acronyms and Abbreviations

3D-VIEW	three-dimensional Virtual Interactive Environmental Worlds		BFI	Biofuels Initiative
AAVP	ARM Aerial Vehicles Program		BIE	Bureau of Indian Education
AAVSHTO	American Association of State Highway Transportation Officials		BLM	Bureau of Land Management
ac	acre		BOR	Bureau of Reclamation
ACRF	Atmospheric Radiation Measurement Climate Research Facility		BRDI	Biomass Research and Development Initiative
ADAGE	Applied Dynamic Analysis of the Global Economy		BTS	Bureau of Transportation Statistics
AEO	*Annual Energy Outlook*		Btu	British thermal unit
AERONET	AErosol RObotic NETwork		°C	degree Centigrade
AFCI	Advanced Fuel Cycle Initiative		C_2F_6	hexafluoroethane
AGAGE	Advanced Global Atmospheric Gases Experiment		C_4F_{10}	perfluorobutane
AIRS	Atmospheric InfraRed Sounder		C_6F_{14}	perfluorohexane
ALU	agriculture and land use		C&D	commercialization and deployment
AMS	American Meteorological Society		CAAFI	Commercial Aviation Alternative Fuels Initiative
ANL	Argonne National Laboratory		CAFE	Corporate Average Fuel Economy
AON	Arctic Observing Network		CALIPSO	Cloud-Aerosol Lidar and Infrared Pathfinder Satellite Observation
APP	Asia-Pacific Partnership on Clean Development and Climate		CAR	*U.S. Climate Action Report*
ARM	Atmospheric Radiation Measurement		CATHALAC	Water Center for the Humid Tropics of Latin America and the Caribbean
ARPA-E	Advanced Research and Projects Agency–Energy		CBI	Net-Zero Energy Commercial Building Initiative
ARRA	American Recovery and Reinvestment Act of 2009		CBM	coalbed methane
ARS	Agricultural Research Service		CCCSTI	Committee on Climate Change Science and Technology Integration
A-Train	Afternoon Train		CCD	Climate Change Division
AZA	Association of Zoos and Aquariums		CCLI	Course, Curriculum, and Laboratory Improvement
BCF	billion cubic feet		CCPI	Clean Coal Power Initiative
BEA	Bureau of Economic Analysis		CCRI	Climate Change Research Initiative
BER	Office of Biological and Environmental Research		CCS	carbon capture and sequestration
			CCSP	U.S. Climate Change Science Program

CCTP	Climate Change Technology Program	COP	Conference of the Parties
CDCP	Clean Development and Climate Program	CPO	Climate Program Office
		CPT	Climate Process and Modeling Teams
CEA	Council of Economic Advisers	CRC–URI	The Coastal Resources Center at the University of Rhode Island
CENR	Committee on Environment and Natural Resources		
		CrIS	Cross-track Infrared Sounder
CEQ	Council on Environmental Quality	CRN	Climate Reference Network
CERES	Clouds and the Earth's Radiant Energy System	CRP	Conservation Reserve Program
		CSLF	Carbon Sequestration Leadership Forum
CESN	Climate Effects Science Network		
CFC	chlorofluorocarbon	CSP	Conservation Stewardship Program
CFP	Climate Friendly Parks	CSREES	Cooperative State Research, Education, and Extension Service
CH_4	methane		
CHE	NSF Division of Chemistry	CTA	conservation technical assistance
CHECK	Climate Change Emission Calculator Kit	CTI	Climate Technology Initiative
		CTI	Coral Triangle Initiative
CHP	combined heat and power	CTIC	Conservation Technology Information Center
CIFs	Climate Investment Funds		
CIG	Climate Impacts Group	DAAC	Distributed Active Archive Center
CIWG	Communications Interagency Working Group	DCA	Development Credit Authority
		DEP	Department of Environmental Protection
CLASS	Comprehensive Large Array-data Stewardship System		
		DERA	Diesel Emissions Reduction Act
CLD	Climate and Large-scale Dynamics	DMR	NSF Division of Materials Research
CLIMAS	Climate Assessment for the Southwest	DMS	NSF Division of Mathematical Science
CLIP	Climate Leadership In Parks	DOC	U.S. Department of Commerce
CMAQ	Congestion Mitigation and Air Quality	DOD	U.S. Department of Defense
CMG	Collaboration in Mathematical Geosciences	DOE	U.S. Department of Energy
		DOI	U.S. Department of the Interior
CMOP	Coalbed Methane Outreach Program	DOS	U.S. Department of State
CNES	Centre National d'Etudes Spatiales	DOT	U.S. Department of Transportation
CNRS	Centre Nationale de Recherche Scientifique	ECO-Asia	Environmental Cooperation–Asia
		ECV	essential climate variable
CO	carbon monoxide	EDA	Economic Development Administration
CO_2	carbon dioxide		
CO_2 Eq.	carbon dioxide equivalents	EDG	EOS Data Gateway
COL	combined license	EECBG	Energy Efficiency and Conservation Block Grant
CommEd	Communications and Education		

EEP	Environmental Exports Program	FRPP	Farm and Ranchlands Protection Program
EFRC	Energy Frontier Research Center	FSA	Farm Service Agency
EFRI	Emerging Frontiers in Research and Innovation	FTA	Federal Transit Administration
EGS	enhanced geothermal systems	FWS	U.S. Fish and Wildlife Service
EHE	excessive heat event	FY	fiscal year
EIA	Energy Information Administration	G8	Group of Eight
EIS	Environmental Impact Statement	GAW	Global Atmospheric Watch
EISA	Energy Independence and Security Act of 2007	GBEP	Global Bioenergy Partnership
		GCCE	Global Climate Change Education
EOS	Earth Observing System	GCCI	*Global Climate Change Impacts in the United States*
EOSDIS	Earth Observing System Data and Information System		
		GCEP	Global Change Education Program
EPA	U.S. Environmental Protection Agency	GCOS	Global Climate Observing System
EPAct	Energy Policy Act of 2005	GCRA	Global Change Research Act
Eq.	equivalents	GDA	Global Development Alliance
EQIP	Environmental Quality Incentives Program	GDP	gross domestic product
		GEF	Global Environment Facility
ESP	early site permit	Gen IV	Generation IV Nuclear Energy Systems
ESSEA	Earth System Science Education Alliance	GEO	Intergovernmental Group on Earth Observations
ETBC	Emerging Topics in Biogeochemical Cycles	GEOSS	Global Earth Observation System of Systems
EUGENE	Ecological Understanding as a Guideline for Evaluation of Nonformal Education		
		Gg	gigagram
EUMETSAT	European Organisation for the Exploitation of Meteorological Satellites	GHG	greenhouse gas
		GIF	Generation IV International Forum
Ex-Im	Export-Import Bank	GIS	geographic information system
°F	degree Fahrenheit	GLOBE	Global Learning and Observations to Benefit the Environment
FAA	Federal Aviation Administration		
FCCA	Forest, Climate & Community Alliance	GLOBEC	Global Ocean Ecosystems Dynamics
FCPF	Forest Carbon Partnership Facility	GLOSS	Global Sea Level Observing System
FEMP	Federal Energy Management Program	GOES	Geostationary Operational Environmental Satellite
FEWS NET	Famine Early Warning Systems Network		
FHA	Federal Housing Administration	GOOS	Global Ocean Observing System
FHWA	Federal Highway Administration	GOSIC	Global Observing System Information Center
FM	frequency modulation		
		GPS	global positioning system

GRACE	Gravity Recovery and Climate Experiment	IPHE	International Partnership for the Hydrogen Economy
GRACEnet	Greenhouse Gas Reduction through Agricultural Carbon Enhancement Network	IPY	International Polar Year
		IRENA	International Renewable Energy Agency
GRP	Grassland Reserve Program	IRG	International Resources Group
GRUAN	GCOS Reference Upper-Air Network	IRI	International Research Institute for Climate and Society
GTOS	Global Terrestrial Observing System		
GWP	global warming potential	ISE	Informal Science Education
ha	hectare	ITAP	International Technical Assistance Program
HCFC	hydrochlorofluorocarbon		
HD	high definition	ITEST	Innovative Technology Experiences for Students and Teachers
HFC	hydrofluorocarbon		
HFC-23	fluoroform	ITP	Industrial Technologies Program
HFEs	hydrofluorinated ethers	IWGEO	Interagency Working Group on Earth Observations
HIV/AIDS	human immunodeficiency virus/ acquired immunodeficiency syndrome		
		JAXA	Japanese Exploration Space Agency
HSPF	Hydrologic Simulation Program–Fortran	K–12	kindergarten through grade 12
HUD	U.S. Department of Housing and Urban Development	kg	kilogram
		km	kilometer
HVAC	heating, ventilation, and air conditioning	km^2	square kilometers
IAC	Industrial Assessment Center	kWh	kilowatt-hour
IASI	Infrared Atmospheric Sounding Interferometer	LDCM	Landsat Data Continuity Mission
		LIDAR	light detection and ranging
ICAA	Initiative for Conservation in the Andean Amazon	LMOP	Landfill Methane Outreach Program
		M2M	Methane to Markets
ICO	Integration and Coordination Office	MCC	Millennium Challenge Corporation
ICS	Incident Command System	MEF	Major Economies Forum on Energy and Climate
IDEA	Integrated Data and Environmental Applications		
		MERITO	Multicultural Education for Resource Issues Threatening Oceans
IEA	International Energy Agency		
IGEM	Intertemporal General Equilibrium Model	MetOp	Meteorological Operational
		mi	mile
IGLO	International Action on Global Warming	mi^2	square miles
		MIT	Massachusetts Institute of Technology
in	inch	mm	millimeter
IOOS	Integrated Ocean Observing System	MMT	million metric tons
IPCC	Intergovernmental Panel on Climate Change	$MMTCO_2$	million metric tons of carbon dioxide

MODIS	Moderate Resolution Imaging Spectroradiometer	NPOESS	National Polar-orbiting Operational Environmental Satellite System
MOU	Memorandum of Understanding	NPS	National Park Service
MPL	Micro Pulse Lidar	NRC	National Research Council
MPLNET	Micro Pulse Lidar Network	NRCS	Natural Resources Conservation Service
MPG	miles per gallon	NSF	U.S. National Science Foundation
MPO	metropolitan planning organization	NSTA	National Science Teachers Association
MSM	Multi-Scale Modeling	NSDL	National STEM Education Distributed Learning
MSW	municipal solid waste		
MW	megawatt	NSTC	National Science and Technology Council
MWe	megawatts electric		
MWh	megawatt-hour	NTI	National Transit Institute
MY	model year	NWCC	National Water and Climate Center
N_2O	nitrous oxide	O_3	ozone
NACP	North American Carbon Program	OAP	Office of Atmospheric Programs
NAS	National Academies of Science	OAS	Organization of American States
NASA	National Aeronautics and Space Administration	OCO	Orbiting Carbon Observatory
		OCS	Outer Continental Shelf
NAST	National Assessment Synthesis Team	ODS	ozone-depleting substance
NCDC	National Clean Diesel Campaign	OECC	White House Office of Energy and Climate Change
NCDC	National Climatic Data Center		
NEMS	National Energy Modeling System	OECD	Organisation for Economic Co-operation and Development
NEON	National Ecological Observatory Network		
		ONMS	Office of National Marine Sanctuaries
NF_3	nitrogen trifluoride	OPEC	Organization of the Petroleum Exporting Countries
NGO	nongovernmental organization		
NHTSA	National Highway Traffic Safety Administration	OPIC	Overseas Private Investment Corporation
		OSTP	Office of Science and Technology Policy
NIDIS	National Integrated Drought Information System	OSVW	ocean surface vector winds
		P2C2	Paleo Perspectives on Climate Change
NIFA	National Institute of Food and Agriculture	PAA	Price-Anderson Act
		PaCIS	Pacific Climate Information System
NMVOC	non-methane volatile organic compound	PANGEA	Partnerships for New GEOSS Applications
NOAA	National Oceanic and Atmospheric Administration	PARTNER	Partnership for AiR Transportation Noise and Emissions Reduction
NO_x	oxides of nitrogen	PES	payment for ecosystem services
NP	nuclear power	PFAN	Private Financing Advisory Network

PFC	perfluorocarbon	SAFETEA-LU	Safe, Accountable, Flexible, Efficient Transportation Equity Act: A Legacy for Users
PHEV	plug-in hybrid electric vehicle	SAPs	Synthesis and Assessment Products
PIH	plug-in hybrid	SAR	Second Assessment Report
PL	Public Law	SCAN	Soil Climate Analysis Network
PM	particulate matter	S'COOL	Students' Cloud Observations On-Line
POES	Polar Operational Environmental Satellites	SeaWiFS	Sea-viewing Wide Field-of-view Sensor
PRICIP	Pacific Region Integrated Climatology Information Products	SEN	Save Energy Now
		SEP	State Energy Program
PTC	production tax credit	SERVIR	Regional Visualization and Monitoring System
QuickSCAT	Quick Scatterometer		
R&D	research and development	SF_6	sulfur hexafluoride
RAMA	Research Moored Array for African-Asian-Australian Monsoon Analysis and Prediction	SFR	Sodium Fast Reactor
		SGCR	Subcommittee on Global Change Research
RANET	Radio and Internet for the Communication of Hydro-Meteorological and Climate-Related Information for Development	SHADOZ	Southern Hemisphere ADditional OZonesondes
		SNAP	Significant New Alternatives Policy program
RCMRD	Regional Center for Mapping of Resources for Development	SNOTEL	SNOpack TELemetry
		SO_2	sulfur dioxide
RD&D	research, development, and demonstration	SOARS	Significant Opportunities in Atmospheric Research and Science
REC	renewable energy credit	SORCE	Solar Radiation and Climate Experiment
REDD	reducing emissions from deforestation and degradation		
		SOS	Science On a Sphere®
REDI	Renewables and Efficiency Deployment Initiative	STAR	Science To Achieve Results
		STEM	science, technology, engineering, and mathematics
REEEP	Renewable Energy and Energy Efficiency Partnership		
		SUV	sport utility vehicle
REPI	Renewable Energy Production Incentive	SWAT	Soil and Water Assessment Tool
RES	renewable energy standard	SWiM	System-Wide Monitoring
RFS	renewable fuels standard	T&D	transmission and distribution
RGGI	Regional Greenhouse Gas Initiative	TAO	Tropical–Atmosphere–Ocean
RISA	Regional Integrated Science and Assessments Program	TBtus	trillion British thermal units
		TFCA	Tropical Forest Conservation Act
RPS	renewable portfolio standard	Tg	teragram
S&A	synthesis and assessment		

Tg CO$_2$ Eq.	teragrams of carbon dioxide equivalents	USDA	U.S. Department of Agriculture
TIGGER	Transit Investment for Greenhouse Gas and Energy Reduction	USDP	U.S. Drought Portal
TIST	International Small Group and Tree Planting Program	USFS	U.S. Forest Service
		USFWS	U.S. Fish and Wildlife Service
UBM	Undergraduates in Biological and Mathematical Sciences	USGCRP	U.S. Global Change Research Program
		USGEO	U.S. Group on Earth Observation
UN	United Nations	USGS	U.S. Geological Survey
UNEP	United Nations Environment Programme	USTDA	U.S. Trade and Development Agency
UNFCCC	United Nations Framework Convention on Climate Change	VCOP	Voluntary Code of Practice for the Reduction of Emissions of HFC & PFC Fire Protection Agents
UNIFEM	United Nations Development Fund for Women	VHTR	Very-High-Temperature Reactor
		WAP	Weatherization Assistance Program
UNPEPP	University-National Park Energy Partnership Program	WCI	Western Climate Initiative
		WHIP	Wildlife Habitat Incentives Program
USAID	U.S. Agency for International Development	WMO	World Meteorological Organization
		WRP	Wetland Reserve Program
USCRN	U.S. Climate Reference Network		
US-CTC	U.S. Climate Technology Cooperation		

1

Executive Summary

"The threat from climate change is serious, it is urgent, and it is growing. Our generation's response to this challenge will be judged by history, for if we fail to meet it—boldly, swiftly, and together—we risk consigning future generations to an irreversible catastrophe."

President Barack Obama
September 22, 2009
United Nations Summit on Climate Change

Throughout the United States, Americans are taking action to address the grave challenge of climate change, and to promote a sustainable and prosperous clean energy future. These efforts are occurring at all levels of government, in the private sector, and through the everyday decisions of individual citizens.

This *U.S. Climate Action Report 2010* (2010 CAR) sets out the major actions the U.S. government is taking at the federal level, highlights examples of state and local actions, and outlines U.S. efforts to assist other countries' efforts to address climate change.

At the federal level, since assuming office in January 2009, President Obama has renewed the U.S. commitment to lead in combating climate change. The Obama administration, together with the U.S. Congress, has taken major steps to enhance the domestic effort to promote clean energy solutions and tackle climate change.

Through the American Recovery and Reinvestment Act (ARRA), signed into law in February 2009, the United States allocated over $90 billion for invest-

ments in clean energy technologies to create green jobs, speed the transformation to a clean, diverse, and energy-independent economy, and help combat climate change.

In May 2009, President Obama announced a commitment to develop the first-ever joint fuel economy and carbon dioxide (CO_2) tailpipe emission standards for cars and light-duty trucks in the United States. These standards will boost fuel efficiency on average 4.3 percent annually and approximately 21.5 percent over the term of the standards, starting in 2012 and ending in 2016.

In September 2009, the U.S. Environmental Protection Agency (EPA) announced its plan to collect greenhouse gas (GHG) emission estimates from facilities responsible for 82.5 percent of the GHG emissions across diverse sectors of the economy, including power generation and manufacturing.

In October 2009, the President issued an Executive Order requiring federal agencies to set and meet strict GHG reduction targets by 2020. President Obama also called for more aggressive efficiency standards for common household appliances and put in motion a program to open the outer continental shelf to renewable energy production.

In December 2009, following an extensive comment and review period, the EPA Administrator issued findings under the U.S. Clean Air Act that the current and projected GHG concentrations in the atmosphere threaten the health and welfare of current and future generations.

The United States is engaged in crafting new laws to provide a comprehensive long-term framework for combating climate change over the coming decades. In February 2009, President Obama announced his intent to work with Congress in seeking new legislation on energy and climate change that would establish a mandatory economy-wide cap on emissions, with emission reductions beginning in 2012 and becoming more stringent annually thereafter, leading to GHG reductions of approximately 83 percent below 2005 levels by 2050. In June 2009, the U.S. House of Representatives took a significant step toward realizing this goal by passing the American Clean Energy and Security Act. The legislation includes major new investments in clean energy technologies that will be needed to achieve a prosperous transformation to a low-carbon economy, help the United States adapt to the effects of climate change, and establish the United States as a leader in creating the skilled green jobs that will help drive economic growth in coming decades.

Leadership involves effective cooperation with countries from all regions of the world. Since entering office, the Obama administration has ramped up U.S. engagement with other countries through the United Nations Framework Convention on Climate Change (UNFCCC) and through complementary efforts in support of a successful global climate agreement.

In April 2009, President Obama launched the Major Economies Forum on Energy and Climate (MEF), establishing an enhanced dialogue among 17 developed and developing economies representing 80 percent of global emissions to help support the multilateral negotiating process and devise new ways to advance the development and deployment of clean energy technologies. Leaders of these nations met in July 2009, and agreed on ways to further consensus on an enhanced future climate regime under the UNFCCC. These leaders also announced the establishment of a new Global Partnership to speed clean energy technology deployment. Experts from these countries have since developed action plans covering 10 key technologies. In December 2009 in Copenhagen, as part of this effort, the United States and other partners announced a new, five-year $350-million Climate Renewables and Deployment Initiative.

In June 2009, the President announced a new Partnership on Clean Energy and Climate of the Americas to promote clean energy technologies across the Western Hemisphere. The United States has accelerated collaboration with key partners, such as China, India, the European Union, Canada, Brazil, Mexico, Russia, and others, to combat climate change, coordinate clean energy research and development, and support efforts to achieve a successful agreement under the UNFCCC.

In September, at the 2009 Group of Twenty (G-20) summit in Pittsburgh, Pennsylvania, President Obama joined other G-20 leaders in committing to phase out fossil fuel subsidies.

In December 2009, at the Fifteenth Conference of the Parties of the UNFCCC, as part of a Copenhagen Accord involving robust GHG mitigation contributions by developed and key developing countries, the Obama administration proposed a U.S. GHG emissions reduction target in the range of 17 percent below 2005 levels by 2020 and approximately 83 percent below 2005 levels by 2050, ultimately aligned with final U.S. legislation. In January 2010, the United States inscribed its near-term proposal—to reduce emissions in the range of 17 percent from 2005 levels by 2020—in the Copenhagen Accord, formally associating itself with the Accord.

The United States announced in Copenhagen that it would increase U.S. climate assistance to ensure a fast start to post-Copenhagen efforts, contributing its share to developed country financing approaching $30 billion for 2010–2012. Also, in the context of meaningful mitigation actions and transparency on implementation, developed countries committed to a goal of mobilizing $100 billion globally by 2020 for countries in need, from various public- and private-sector sources. The fast-start effort will begin in 2010, with U.S. financing increased approximately three times that of 2009, including a substantial increase in adaptation financing for the most vulnerable countries and communities.

This 2010 CAR, submitted as a formal national communication, in accordance with Articles 4 and 12 of the UNFCCC, documents the actions the United States is taking to address climate change. This review accounts for current and proposed activities up to 2010. This report cites information and data available through 2009, except under very special circumstances, where more recent data were available. It explains how U.S. social, economic, and geographic circumstances affect U.S. GHG emissions, summarizes U.S. GHG emission trends from 1990 through 2007, identifies existing and planned U.S. policies and measures to reduce GHGs, and reports, wherever possible, measurable and verifiable emission reduction estimates for those policies and measures. The report also indicates future trends for U.S. GHG emissions, outlines the impacts of climate change on the United States and the adaptation measures the nation is taking to address those impacts, provides information on climate-related financial resources and technology transfer, details U.S. research and systematic observation efforts, and describes U.S. climate education, training, and outreach initiatives.

The activities in this report outline a set of initiatives in an ambitious, sustained effort that will be required to fully address climate change. The United States, in close cooperation with other nations of the world, is ready to build on its actions to date and assume a leadership role in this effort. The United States will continue to vigorously develop and build on its domestic and international efforts in coming months and years.

NATIONAL CIRCUMSTANCES

Chapter 2 of this report outlines the national circumstances of the United States and how they affect U.S. GHG emissions. The United States is a large country with a diverse geography that encompasses a full range of tropical, temperate, and Arctic ecosystems, stretching across seven time zones, from the Atlantic seaboard to the Hawaiian Islands. The total U.S. land area is 3,548,112 square miles (mi^2) (9,192,000 square kilometers [km^2]); about 28 percent of that land is owned and managed by the federal government in a system of parks, forests, wilderness areas, wildlife refuges, and other public lands.

The United States is a federal republic, and its government is divided into three distinct branches: executive, legislative, and judicial. Each branch plays a distinct role in the creation, implementation, and adjudication of America's laws. In addition, the governments of U.S. states and localities are responsible for developing environmental and energy laws and policies that collectively have a substantial influence on the U.S. climate response.

As of 2009, the United States is the third most populous country in the world, with an estimated population of 308 million. From 1990 to 2009, the U.S. population grew by about 59 million, at an annual rate of about 1 percent. This growth rate is relatively high compared to the approximately 0.4 percent annual growth rates of Organisation of Economic Co-operation and Development (OECD) member countries.

The United States has the highest real gross domestic product (GDP) in the world. Between 1990 and 2008, U.S. GDP grew by over $5.78 trillion (in constant 2008 dollars) or 66.9 percent, to reach $14.4 trillion (2008 dollars). Per capita income on a purchasing power parity basis was $46,716 in 2008—the fourth highest in the world behind Luxembourg, Norway, and Singapore.

The United States is the world's largest producer and consumer of energy. The nation currently consumes energy from petroleum, natural gas, coal, nuclear, conventional hydropower, and other renewable energy sources. For baseload electricity and most energy needs in the transportation sector, the fuels used most often are fossil fuels, accounting for approximately 79 percent of all U.S. energy consumption from 2005 through 2008. Petroleum remains the largest single source of U.S. energy consumption; in 2008 it accounted for 37.7 percent of total U.S. energy demand, down from 41 percent in 2005. Natural gas accounts for 24.4 percent, coal for 22.4 percent, nuclear for 8.1 percent, conventional hydro for 2 percent, and other renewables for 3 percent (U.S. DOE/EIA 2009d).

Annual U.S. energy consumption has been variable over the last decade, closely tracking economic growth rates and trends in energy efficiency in the residential, commercial, industrial, and transportation sectors. Between 2005 and 2007, total U.S. primary energy consumption grew by slightly over 1 percent. However, total primary energy consumption fell by 2.2 percent in 2008, as the economy weakened. Also, overall U.S. energy intensity has continually decreased, indicating an overall trend toward increasing energy efficiency in the economy. The decline of the U.S. economy's energy intensity has a direct effect on U.S. CO_2 emissions, 94 percent of which derive directly from the burning of fossil fuels according to 2007 data (U.S. EPA/OAP 2009).

The U.S. transportation system has evolved to meet the needs of a highly mobile, dispersed population and a large economy. Automobiles and light trucks dominate the passenger transportation system, and the highway share of passenger miles traveled by these vehicles in 2006 (the most recent year of available data) was 89 percent of total passenger miles. Air travel accounted for slightly over 10 percent, and mass transit and rail travel combined accounted for only about 1 percent of passenger miles traveled.

Many of the long-term trends identified in the 2006 *U.S. Climate Action Report* (2006 CAR) continue today, but recent events have significantly affected U.S. national circumstances. In particular, the economic slowdown in 2008 and early 2009 had a substantial impact on energy use and, correspondingly, GHG emissions. Technological change, energy efficiency improvements in transportation, buildings, and other sectors, private and public investment in low-carbon energy infrastructure, and a shift to less energy-intensive economic activity have continued to slow the growth of energy demand.

GREENHOUSE GAS INVENTORY

Chapter 3 summarizes U.S. anthropogenic GHG emission trends from 1990 through 2007. The estimates presented in the report were calculated using methodologies consistent with those recommended by the Intergovernmental Panel on Climate Change (IPCC). A complete accounting of GHGs in the United States is referenced in Chapter 3 of this report in Figure 3-1 and Table 3-2. In 2007, total U.S. GHG emissions were 7,150.1 teragrams of carbon dioxide equivalents (Tg CO_2 Eq.). Overall, total U.S. emissions rose by 17 percent from 1990 through 2007. Over that same time period, the U.S. GDP increased by 65 percent and population increased by 21 percent. CO_2 accounted for approximately 85 percent of total U.S. GHG emissions in 2007.

As the largest source of U.S. GHG emissions, CO_2 from fossil fuel combustion has accounted for approximately 79 percent of global warming potential-weighted emissions since 1990. Emissions of CO_2 from fossil fuel combustion increased at an average annual rate of 1.3 percent from 1990 through 2007. The fundamental factors influencing this trend include general domestic economic growth over the last 17 years, and significant growth in emissions from transportation activities and electricity generation. Between 1990 and 2007, CO_2 emissions from fossil fuel combustion increased from 4,708.9 Tg CO_2 Eq. to 5,735.8 Tg CO_2 Eq., a 21.8 percent total increase over the 17-year period. Historically, changes in emissions from fossil fuel combustion have been the dominant factor affecting U.S. emission trends.

Methane (CH_4) accounted for approximately 8 percent of total U.S. GHG emissions in 2007, with enteric fermentation (methane produced by livestock) being the largest source of CH_4 emissions. U.S. emissions of CH_4 declined by 5 percent from 1990 through 2007, mostly due to increased collection and combustion of landfill gas, as well as improvements in technology and management practices at natural gas plants.

Nitrous oxide (N_2O) accounted for approximately 4.4 percent of total U.S. GHG emissions in 2007. The main U.S. anthropogenic activities producing N_2O are agricultural soil management and fuel combustion in motor vehicles. Overall, U.S. emissions of N_2O declined by 1 percent from 1990 to 2007, largely due to the installation of newer N_2O control technologies in motor vehicles throughout the past decade.

Fluorinated substances—hydrofluorocarbons (HFCs), perfluorocarbons (PFCs), and sulfur hexafluoride (SF_6)—accounted for 2 percent of total U.S. GHG emissions in 2007. The increasing use of these compounds since 1995 as substitutes for ozone-depleting substances has been largely responsible for their upward emission trends.

Net CO_2 sequestration from land use, land-use change, and forestry increased by 221.1 Tg CO_2 Eq. (26 percent) from 1990 through 2007. This increase was primarily due to growth in the rate of net carbon accumulation in forest carbon stocks, particularly in above-ground and below-ground tree biomass.

POLICIES AND MEASURES

Chapter 4 of this report outlines near-term policies and measures undertaken by the U.S. government to mitigate GHG emissions. In addition to the major new 2009 initiatives highlighted earlier in this chapter, the U.S. government is making important progress toward reducing GHG emissions through some 80 energy policies and measures that promote increased investment in end-use efficiency, clean energy development, and reductions in agricultural GHG emissions. The U.S. government is also committed to reducing emissions from the most potent GHGs; more than a dozen initiatives across five executive agencies target these potent gases. These activities form a foundation for a comprehensive approach for achieving a transformation in the way the United States will use energy over coming decades.

In addition, a large number of U.S. states and localities are implementing clean energy incentives and clean energy targets—from voluntary emission goals and green building standards to mandatory cap-and-trade laws.

PROJECTED GREENHOUSE GAS EMISSIONS

Chapter 5 provides projections of U.S. GHG emissions through 2020 and beyond. These projections incorporate national estimates of population growth, economic growth, technology improvement, and normal weather patterns. Based on anticipated trends in technology development and adaptation, demand-side efficiency gains, and fuel switching, the projections represent a "business-as-usual" scenario that incorporates major policies in place as of March 31, 2009, including the Energy Independence and Security Act of 2007 (EISA) and ARRA, as well as a number of other federal and state measures outlined in Chapter 4. In this sense, the "business-as-usual" projections equal the "with measures" scenario called for under the UNFCCC Guidelines for Annex 1 Communications.

Despite the recent global economic turmoil, the U.S. economy is expected to recover and emissions are expected to grow in the long term in a business-as-usual/"with measures" case. Though absolute emissions are expected to grow in a "with measures" scenario, emissions per unit of GDP are expected to decline.

Projected GHG emissions under the "with measures" scenario presented in this report are significantly lower than emission estimates in the 2006 CAR. This shift is a result of changes in expectations regarding energy prices and economic growth and new policies taken since 2006, including EISA and ARRA, as well as various actions at the state level.

Projections are provided by gas and by sector from the present to 2020. Gases included in this report are CO_2, CH_4, N_2O, HFCs, PFCs, and SF_6. Sectors reported include electric power generation and residential, commercial, industrial, and transportation end use. Proposed or planned policies that had not been implemented as of March 31, 2009, as well as sections of existing legislation that require implementing regulations or funds that have not been appropriated, are not included in the projections. The projections include ARRA provisions, but do not include, for example, the vehicle fuel economy and emission standards announced by the President in May 2009.

The GHG emission projections in this chapter generally extend to 2020. Because of the extensive discussion of 2050 emission reduction goals in the United States and internationally, this chapter also includes a brief discussion of projected GHG emissions out to 2050.

Additional measures will be needed beyond those currently in place to ensure that the United States plays its part in the global effort to avoid the most damaging effects of climate change. The Obama administration supports the implementation of a market-based cap-and-trade program to spur growth in a low-carbon economy and reduce GHG emissions to approximately 83 percent below 2005 levels by

2050. In June 2009, the U.S. House of Representatives passed the landmark American Clean Energy and Security Act, which includes economy-wide GHG reduction goals of 3 percent below 2005 levels in 2012, 17 percent below 2005 levels in 2020, and 83 percent below 2005 levels in 2050.

As this report was finalized, the U.S. Senate was considering its own legislation, with similarly bold targets to promote clean energy and reduce GHG emissions. The target trajectories stipulated in those bills—aligned with the Obama administration's goals and with the U.S. inscription in the Copenhagen Accord—are presented against a long-term business-as-usual/"with measures" reference scenario in Figure 5-1.

From 2005 through 2020, total GHG emissions are projected to rise by 4 percent under a "with measures" scenario, from 7,109 Tg CO_2 Eq. to 7,416 Tg CO_2 Eq., while the U.S. GDP is projected to grow by 40 percent. Over that period, CO_2 emissions in the baseline projection are estimated to increase by 1.5 percent, although CH_4, N_2O, and PFC emissions are expected to grow more rapidly by 8 percent, 5 percent, and 4 percent, respectively. A large portion of emissions growth is driven by HFCs, which are projected to more than double between 2005 and 2020, as they are more extensively used as a substitute for ozone-depleting substances. The relatively slow growth forecast for CO_2 emissions is attributable to increasing use of renewable energy and policies implemented to increase efficiency.

With additional mitigation measures, such as those that would be implemented under the American Clean Energy and Security Act of 2009, the United States would have a GHG reduction goal of 17 percent by 2020, though U.S. GDP is projected to grow by 40 percent in that time. Over that period, with additional mitigation measures, CO_2, CH_4, N_2O, and PFC emissions would decrease significantly, and HFCs—among the most potent of GHGs—would be subject to a targeted cap and phase-down process.

More rapid improvements in technologies that emit fewer GHGs, new GHG mitigation requirements, or more rapid adoption of voluntary GHG emission reduction programs could result in lower GHG emission levels than in the baseline scenario projection.

IMPACTS AND ADAPTATION

Chapter 6 of this report highlights actions taken in the United States to better understand and respond to vulnerabilities and impacts associated with climate change. The U.S. government is involved in a wide array of climate assessments, research, and other activities to understand the potential impacts of climate change on the environment and the economy, and to develop methods and tools to enhance adaptation options. The U.S. government sponsors and adheres to some of the world's most advanced scientific research on climate change.

Chapter 6 outlines a set of studies by the U.S. Global Change Research Program, *Global Climate Change Impacts in the United States*, released in July 2009, which highlight key vulnerabilities in the United States associated with climate change (Karl et al. 2009). These key vulnerabilities, shared by the United States and many nations around the world, include the potential for water scarcity, unreliable energy production and transmission, damage to transportation infrastructure, public health problems, damage to ecosystems, and catastrophic harm to coasts and coastal communities.

Through the creation of special funds and programs related to climate adaptation, the U.S. government is working to address these vulnerabilities. States and localities have a major role to play in vulnerability assessment and adaptation. Chapter 6 includes examples of these efforts. The United States is committed to establishing and maintaining climate adaptation assistance for both domestic and international communities.

FINANCIAL RESOURCES AND TRANSFER OF TECHNOLOGY

Chapter 7 outlines U.S. government initiatives and partnerships and U.S. agency roles in climate-related international assistance and technology transfer. Chapter 7 also notes the U.S. offer of contributing to total developed country climate financing approaching $30 billion by 2012, with a goal of mobilizing $100 billion by 2020, in the context of robust mitigation efforts by all Parties to the UNFCCC.

The United States is committed to helping countries in need of addressing climate change, and is scaling up its efforts to promote clean energy, reduce emissions in the land-use sector, and adapt to climate change in a manner consistent with intellectual property rights. The fiscal year (FY) 2010 budget provides more than a threefold increase in bilateral and multilateral funding for climate-related activities, relative to the enacted funding provided in the previous fiscal year. It increases U.S. Agency for International Development (USAID) climate programs by 70 percent, with significant new investments in mitigation and adaptation strategies that will build on extensive USAID experience in promoting clean and climate-resilient development. It also provides first-ever contributions to multilateral climate funds, including $375 million to the World Bank Climate Investment Funds, and $50 million total to the UNFCCC Special Climate Change Fund and the UNFCCC Least Developed Country Fund. The President's FY 2011 budget, submitted to Congress in February 2010, proposes a further increase in core climate change funding of almost 40 percent, as well as substantial increases in funding for activities in such areas as food security, water, and health that will have significant climate co-benefits.

These investments—combined with U.S.-based foundation grants, nongovernmental organization (NGO) resources, private-sector commercial sales, commercial lending, foreign direct investment, market-enabled GHG reductions, private equity investment, and dozens of already-existing government programs that distribute aid, fund clean energy research, and conserve natural resources—support a robust contribution by the United States in organizing and supporting the world's response to climate change.

The U.S. government leads or is involved in a number of bilateral and multilateral clean energy partnerships, including new partnerships established in 2009. The United States will join a number of other countries in the Climate Renewables and Efficiency Deployment Initiative (Climate REDI), which will channel $350 million to fund programs over five years, including $85 million in U.S. funding. The United States announced in Copenhagen that it will partner with other key donors to channel $3.5 billion, including $1 billion in U.S. funding, to reduce emissions from deforestation, land degradation, and other activities through 2012. The United States will also join member countries of the Arctic Council to advance efforts to reduce black carbon from sources leading to reduced albedo in Arctic polar ice. The United States has made combating climate change a central part of its bilateral and regional strategic relationships, and recent climate partnerships have followed from leader summits with China, India, and countries across the Western Hemisphere.

U.S. agencies are engaged in extensive cooperative activities at the technical level. The National Oceanic and Atmospheric Administration (NOAA), Department of Energy, EPA, and other agencies lead U.S. engagement in specific climate partnerships, such as the Global Earth Observation System of Systems, the Carbon Sequestration Leadership Forum, and the Methane to Markets Partnership. These agencies also have extensive bilateral relationships designed to further climate protection in their areas of responsibility.

Agencies that promote international trade, including the U.S. Trade and Development Agency, the U.S. Export-Import Bank, and the Overseas Private Investment Corporation, are also enhancing efforts to promote clean investments. These and other U.S. government efforts will be ramped up over time as part of a successful global agreement on climate change.

RESEARCH AND SYSTEMATIC OBSERVATION

Chapter 8 describes how the United States is laying a strong scientific and technological foundation to reduce uncertainties, clarify risks and benefits, and develop effective mitigation options for addressing the impacts of climate change. It outlines the lead role of the interagency U.S. Global Change Research Program (USGCRP, formerly known as the Climate Change Science Program) in U.S. climate research. The chapter describes U.S. efforts to collect scientific observations about climate, archive climate-related data, and provide access to the data. This chapter also details how the U.S. government is supporting clean energy technology and climate change mitigation technologies.

The essential capacities for research and observations are widely distributed across U.S. government agencies, and are brought together into a single interagency program through the USGCRP. Growing out of interagency activities and planning that began in 1988, the creation of the USGCRP energized cooperative interagency activities, with each agency bringing its strengths to the collaborative effort. The FY 2010 budget provides over $2 billion for programs under the USGCRP—an increase of $46 million, or about 3 percent, over the 2009 level (excluding ARRA funds) (OMB 2009). The Office of Science and Technology Policy and Office of Management and Budget work closely with the Integration and Coordination Office and the working groups to establish research priorities and funding plans to ensure the program is aligned with the Obama administration's priorities and reflects agency planning.

The USGCRP helps organize and analyze critical climate data, which form the foundation of climate science. Long-term, high-quality observations of the global environmental system are essential for defining the current state of the Earth's system, its history, and its variability. This task requires both space- and surface-based observation systems.

The United States has committed not only to improving the scientific understanding of global climate change, but also to accelerating the development and deployment of technologies to reduce GHG emissions.

These efforts are targeted at increasing energy end-use efficiency and supplying energy with greatly reduced GHG emissions to meet the nation's goals of reducing GHG emissions and stabilizing GHG atmospheric concentrations at a level that avoids dangerous human interference with the climate system. To address these challenges, the Obama administration and Congress are working together to spur a revolution in clean energy technologies. An example of that commitment is the creation of the Advanced Research and Projects Agency–Energy (ARPA-E). Modeled on the Defense Department's Defense Advanced Research Projects Agency, which produced the predecessor to the Internet, ARPA-E was brought to life in 2009 by Congress with funding from the federal stimulus bill for the purpose of overcoming the long-term, high-risk technological barriers to the development of clean energy technologies.

EDUCATION, TRAINING, AND OUTREACH

Chapter 9 outlines how U.S. climate change education, training, and outreach efforts have expanded significantly since the publication of the 2006 CAR. U.S. federal agencies—including USAID; the Departments of Agriculture, Energy, the Interior, and Transportation; EPA; the National Aeronautics and Space Administration; NOAA; and the National Science Foundation—work on a wide range of climate change education, training, and outreach programs. A Climate Change Education Interagency Working Group was formed in 2008 to coordinate these efforts and develop an integrated national approach to climate change. Efforts by industry, states, local governments, universities, schools, and NGOs are essential complements to more than 100 federal programs that educate industry and the public regarding climate change. The combined efforts of the U.S. federal, state, and local governments and private entities are ensuring that the American public is better informed about climate change and more aware of the impact the nation's choices may have on the sustainability of the planet.

2

National Circumstances

Greenhouse gas (GHG) emissions in the United States are influenced by a multitude of factors. These include population and density trends, economic growth, energy production and consumption, technological development, use of land and natural resources, as well as climate and geographic conditions. This chapter focuses on both current circumstances and departures from historical trends since the 2006 *U.S. Climate Action Report* (2006 CAR) was submitted to the United Nations Framework Convention on Climate Change (UNFCCC) in 2007, and the impact of these changes on GHG emissions and removals (U.S. DOS 2007).

GOVERNMENT STRUCTURE

The United States is a federal republic. As such, local, state, and federal governments share responsibility for the nation's economic development, energy, natural resources, and many other issues. At the federal government level, a number of federal agencies, commissions, and advisory offices to the President are involved in developing, coordinating, and implementing nationwide policies to act on climate change.

The United States government is divided into three distinct branches: executive, legislative, and judicial. Each branch possesses distinct powers, but each also affects the other two, which creates a system of "checks and balances" and separates the powers to create, implement, and adjudicate laws.

Executive Branch

The executive branch is charged with implementing and enforcing the laws of the United States. The President of the United States is the U.S. Head of State and

oversees the executive branch. The President is advised by a Cabinet that includes a Vice President and the heads of 15 executive agencies—the Departments of State, Treasury, Defense, Justice, Interior, Agriculture, Commerce, Labor, Health and Human Services, Housing and Urban Development, Transportation, Energy, Education, Veterans Affairs, and Homeland Security. Other positions with Cabinet rank include the President's Chief of Staff, the Administrator of the Environmental Protection Agency, the Director of the Office of Management and Budget, the U.S. Trade Representative, the Chair of the Council of Economic Advisers, and the U.S. Ambassador to the United Nations. The Executive Office of the President, overseen by the President's Chief of Staff, includes a number of offices that play important roles in U.S. climate policy, such as the the Office of Energy and Climate Change, the Office of Science and Technology Policy, the Council on Environmental Quality, and the National Security Council. The executive branch also includes a number of independent commissions, boards, and agencies, such as the Federal Energy Regulatory Commission and the Export-Import Bank. Collectively, executive branch institutions cover a wide range of responsibilities, such as serving America's interests overseas, developing and maintaining the federal highway and air transit systems, researching the next generation of energy technologies, and managing the nation's abundant public lands.

Legislative Branch

The legislative branch consists of the two bodies in the U.S. Congress—the House of Representatives and the Senate—which are the primary lawmaking bodies of the government. This branch represents the U.S. citizenry through a bicameral system intended to balance power between representation based on population and representation based on statehood. The Senate is composed of 100 members, two from each of the 50 U.S. states. The House is composed of 435 members; each represents a single congressional district of approximately 650,000 people. In Congress, climate change is addressed by committees that are charged with developing legislation on energy and other relevant issues. In the House, the Committees on Agriculture, Labor, Ways and Means, and Energy and Commerce, among others, play vital roles in developing legislation. In the Senate, the Committees on Environment and Public Works; Finance; Foreign Relations; Agriculture; Commerce, Science, and Transportation; and Energy and Natural Resources are critical venues for debate.

Each body of Congress has the authority to develop legislation. A completed bill must receive a majority of votes in both the House and the Senate, and any differences between the House and the Senate versions must be reconciled before that bill can be sent to the President to be signed into law. The legislation becomes effective upon the President's signature.

Because the legislative process requires the support of both chambers of Congress and also involves the executive branch, a strong base of support is necessary to enact new legislation. As climate legislation is developed, this high threshold will remain very relevant.

Judicial Branch

The third branch, the judicial branch, serves as the government's court system responsible for interpreting the U.S. Constitution. It includes the Supreme Court, which is the highest court in the United States. The judicial branch in particular plays a significant role in defining the jurisdiction of the executive departments and interpreting the application of climate and energy policies under existing laws.

Governance of Energy and Climate Change Policy

Jurisdiction for addressing climate change within the federal government cuts across each of the three branches. Within the executive branch alone, some two dozen federal agencies and executive offices work together to advise, develop, and implement policies that help the U.S. government understand the workings of the Earth's climate system, reduce GHG emissions and U.S. dependence on oil, promote a clean energy economy, and assess and respond to the adverse effects of climate change. Chapters 4, 6, 7, 8, and 9 of this report describe the activities of these agencies related to these policies.

As with many other policy areas, jurisdiction for energy policy is shared by federal and state governments. Economic regulation of the energy distribution segment is a state responsibility, with the Federal Energy Regulatory Commission regulating wholesale sales and transportation of natural gas and electricity. In the absence of comprehensive federal climate change legislation, U.S. states have increasingly enacted climate change legislation or other policies designed to promote clean energy. Examples of these policies are described in Chapter 4 of this report. Similarly, land-use oversight is subject to mixed jurisdiction, with localities playing strong roles as well, and many areas related to adaptation policy are taken up by state and local entities. Examples of these activities are provided in Chapter 6.

POPULATION PROFILE

Population changes and growth patterns are fundamental drivers of trends in energy consumption, land use, housing density, and transportation, all of which have a significant effect on U.S. GHG emissions. The United States is the third most populous country in the world, with an estimated population of 308 million. From 1990 to 2009, the U.S. population grew by 55.3 million, at an average annual rate of just over 1 percent, for a total growth of approximately 22 percent since 1990. This growth rate is one of the highest among advanced economies, and is more than three times that of the European Union during this time period. The U.S. Census Bureau projects that the annual growth rate will remain relatively constant at nearly 1 percent through 2020, when the U.S. population is projected to be 341 million. Compared to 2008 levels, the U.S. population is expected to grow to 373 million (22 percent) in 2030 and to 439 million (44 percent) in 2050. Although the projections show increasing population, between 2020 and 2050, the steady increase of immigration is expected to be balanced by decreasing U.S. population growth rates. Overall, U.S. population growth will slow slightly to 0.87 percent in 2030 and will decline to 0.79 percent in 2050 (U.S. DOC/Census 2008b).

Population density trends show that more Americans are moving into cities and metropolitan areas. From 2000 to 2007, the U.S. population living in metropolitan areas grew from 80.2 percent to 83.5 percent. In general, increasing urbanization changes commuter patterns and reduces GHG emissions from the transportation sector. However, compared to cities in many other industrialized countries, major U.S. cities have relatively low population densities, and U.S. urban commuters use more energy for transportation and generate higher GHG emissions per person (U.S. DOE/EIA 2009c).

In addition, within any metropolitan region, the population density, walkability of neighborhoods, and access to public transit vary substantially. As a result, the average GHG emissions from household transportation vary significantly. For example, monitoring surveys at Atlantic Station, a redeveloped steel mill in midtown Atlanta, Georgia, reveal residents drive 14 miles per day compared to the regional average of 33 miles per day (U.S. EPA and Jacoby 2008).

GEOGRAPHIC PROFILE

The United States is one of the largest countries in the world, with a total area of 9,192,000 square kilometers (km²) (3,548,112 square miles [mi²]) stretching over seven time zones. The topography is diverse, featuring deserts, lakes, mountains, plains, and forests. The federal government owns and manages the natural resources on about 28 percent of U.S. land, most of which is managed as part of the national systems of parks, forests, wilderness areas, wildlife refuges, and other public lands. More than 60 percent of land area is privately owned, 9 percent is owned by state and local governments, and 2 percent is held in trust by the Bureau of Indian Affairs (Lubowski et al. 2006b).

CLIMATE PROFILE

The climate of the United States is highly diverse, ranging from tropical conditions in south Florida and Hawaii to arctic and alpine conditions in Alaska and across the Rocky Mountains. Temperatures for the continental United States show a strong gradient across regions and seasons, from very high temperatures in southern coastal states where the annual average temperatures exceed 21°C (70°F), to much cooler conditions in the northern parts of the country along the Canadian border, and from seasonal differences as great as 50°C (90°F) between summer and winter in the northern Great Plains. Similarly, precipitation also varies across the country and by seasons, measuring more than 127 centimeters (50 inches) per year along the Gulf of Mexico, while annual precipitation can be less than 30 cm (12 in) in the Intermountain West and Southwest (Figure 2-1). The peak rainfall season also varies by region. Many parts of the Great Plains and Midwest experience late-spring peaks, West Coast

Figure 2-1 Observed Change in Annual Average Precipitation: 1958–2008

While U.S. annual average precipitation has increased about 5 percent over the past 50 years, there have been important regional differences.

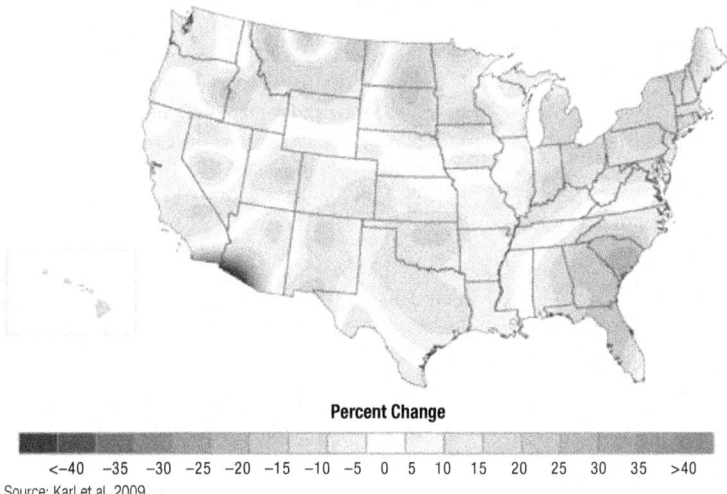

Percent Change

<−40   −35   −30   −25   −20   −15   −10   −5   0   5   10   15   20   25   30   35   >40

Source: Karl et al. 2009.

states have a distinct rainy season during winter, the Desert Southwest is influenced by summer's North American Monsoon, and many Gulf and Atlantic coastal regions experience summertime peaks.

The United States is subject to almost every kind of weather extreme, including severe thunderstorms, almost 1,500 tornadoes per year, and an average of 17 hurricanes that make landfall along the Gulf and Atlantic coasts each decade. At any given time, approximately 20 percent of the country experiences drought conditions. Differing U.S. climate conditions can be expressed by the number of annual heating and cooling degree-days.[1] Heating and cooling degree-days represent the number of degrees that the daily average temperature—the mean of the maximum and the minimum temperatures for a 24-hour period—is below (heating) or above (cooling) 18.3°C (65°F). For example, a weather station reporting a mean daily temperature of 4°C (40°F) would report 25 heating degree-days. From 2001 to 2008, the number of heating degree-days averaged 4,259, which was 3.8 percent below the 20th-century average. Over the same period, the annual number of cooling degree-days averaged 1,335, which was 5.4 percent above the long-term average.

[1] See http://www.whitehouse.gov/issues/economy.

Figure 2-2 **Energy Flow Through the U.S. Economy in 2008** (Quadrillion Btus)

The U.S. energy system is the world's largest, and it uses a diverse array of fuels from many different sources. The United States is largely self-sufficient in most fuels, except for petroleum. In 2008, net imports of crude oil and refined products accounted for about 57 percent of U.S. petroleum consumption on a Btu basis (U.S. DOE/EIA 2009m).

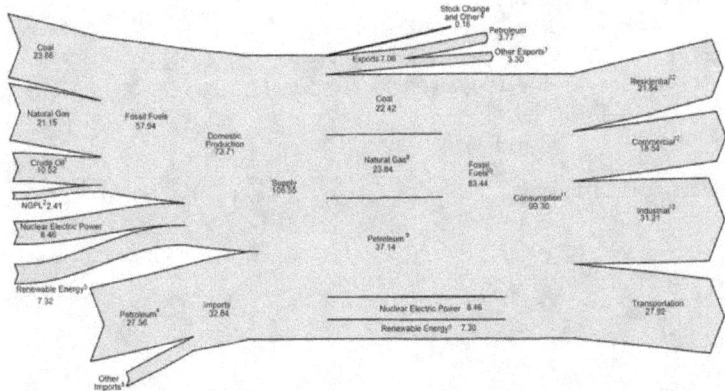

[1] Includes lease condensate.

[2] Natural gas plant liquids.

[3] Conventional hydroelectric power, biomass, geothermal, solar/photovoltaic, and wind.

[4] Crude oil and petroleum products. Includes imports into the Strategic Petroleum Reserve.

[5] Natural gas, coal, coal coke, fuel ethanol, and electricity.

[6] Adjustments, losses, and unaccounted for.

[7] Coal, natural gas, coal coke, and electricity.

[8] Natural gas only; excludes supplemental gaseous fuels.

[9] Petroleum products, including natural gas plant liquids, and crude oil burned as fuel.

[10] Includes 0.04 quadrillion British thermal units (Btus) of coal coke net imports.

[11] Includes 0.11 quadrillion Btus of electricity net imports.

[12] Primary consumption, electricity retail sales, and electrical system energy losses, which are allocated to the end-use sectors in proportion to each sector's share of total electricity retail sales.

See Note, "Electrical Systems Energy Losses," at end of Section 2 of U.S. DOE/EIA 2009b.

Notes:
• Data are preliminary.
• Values are derived from source data prior to rounding for publication.
• Totals may not equal sum of components due to independent rounding.

Sources: U.S. DOE/EIA 2009b, Tables 1.1, 1.2, 1.3, 1.4, and 2.1a.

ECONOMIC PROFILE

The U.S. economy is the largest national economy in the world, with a nominal gross domestic product (GDP) of $14.4 trillion in 2008, slightly smaller than that of the European Union (U.S. DOC/BEA 2009e; Eurostat 2009). The U.S. per capita GDP in 2007 was just over $47,000. Between 1990 and 2008, the U.S. economy grew by over 60 percent (in constant 2005 dollars), one of the highest growth rates among advanced economies in this time frame. Economic growth had a significant impact on GHG emissions during this period, though declining energy intensity in the U.S. economy resulted in significantly lower GHG emission increases in the past decade than in the 1990s.

Between 2005 and 2008, the U.S. GDP grew by $674 billion (in constant 2005 dollars), or 5.3 percent (U.S. DOC/BEA 2009e). Much of this growth was driven by strong consumer demand related to asset appreciation and easy credit. In the second half of 2008, substantial imbalances in the financial sector and in the economy generally gave rise to a deep global recession. The U.S. economy contracted 3.8 percent from the second quarter of 2008 through the second quarter of 2009 (U.S. DOC/BEA 2009e), and unemployment rose precipitously as a result, making job retention and creation the highest domestic policy priority. As one of his first acts, President Obama worked with Congress to enact the American Recovery and Reinvestment Act of 2009 (ARRA), designed to save or create about 3.5 million jobs, while making long-term investments to put the U.S. economy on a sound footing in coming years and decades. ARRA incorporates measures to increase production of alternative energy, modernize and weatherize buildings and homes, expand broadband technology across the country, and create a more efficient health care system.[1]

ENERGY RESERVES AND PRODUCTION

The United States is the world's largest producer and consumer of energy. The United States has large reserves of energy sources currently used for energy production, including fossil fuels, uranium ore, renewable biomass, and hydropower. Other renewable energy sources like solar and wind power, though currently a small portion of the energy resources used in the United States, are growing rapidly, with wind energy in the lead of other non-hydro renewable resources.

Figure 2-2 provides an overview of energy flows through the U.S. economy in 2008. This section focuses on changes in U.S. energy supply and demand since the 2006 CAR, which covered changes through 2005.

Fossil Fuels

The current base of U.S. energy resources used is fossil fuels, accounting for approximately 79 percent of all U.S. energy production from 2005 through 2008 (U.S. DOE/EIA 2009d).

Coal

Coal, the fuel most frequently used for power generation and supplying over 48 percent of the total electricity generated in the United States, also has the highest emissions of carbon dioxide (CO_2) per unit of energy. The United States uses around 1.1 billion tons of coal per year. Current recoverable coal reserves would supply the U.S. demand for energy, assuming constant 2007 rates of consumption, for approximately 232 years (U.S. DOE/EIA 2009f). Coal is particularly plentiful, yet due to property rights, land-use conflicts, and physical and environmental restrictions, only about 50 percent of the world's coal reserves (equal to about 489 billion short tons) may be available or accessible for mining. Of the estimated recoverable coal reserves, the United States holds the world's largest share (27 percent), followed by Russia (17 percent), China (13 percent), and Australia (9 percent) (U.S. DOE/EIA 2009e).

Oil

The trends in oil reserves and production identified in the 2006 CAR have changed very little. Both peaked in 1970, when Alaskan North Slope oil fields were discovered and developed, and generally have declined since then. Proven domestic reserves of crude oil stand at about 21.3 billion barrels. U.S. domestic crude oil production at the end of 2008 was estimated at 4.95 million barrels per day, of which only 0.5 percent was exported. Crude oil imports in 2008 actually averaged 9.95 million barrels per day, with another 3.18 million barrels of petroleum products imported.[2] Since around 1985, net crude oil imports have generally increased, while U.S. production has declined and consumption has grown. The United States relies on net petroleum imports to meet 56.9 percent of its petroleum needs, a decrease of 3.1 percent since the 2006 CAR, but an increase of 3.9 percent since 2000. The countries from which the United States imports the largest shares of crude oil and petroleum products include Canada (18.2 percent), Mexico (11.4 percent), Saudi Arabia (11.0 percent), Venezuela (10.1 percent), and Nigeria (8.4 percent) (U.S. DOE/EIA 2009g).

Natural Gas

Natural gas, the fossil fuel with the lowest emissions of CO_2 per unit of energy, has become an increasingly prominent fuel source in the United States in recent years. The increase in natural gas use for electricity generation, which has grown from 17.8 percent of total electricity generation in 2004 to 21.3 percent in 2008, has boosted demand for natural gas resources. Proven U.S. reserves of dry natural gas are rapidly increasing, growing by 33.6 percent from 177,427 billion cubic feet in 2000 to 237,726 billion cubic feet in 2007. Notably, the 2007 increase in proven dry natural gas reserves represented 237 percent of the total dry gas production for that year.

In 2008, the United States produced 20,561 billion cubic feet of dry natural gas. Imports totaled 3,980 billion cubic feet, of which 91.1 percent was imported by pipeline and 8.9 percent by liquid natural gas terminal. The United States imports natural gas by pipeline from Mexico and Canada, and by liquid natural gas terminal from Canada, Japan, Mexico, and Russia. Since 2004, pipeline imports from Mexico have grown from 0 percent to 1.2 percent, with the remainder of pipeline imports coming from Canada.

Nuclear Energy

In 2008, nuclear energy from 104 operating reactor units accounted for 19.6 percent of all electricity generated in the United States. The U.S. supply of uranium, the fuel used for nuclear fission, is mostly imported from other countries, with only 14.5 percent of the uranium used in 2008 being supplied by the United States. Most of these reserves can be found in Wyoming, New Mexico, Arizona, and Texas (U.S. DOE/EIA 2004). The average yearly U.S. uranium concentrate production in 2006–2008 was 4 million pounds, up from an average yearly production of 2 million pounds during 2003–2005.

In July 2007, the first application in over three decades was filed to build and operate a commercial nuclear reactor in the United States. By the end of 2007, 5 combined license (COL) applications were on file with the Nuclear Regulatory Commission (NRC). In 2008, the number of COL applications doubled; and as of November 2009, 17 applications were on file with the NRC.[3]

Renewable Energy

Renewable energy represents a rapidly growing source of U.S. energy production, currently accounting for 3 percent of U.S. electric generation excluding conventional hydro, or 9 percent including conventional hydro (U.S. DOE/EIA 2009i). Though there is currently no federally mandated standard for the use of renewable energy sources, as of 2009, 35 states and the District of Columbia had legislatively mandated a renewable energy portfolio standard (RPS). The RPS requirements vary by state, though many states have mandated that 15–25 percent of electricity sales come from renewable sources by 2020 or 2025. With the passing of the Energy Policy Act of 2005 (EPAct), federally mandated investment tax credits were established for those investing in residential, commercial, and industrial renewable energy, and the production tax credit for renewable energy electricity generation was extended. These provisions were extended until 2016 with the Energy Improvement and Extension Act of 2008.

These policies have played a primary role in the rapid expansion of electricity generated from renewable resources, such as wind, from 2006 to 2009. Conven-

[2] U.S. Department of Energy, Energy Information Adminnistration. See http://tonto.eia.doe.gov/dnav/pet/hist/LeafHandler.x?n=PET&s=MPEIMUS1&f=M

[3] The Energy Policy Act of 2005 (EPAct) provides a production tax credit for new nuclear reactors brought on line through 2020. The credit will pay producers 1.8 cents per kilowatt-hour of nuclear-generated electricity during the first 8 years of operation (U.S. DOE/EIA 2005). In addition, EPAct authorizes the Department of Energy (DOE) to provide up to $18.5 billion in loan guarantees for new reactors. DOE has selected four projects for final due diligence and detailed negotiations to receive the loan guarantees.

tional hydro remains by far the largest renewable source of electricity generation, generating 248 billion kilowatt-hours (kWh) in 2008, slightly more than double that of all other renewable sources combined. However, due to drought conditions on the West Coast that began in 2006 (U.S. DOE/EIA 2008d), this represents an 8 percent decrease from generation levels in 2005. Electricity production from renewable sources, excluding conventional hydro, totaled 123 billion kWh in 2008, which represents a 41 percent increase in production from 2005. Major growth is visible in the wind power industry alone, with electricity generation from wind increasing by 192 percent from 2005 levels to reach 52 billion kWh in 2008 (U.S. DOE/ EIA 2009j). Notably, between 2004 and 2008, total installed wind capacity increased by 369 percent to reach 23,847 megawatts (U.S. DOE/EIA 2008d).

Electricity

Total U.S. electricity generation reached 4,110 billion kWh in 2008, up by 8.1 percent compared to generation levels in 2000 and 1.3 percent compared to levels in 2005 (U.S. DOE/EIA 2009i). The Energy Information Administration (EIA) projects that U.S. electricity demand will continue to rise by 26 percent between 2007 and 2030 (U.S. DOE/EIA 2009k).[4]

[4] It is important to note that EIA projections used here and elsewhere in this report represent "business as usual" and do not include climate change and energy legislation or other policies currently under consideration.

[5] All EIA projections assume a business-as-usual scenario and do not incorporate any changes from current or future legislative action to curb GHGs supported strongly by the Obama administration.

In 2009, U.S. electricity generation was largely powered by coal-fired power plants, at 48 percent of total generation. Compared to previous years, the share of electricity generated from coal is declining, down from 51.7 percent in 2000 and 49.6 percent in 2005. This trend is due to rapid growth in natural gas-fired generation, which has risen from 15 percent of total electric generation in 2000 to 21 percent in 2008 (U.S. DOE/EIA 2009k). Figure 2-3 shows electricity flow through the U.S. economy in 2008.

ENERGY CONSUMPTION

The United States currently consumes energy from petroleum, natural gas, coal, nuclear, conventional hydro, and renewables. Petroleum remains the largest single source of U.S. primary energy consumption; in 2008 it accounted for 37.7 percent of total U.S. energy demand, down from 41 percent in 2005. Natural gas accounts for 24.4 percent, coal for 22.4 percent, nuclear for 8.1 percent, conventional hydro for 2 percent, and other renewables for 3 percent.

U.S. energy consumption has shifted over the last decade with economic growth rates and trends in energy efficiency in the residential, commercial, industrial, and transportation sectors. Between 2005 and 2007, total U.S. primary energy consumption grew by slightly over 1 percent. However, along with the current decrease in economic growth, total primary energy consumption fell by 2.2 percent in 2008 alone. The 99.3 quadrillion British thermal units (Btus) of energy consumed in 2008 represent a 1.1 percent decrease in consumption from 2005 levels and a 0.3 percent increase from 2000 levels.

The effects of the economic crisis on energy consumption are visible on a per capita basis as well. With an annual population growth rate of around 1 percent, per capita energy consumption between 2000 and 2007 declined to nearly 0.3 percent per year (Figure 2-4). In 2008, however, per capita energy use dropped sharply, down by 2.9 percent to 327 million Btus per person (U.S. DOE/EIA 2009b). EIA projects that high oil prices, increasing energy efficiency due to fuel efficiency standards, and increasing regulations on carbon intensity between 2008 and 2020 will contribute to declining per capita energy use, which is expected to reach 310 million Btus per person in 2020 and decline thereafter (U.S. DOE/EIA 2009h).[5]

In addition to the trend of decreasing energy consumption in recent years, overall energy intensity in the United States has decreased, indicating an overall trend toward increasing energy efficiency in the economy. Between 2000 and 2005, the energy intensity (energy consumed per dollar of GDP) of the U.S. economy declined on average by 1.9 percent per year. Between 2006 and 2008, energy intensity fell by 2.3 percent per year, from 9,140 Btus per dollar in 2005

Figure 2-3 Electricity Flow Through the U.S. Economy in 2008 (Quadrillion Btus)
In 2008, U.S. electricity generation was mostly powered by coal-fired power plants, at 48.5 percent of total generation.

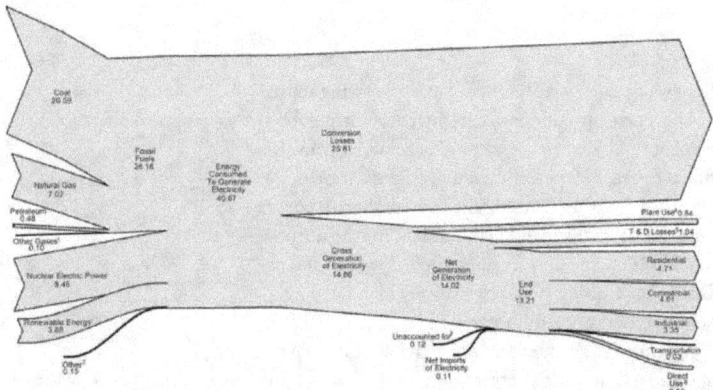

[1] Blast furnace gas, propane gas, and other manufactured and waste gases derived from fossil fuels.

[2] Batteries, chemicals, hydrogen, pitch, purchased steam, sulfur, miscellaneous technologies, and non-renewable waste (municipal solid waste from non-biogenic sources, and tire-derived fuels).

[3] Data collection frame differences and nonsampling error. Derived for the diagram by subtracting the "T&D (transmission and distribution) Losses" estimate from "T&D Losses and Unaccounted for" derived from Table 8.1.

[4] Electric energy used in the operation of power plants.

[5] Transmission and distribution losses (electricity losses that occur between the point of generation and delivery to the customer) are estimated as 7 percent of gross generation.

[6] Use of electricity that is (1) self-generated, (2) produced by either the same entity that consumes the power or an affiliate, and (3) used in direct support of a service or industrial process located within the same facility or group of facilities that house the generating equipment. Direct use is exclusive of station use.
Notes:
• Data are preliminary.
• See Note, "Electrical System Energy Losses," at the end of Section 2 of U.S. DOE/EIA 2009b.
• Values are derived from source data prior to rounding for publication.
• Totals may not equal sum of components due to independent rounding.
Sources: U.S. DOE/EIA 2009b, Tables 8.1, 8.4a, 8.9, A6 (column 4), and U.S. DOE/EIA 2008c.

to 8,520 Btus per dollar in 2008 (U.S. DOE/EIA 2009b). These data reflect a trend of advances in energy technologies and efficiency, and the growing importance of service industries and declining contribution of energy-intensive industries to the GDP.

The decline of the U.S. economy's energy intensity has a direct effect on U.S. CO_2 emissions, 94 percent of which derived directly from the burning of fossil fuels in 2007 (U.S. EPA/OAP 2009). As a result, the carbon intensity—measured as the ratio of metric tons of CO_2 emitted from energy consumption per million dollars of real GDP—also declined steadily in the past few years, falling by 4.5 percent from 544 in 2005 to 520 in 2007 (U.S. DOE/EIA 2009b).

Residential Sector

The residential sector's energy base fluctuates according to season, region, and year, and petroleum and natural gas demand varies much more than electricity demand. Consumption of petroleum, as fuel oil or liquefied petroleum gas, has been on a relative decline

after a peak of 885,000 barrels per day in 2003, dropping to 684,000 barrels per day in 2008. Consumption of natural gas has fluctuated as well in recent years, declining after a 2003 peak of 5,079 billion cubic feet to 4,368 billion cubic feet in 2006—a level not seen since 1987—and then increasing again to 4,866 billion cubic feet in 2008 (U.S. DOE/EIA 2009b).

The residential sector, made up of living quarters for private households, uses energy through a variety of sources: space heating, water heating, air conditioning, lighting, refrigeration, cooking, appliances, and electronics. In 2008, residential energy consumption, including electricity losses, totaled 21.6 quadrillion Btus, representing 21.7 percent of U.S. consumption. Residential buildings accounted for nearly 21 percent of GHG emissions from the use of fossil fuels.

The residential sector's energy use has also declined. Residential electricity consumption grew an average of 2.9 percent annually between 2000 and 2005, but slowed to an average 0.5 percent growth between 2006 and 2008.

Figure 2-4 **Energy Consumption and Expenditures Indicators**

Over the last decade, the U.S. demand for energy has plateaued, while the total relative cost of energy for individual Americans has risen dramatically.

Energy Consumption: 1949–2008

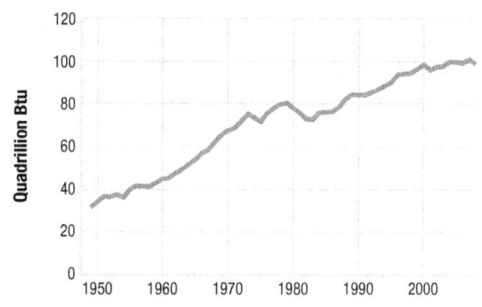

Energy Consumption per Person: 1949–2008

Energy Expenditures as a Share of
Gross Domestic Product: 1970–2006

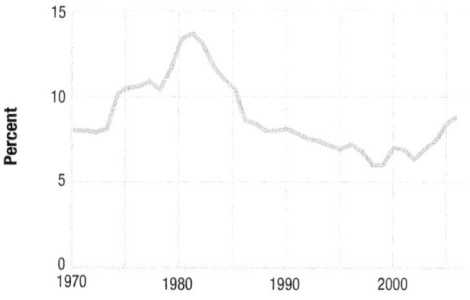

Energy Expenditures per Person: 1970–2006

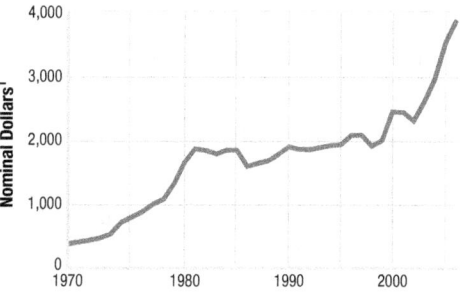

[1] See "Nominal Dollars" in Glossary of U.S. DOE/EIA 2009b.
Source: U.S. DOE/EIA 2009b, Table 1.5.

Commercial Sector

The commercial sector is made up of service facilities and equipment used by businesses, governments, private and public organizations, institutional living quarters, and sewage treatment plants. The most common uses of energy in this sector include space ventilation and air conditioning, water heating, lighting, refrigeration, cooking, and the operation of office and other equipment. Less common uses of energy include transportation. In 2008, the total energy consumed in the commercial sector was 3.7 percent higher than in 2005. At 18.5 quadrillion Btus, the commercial sector's energy use represented 18.6 percent of total U.S. energy demand in 2008. When GHG emissions from electricity generation are assigned to end-use sectors, the commercial sector was responsible for approximately 17 percent of total U.S. GHG emissions in 2007 (U.S. DOE/EIA 2009b; U.S. EPA/OAP 2008).

Electricity accounts for 78 percent of the commercial sector's energy use, followed by natural gas, which is close to 17 percent. Demand responds largely to a combination of prices and weather, although the impact of weather is less significant in the commercial sector than in the residential sector. Since the period covered by the 2006 CAR, demand for electricity increased annually at 1.96 percent; demand for natural gas fluctuated between 2005 and 2008, but increased by 6.6 percent in 2007 and by 3.3 percent in 2008.

EIA projects that between 2008 and 2030, greater energy efficiency will offset growth in commercial energy demand and thereby lead to an overall decline in the energy intensity of the commercial sector over the next two decades.

Industrial Sector

The U.S. industrial sector consists of all facilities and equipment used for producing, processing, or assembling goods, including manufacturing, mining, agriculture, and construction. The sector depends largely on coal, natural gas, and petroleum for its energy use. Electricity use, including system losses, represents around one-third of all energy consumed in the industrial sector, while petroleum and natural gas account for 27 percent and 26 percent, respectively. Use of coal in industrial energy consumption, falling from about 6.4 percent in 2000 to 5.7 percent in 2008, is slowly being replaced by increasing use of renewable energy sources, notably biomass.

Industrial sector energy consumption has declined steadily since 1973, falling from 43 percent of total energy consumption to 35 percent in 2000. This trend has steadily continued in recent years, dropping from 32.3 percent in 2005 to 31.4 percent in 2008. Within the industrial sector, total energy consumption has been decreasing—slightly from 2006 to 2008, and more quickly in 2008, when it fell by 4.1 percent. In 2008 energy consumption in the industrial sector had

dropped by 10.2 percent compared to 2000 levels, largely due to the U.S. economic downturn, which drastically reduced demand for manufactured goods. When emissions from electricity generation are distributed to end-use sectors, fossil fuel-related CO_2 emissions from the industrial sector have risen by 0.2 percent since 2006 and 2.4 percent since 1990 (U.S. DOE/EIA 2009n). As of 2007, using the same assumptions, the industrial sector accounted for 29.1 percent of total U.S. GHG emissions (U.S. EPA/OAP 2009).

Approximately four-fifths of the total energy used in the industrial sector is for manufacturing, with chemicals and allied products, petroleum and coal products, paper and nonmetallic minerals, and primary metals accounting for most of this share. The top five energy-consuming industries—bulk chemicals, refining, paper, steel, and food—account for around 60 percent of total industrial energy use but comprise only 20 percent of the total shipments. Projected slow growth in these energy-intensive industries is likely to result from increased foreign competition, reduced domestic demand for raw materials and basic goods they produce, and movement of investment capital to more profitable areas (U.S. DOE/EIA 2009h). Therefore, EIA estimates that industrial energy consumption will grow by only 4 percent between 2007 and 2030, primarily as a result of declines in output from most of the energy-intensive manufacturing industries.

Transportation Sector

Energy consumption in the transportation sector includes all energy used to move people and goods: automobiles, trucks, buses, and motorcycles; trains, subways, and other rail vehicles; aircraft; and ships, barges, and other waterborne vehicles.[6] Transportation is responsible for about 70 percent of all the petroleum used. In 2008, petroleum supplied 94 percent of the energy used in the transportation sector, down slightly from 96 percent in 2005, and the passage of the Energy Independence and Security Act of 2007 (EISA) (U.S. DOE/EIA 2009b). Energy consumption of biomass has grown significantly in recent years. High world oil prices, EPAct's passage in 2005, and EISA's passage in 2007 have encouraged the use of agriculture-based ethanol and biodiesel in the transportation sector. Therefore, between 2007 and 2008, for example, biomass consumption grew by over 35 percent, while petroleum consumption fell by 5.1 percent. In comparison, biomass consumption grew by an annual average of 20 percent from 2000 to 2004 and by 25 percent from 2005 to 2008 (U.S. DOE/EIA 2009b).

Demand in the transportation sector accounted for 28 percent of total U.S. energy demand in 2008 and approximately 28 percent of total U.S. GHG emissions. Slowing economic growth and high oil prices in 2008 were primary factors affecting the change in energy use

---

[6] Transportation does not include such vehicles as construction cranes, bulldozers, farming vehicles, warehouse tractors, and forklifts, whose primary purpose is not transportation.

from the transportation sector during the period since the 2006 CAR. Between 2000 and 2004, transport-related energy use grew by around 5 percent, at an average annual rate of 1.4 percent, continuing at nearly the same rate of annual growth between 2005 and 2007. However, in 2008 transport-related energy consumption declined by over 4 percent (U.S. DOE/EIA 2009b). Crude oil prices had increased rapidly in mid-2007, due to several factors, including the inability of non-Organization of the Petroleum Exporting Countries (OPEC) production rates to meet increasing global demand, OPEC members' production decisions, lack of spare production capacity, and geopolitical instability in key oil-producing regions. Therefore, crude oil prices increased from around $60 per barrel in 2007 to a peak of $145.16 per barrel in July 2008 (U.S. DOE/EIA 2009l), in stark contrast to 2005–2006 average prices of approximately $53 per barrel. With the decline in energy use, 2008 transportation-related energy consumption returned to 2005 levels, at approximately 27.94 quadrillion Btus.

Federal Government

The U.S. government remains the nation's largest single user of energy. Facilitated and coordinated by DOE's Federal Energy Management Program, federal agencies have invested in energy efficiency over the past two decades. As of 2007, the last year for which final data are available, the U.S. government's total primary energy consumption—excluding energy consumed to produce electricity and enrich uranium—was 1.085 quadrillion Btus, about 1.0 percent of total U.S. energy consumption. This is down by 6.6 percent compared to 2005 levels and 24.5 percent from 1990 levels. Compared to 2000 levels, however, the U.S. government's total primary energy consumption is up by 9.2 percent.

EISA set new goals for federal government energy use, including: a requirement to reduce energy intensity by 30 percent in 2030 compared to 2003 levels; a requirement to reduce the fossil-fuel-based energy consumption of new federal buildings, beginning with a 55 percent reduction in 2010 compared to 2003 levels and reaching a 100 percent reduction in 2030; a prohibition against purchasing light- or medium-duty vehicles that are not low-GHG-emitting vehicles; a requirement to reduce annual petroleum consumption by 20 percent in 2015 compared to 2005; and many other requirements regarding procurement of energy-efficient products, improved metering, and reporting (U.S. DOE/FEMP 2007).

Federal agencies are also subject to requirements for the use of electricity from renewable sources. Section 203 of EPAct requires the federal government to consume:

- not less than 3 percent in FY 2007–2009,
- not less than 5 percent in FY 2010–2012, and

- not less than 7.5 percent in FY 2013 and each fiscal year thereafter.

Preliminary data for FY 2008 indicate that federal agencies purchased or produced 1.9 terawatt-hours of renewable electric energy in 2008, which is equivalent to 3.4 percent of the federal government's electricity use of 56.1 terawatt-hours.

Executive Order 13123 of June 3, 1999 (later superseded by Executive Order 13423) established a GHG reduction goal for federal government facilities at 30 percent below 1990 levels by 2010.[7] Data for FY 2005 (the last year for findings under this reporting framework) show emissions from these facilities decreased by 22.1 percent since FY 1990, from 54.8 teragrams of CO_2 equivalent (Tg CO_2 Eq.) in FY 1990 to 42.7 Tg CO_2 Eq. in FY 2005.[8]

On October 5, 2009, President Obama signed Executive Order 13514, *Federal Leadership in Environmental, Energy, and Economic Performance*. The order expands the energy reduction and environmental requirements of Executive Order 13423 by making GHG management a priority for the federal government.[9] Under the new order, federal agencies are required to measure, manage, and reduce GHG emissions toward agency-defined targets. Provisions of the order require agencies to:

- Set scope 1, 2, and 3 GHG emission reduction targets for 2020 relative to a 2008 baseline.

- Reduce vehicle fleet petroleum use by 30 percent by 2020 compared to 2005.

- Improve water efficiency by 26 percent by 2020 compared to 2007.

- Implement the 2030 net-zero-energy building requirement.

- Meet sustainability requirements across 95 percent of all applicable contracts.

- Develop and carry out an integrated strategic sustainability performance plan.

TRANSPORTATION

The U.S. transportation system has evolved to meet the needs of a highly mobile, dispersed population and a large, dynamic economy. While the transportation system supports the movement of people and goods and the economic vitality of the country, efforts are underway to ensure that it is also as sustainable as possible.

Over the years, the United States has developed an extensive multimodal system that includes road, air, rail, and water transport capable of moving large volumes of people and goods long distances. Automobiles and light trucks still dominate the passenger transportation system, and the highway share of passenger miles traveled in 2006, the most recent year of avail-

[7] See http://www1.eere.energy.gov/femp/pdfs/eo13123.pdf.

[8] See http://www1.eere.energy.gov/femp/pdfs/annrep05.pdf.

[9] See http://www.eere.energy.gov/femp/pdfs/eo13514.pdf.

able data, was about 89 percent of the total, down by around 1 percent from the 2006 CAR. Air travel accounted for slightly over 10 percent (up 1 percent from the 2006 CAR), and mass transit and rail travel combined accounted for only about 1 percent of passenger miles traveled. For-hire transport services, as a portion of GDP, have barely changed since the 2006 CAR, accounting for 2.9 percent of GDP in 2007 (U.S. DOT/BTS 2010).

Highway Vehicles

The trends in highway vehicles described in the 2006 CAR have not changed appreciably. Between 2004 and 2007, the number of passenger vehicles rose steadily by an annual average of 1.8 percent to reach 254.3 million. This high degree of vehicle ownership is a result of population distribution, land-use patterns, location of work and shopping, and public preferences for personal mobility. Single-occupant passenger automobiles dominated daily traveling between home and work place in 2007, with 76 percent of the workforce (105 million of 139 million) driving themselves, down slightly from almost 80 percent in 2004. Just over 10 percent of workers commuted in carpools of two or more people, around 5 percent used public transportation, and the rest of the workforce used other means (biking, walking, taxis, etc.) (U.S. DOT/BTS 2010).

Passenger cars account for over 50 percent of highway vehicles and over one-third of all the energy consumed in the transportation sector. Light trucks, sport utility vehicles (SUVs), and vans comprise almost 40 percent of all highway vehicles and consume around 39 percent of energy in the sector. Between 2004 and 2006, the number of registered light trucks, SUVs, and vans increased by an annual average of 4.3 percent, continuing the upward trend seen in the previous years. This growth declined to 2.8 percent in 2007, as crude oil and gasoline prices began to rise steadily in mid-2007 and consumer preferences changed due to the increased costs of operating less fuel-efficient vehicles.

The number of miles driven is another major factor affecting energy use in the highway sector. From 2004 to 2006, the total number of vehicle miles driven each year by all on-road vehicles grew by an average of 0.8 percent to reach around 3 trillion miles, compared to around 1.8 percent average annual growth between 2000 and 2003. During these periods, total energy consumed by on-road vehicles grew by 1.02 percent and 1.1 percent, respectively (U.S. DOT/BTS 2010).

The fuel efficiency of passenger cars, light trucks, SUVs, and vans plays a large role in determining energy consumption and GHG emissions from the highway transport sector. The average fuel efficiency of passenger cars in the United States was relatively unchanged in the period since the 2006 CAR, increasing from an average of 22.1 miles per gallon (MPG) for 2000–2003 to an average of 22.3 MPG for 2004–2006. In 2008,

the average fuel efficiency for new passenger cars and light trucks was 24.3 MPG and 18.2 MPG, respectively (U.S. DOT/BTS 2010). Standards for new vehicles, known as Corporate Average Fuel Economy (CAFE) standards, play an integral role in determining the fuel efficiency of passenger cars and light trucks in the United States. New laws and policies outlined in Chapter 4 of this report will result in substantial increases in fuel efficiency over the next six years, with all new fleets of passenger vehicles and light trucks required to reach an average fuel efficiency of 35.5 MPG and a fleet average CO_2 emissions level of 250 grams per mile by 2016.

Air Carriers

Despite significant economic turbulence in 2008 and 2009, total U.S. air passenger miles remain near historic highs. Fueled in part by economic growth, airlines booked more revenue-paying U.S. passengers in 2007—more than 835 million—than in any other year. In 2008, total passengers (810 million) and total passenger miles (824 billion) declined by 3 and 2 percent, respectively, but those figures were still 10 and 18 percent above 2000 levels, respectively (U.S. DOT/BTS 2009b, 2009c).

Overall air passenger miles traveled are still declining in sync with the recent economic recession. Energy consumed from jet fuel of certified air carriers is decreasing more rapidly, spurred by the rising cost of fuel, increasing passenger load factors, and, to a greater degree, improvements in energy efficiency. In 2008, fuel use in aviation declined by 2.5 percent from 2007 levels, as the average price of a gallon of jet fuel jumped by 46 percent over 2007 prices (U.S. DOT/BTS 2009a).

Freight

Between 2004 and 2007 (the latest year for which freight data are available), U.S. freight transportation grew by 1.5 percent to 4.61 trillion ton miles, representing an average annual growth rate of 0.5 percent compared to 0.7 percent average annual growth between 2000 and 2003. Rail accounts for the largest share of total freight ton miles (39 percent), followed by trucks (29 percent), pipelines (20 percent), waterways (12 percent), and air (less than 1 percent) (U.S. DOT/BTS 2010).

INDUSTRY

The U.S. industrial sector boasts a wide array of light and heavy industries in manufacturing and nonmanufacturing subsectors, the latter of which include mining, agriculture, and construction. Private goods-producing industries accounted for slightly less than 19 percent of total GDP in 2008, and utilities accounted for another 2.1 percent of GDP. The industrial sector as a whole represents just over 29 percent of total U.S. GHG emissions (2007 data). Compared to the period covered under the 2006 CAR, declines in the portion of the economy made up by manufacturing have slowed, with the average change in manufacturing as a percentage of total GDP slowing from a decline of 3.7 percent between 2000 and 2004 to a decline of 1.5 percent between 2005 and 2008 (U.S. DOC/BEA 2009c).

The decline in housing prices, which began in 2007, took a large toll on the construction industry, which saw decreases in total value-added output of 5.4 percent and 4.8 percent in 2007 and 2008, respectively. Mining continued to rise as a share of GDP, growing from 1.8 percent in 2005 to 2.2 percent in 2008.

WASTE

In 2007, the United States generated approximately 230 million metric tons (254 million tons) of municipal solid waste (MSW), about 45 million metric tons (nearly 50 million tons) more than in 1990. Paper and paperboard products made up the largest component of MSW generated by weight (33 percent), and yard trimmings comprised the second-largest material component (more than 13 percent). Glass, metals, plastics, wood, and food each constituted between 5 and 13 percent of the total MSW generated, while rubber, leather, and textiles combined made up about 8 percent of the MSW.

Recycling has been the most significant change in waste management from a GHG perspective. From 1990 to 2007, the recycling rate increased from just over 16 percent to about 33 percent. Of the remaining MSW generated, about 13 percent was combusted and 54 percent was disposed of in landfills. The number of operating MSW landfills in the United States has decreased substantially over the past 20 years, from about 8,000 in 1988 to about 1,750 in 2007, while the average landfill size has increased (U.S. EPA/OSW 2008).

Landfills are the second-largest U.S. source of anthropogenic methane emissions, accounting for 23 percent of the total. Present data suggest a marked increase in the amount of methane recovered for either gas-to-energy or flaring purposes in recent years (U.S. EPA/ OAP 2009).

BUILDING STOCK AND URBAN STRUCTURE

Buildings are large users of energy. Their number, size, and distribution and the appliances and heating and cooling systems that go into them influence energy consumption and GHG emissions. Buildings account for about 37 percent of total U.S. energy consumption and about 70 percent of total electricity consumption.

Residential Buildings

Growth in the U.S. housing market has slowed significantly since the advent of the U.S. economic slowdown in 2007. Prior to the economic slump, the housing market had been relatively strong since 2000.

Between 1997 and 2003, the number of residences in the United States grew by 8.3 percent to approximately 121 million households, 62 percent of which were single, detached dwellings. However, since 2007, the number of new residential units under construction fell from an all-time high of two million new housing units per year in 2005 to less than one million by 2008. During the downturn, the amount of new multi-family rental units (i.e., apartment buildings) has held steady, and they now account for 24 percent of new residential construction (U.S. DOC/Census 2009a).

As the growth of the U.S. housing market has slowed, the growth of energy use in residential buildings has also slowed. The housing boom of 2000–2005 had been especially strong in the Sunbelt, where almost all new homes have air conditioning. Overall, the growth in residential buildings' energy use has been mitigated by recent economic events.

While new homes are larger and more plentiful, their energy efficiency has increased significantly. In 2004, 8 percent of all new single-family homes were certified as ENERGY STAR compliant, implying at least a 30 percent energy savings for heating and cooling relative to comparable homes built to current code (U.S. DOE/EIA 2006). New homes are on average about 13 percent larger than the stock of existing homes, and thus have greater requirements for heating, cooling, and lighting. Nevertheless, under current building codes and appliance standards for heat pumps, air conditioners, furnaces, refrigerators, and water heaters, the energy requirement per square foot of a new home is typically lower than of an existing home (U.S. DOE/EIA 2005).

Commercial Buildings

Between 2000 and 2003 (the latest data available), commercial floorspace rose an estimated 1.8 percent per year. By 2003 there were nearly 4.9 million commercial buildings and more than 6.7 billion square meters (71.7 billion square feet) of floorspace. Much of this growth has been related to the rapidly expanding information, financial, and health services sectors.

More than half of commercial buildings are 465 square meters (5,000 square feet) or smaller, and nearly three-fourths are 929 square meters (10,000 square feet) or smaller. Just 2 percent of buildings are larger than 9,290 square meters (100,000 square feet), but these large buildings account for more than one-third of commercial floorspace (U.S.DOE/EIA 2003).

Electricity and natural gas are the two largest sources of energy used in commercial buildings. Over 85 percent of commercial buildings are heated, and more than 75 percent are cooled. The use of computers and other office electronic equipment continues to grow and will have an impact on the demand for electricity (U.S. DOE/EIA 2006).

AGRICULTURE AND GRAZING

Agriculture in the United States is highly productive. U.S. croplands produce a wide variety of food and fiber crops, feed grains, oil seeds, fruits and vegetables, and other agricultural commodities for both domestic and international markets. Although the United States harvests roughly the same area as it did in 1910, U.S. agriculture feeds a population three times larger, with crops still available for export. Technological changes account for most of the increased productivity. In 2002, U.S. cropland was 179 million hectares (ha) (442 million acres[ac]), about 3 percent lower than in 1997 (Lubowski et al. 2006a, 2006b; USDA 2008).

Soils vary across the landscape in response to the effects of climate, topography, vegetation, and other organisms (including humans) on the rate and direction of soil development processes acting on parent materials over time. In the United States, the wide range and endless combinations of these factors have resulted in a great range of soils with widely varying properties. All soils provide an effective natural filter that protects ground and surface water by removing potential contaminants applied on or in the soil. Soils across the United States have the potential to sequester substantial amounts of organic and inorganic carbon to help reduce atmospheric CO_2 levels. Although soils vary in their resistance and resilience, all are subject to degradation through erosion, salinization, and other mechanisms without proper management.

Conservation is an important objective of U.S. farm policy. The U.S. Department of Agriculture administers a set of conservation programs that have been highly successful at removing environmentally sensitive lands from commodity production and encouraging farmers to adopt conservation practices on working agricultural lands. The largest of these programs, the Conservation Reserve Program (CRP), seeks to reduce soil erosion, improve water quality, and enhance wildlife habitat by retiring environmentally sensitive lands from crop production. About 12.6 million ha (31.2 million ac) of land are enrolled in CRP.

Improved tillage practices also have helped reduce soil erosion and conserve and build soil carbon levels. From 1998 to 2004, the amount of cropland managed with no-till systems increased by 31 percent to 25.4 ha (62.7 ac), in part because of the widespread adoption of herbicide-tolerant crops developed using biotechnology. Land managed using all conservation tillage systems has fluctuated between about 40 and 46 million ha (98.8 and 113.6 million ac) (CTIC 2004).

Sources of GHG emissions from U.S. croplands include nitrous oxide from nitrogen fertilizer use and methane from farm animals' enteric fermentation and manure management. Nitrous oxide from agricultural soil management is the largest source of GHG emissions from the agricultural sector, representing just over 35 percent of that sector's emissions in 2007 (USDA 2008).

Grasslands account for slightly more than one-third of U.S. land uses. Pasture and range ecosystems can include a variety of different flora and fauna communities. They are generally managed by varying grazing pressure, using fire to shift species abundance, and occasionally disturbing the soil surface to improve water infiltration. In 2002, grasslands totaled about 238 million ha (587 million ac), about the same as in 1997. Since 1949, grassland acreage has declined by about 9 percent, reflecting improved productivity of grazing lands, land-use changes, and a decline in the number of domestic animals raised on grazing lands (Lubowski et al. 2006b).

FORESTS

U.S. forests are predominately natural stands of native species, and vary from the complex hardwood forests in the East to the highly productive conifer forests of the Pacific Coast. Planted forest land is most common in the East, and planted stands of native pines are common in the South. In 1630, forest land comprised an estimated 46 percent of the total U.S land area, whereas in 2007, forests covered about one-third of the total area. Historically, most of the forest land loss was due to agricultural conversions in the late 19th century, but today most losses are due to such intensive uses as urban development.

Of the 305 million ha (751 million ac) of U.S. forest land, nearly 208 million ha (514 million ac) are timberland, most of which is privately owned in the conterminous United States. However, a significant area of forest land is reserved forests, which in 2007 accounted for 10 percent of all forest land, or about 30 million ha (75 million ac) (Smith et al. 2009).

Most timber removals come from private lands, with the South providing nearly two-thirds of all domestic timber. Management inputs over the past several decades have been gradually increasing the production of marketable wood in U.S. forests, especially on the private lands and in the South. The United States currently grows more wood than it harvests, with a growth-to-harvest ratio of nearly 2 to 1. As the average age of U.S. forests continues to rise and growth continues to exceed removals, standing volume has increased by 37 percent since 1953 to a level of nearly 33 billion cubic meters.

Existing U.S. forests are an important net sink for atmospheric carbon. Improved forest management practices, the regeneration of previously cleared forest areas, as well as timber harvesting and use have resulted in net sequestration of CO_2 every year since 1990 (U.S. EPA/OAP 2009). In 2007, the land use, land-use change, and forestry sector absorbed a net of 1,062.6 Tg of CO_2. This sequestration represents an offset of 14.9 percent of U.S. GHG emissions on a global warming potential-weighted basis.

3

Greenhouse Gas Inventory

An emissions inventory that identifies and quantifies a country's primary anthropogenic[1] sources and sinks of greenhouse gases (GHGs) is essential for addressing climate change. *The Inventory of U.S. Greenhouse Gas Emissions and Sinks: 1990–2007* (U.S. EPA/OAP 2009) adheres to both (1) a comprehensive and detailed set of methodologies for estimating sources and sinks of anthropogenic GHGs, and (2) a common and consistent mechanism that enables Parties to the United Nations Framework Convention on Climate Change (UNFCCC) to compare the relative contribution of different emission sources and GHGs to climate change.

In 1992, the United States signed and ratified the UNFCCC. As stated in Article 2 of the UNFCCC, "The ultimate objective of this Convention and any related legal instruments that the Conference of the Parties may adopt is to achieve, in accordance with the relevant provisions of the Convention, stabilization of GHG concentrations in the atmosphere at a level that would prevent dangerous anthropogenic interference with the climate system. Such a level should be achieved within a time-frame sufficient to allow ecosystems to adapt naturally to climate change, to ensure that food production is not threatened and to enable economic development to proceed in a sustainable manner."[2]

Parties to the Convention, by ratifying, "shall develop, periodically update, publish and make available... national inventories of anthropogenic emissions by sources and removals by sinks of all GHGs not controlled by the *Montreal Protocol*, using comparable

[1] The term "anthropogenic," in this context, refers to greenhouse gas emissions and removals that are a direct result of human activities or are the result of natural processes that have been affected by human activities (IPCC/UNEP/OECD/IEA 1997).

[2] Article 2 of the United Nations Framework Convention on Climate Change, published by the United Nations Environment Programme/World Meteorological Organization Information Unit on Climate Change. See http://unfccc.int/essential_background/convention/background/items/1353.php.

methodologies...."[3] The United States views the *Inventory of U.S. Greenhouse Gas Emissions and Sinks: 1990–2007* (U.S. EPA/OAP 2009) as an opportunity to fulfill these commitments.

This chapter summarizes the latest information on U.S. anthropogenic GHG emission trends from 1990 through 2007. To ensure that the U.S. emissions inventory is comparable to those of other UNFCCC Parties, the estimates presented here were calculated using methodologies consistent with those recommended in the *Revised 1996 IPCC Guidelines for National Greenhouse Gas Inventories* (IPCC/UNEP/OECD/IEA 1997), the *IPCC Good Practice Guidance and Uncertainty Management in National Greenhouse Gas Inventories* (IPCC 2000), and the *IPCC Good Practice Guidance for Land Use, Land-Use Change and Forestry* (IPCC 2003). Additionally, the U.S. emissions inventory has begun to incorporate new methodologies and data from the 2006 IPCC Guidelines for National Greenhouse Gas Inventories (IPCC 2006). The structure of the *Inventory of U.S. Greenhouse Gas Emissions and Sinks: 1990–2007* (U.S. EPA/OAP 2009) is consistent with the UNFCCC guidelines for inventory reporting (UNFCCC 2003).[4] For most source categories, the Intergovernmental Panel on Climate Change (IPCC) methodologies were expanded, resulting in a more comprehensive and detailed estimate of emissions (Box 3-1).

BACKGROUND INFORMATION

Naturally occurring GHGs include water vapor, carbon dioxide (CO_2), methane (CH_4), nitrous oxide (N_2O), and ozone (O_3). Several classes of halogenated substances that contain fluorine, chlorine, or bromine are also GHGs, but they are, for the most part, solely a product of industrial activities. Chlorofluorocarbons (CFCs) and hydrochlorofluorocarbons (HCFCs) are halocarbons that contain chlorine, while halocarbons that contain bromine are referred to as bromofluorocarbons (i.e., halons). As stratospheric ozone-depleting substances (ODS), CFCs, HCFCs, and halons are covered under the *Montreal Protocol on Substances That Deplete the Ozone Layer*. The UNFCCC defers to this earlier international treaty. Consequently, Parties to the UNFCCC are not required to include these gases in their national GHG emission inventories.[5] Some other fluorine-containing halogenated substances—hydrofluorocarbons (HFCs), perfluorocarbons (PFCs), and sulfur hexafluoride (SF_6)—do not deplete stratospheric ozone but are potent GHGs (Box 3-2). These latter substances are addressed by the UNFCCC and accounted for in national GHG emission inventories.

There are also several gases that do not have a direct global warming effect but indirectly affect terrestrial and/or solar radiation absorption by influencing the formation or destruction of GHGs, including

---

Box 3-1 **Recalculations of Inventory Estimates**

Each year, emission and sink estimates are recalculated and revised for all years in the *Inventory of U.S. Greenhouse Gas Emissions and Sinks*, as attempts are made to improve both the analyses themselves, through the use of better methods or data, and the overall usefulness of the report. In this effort, the United States follows the *IPCC Good Practice Guidance and Uncertainty Management in National Greenhouse Gas Inventories* (IPCC 2000), which states, regarding recalculations of the time series, "It is good practice to recalculate historic emissions when methods are changed or refined, when new source categories are included in the national inventory, or when errors in the estimates are identified and corrected." In general, recalculations are made to the U.S. GHG emission estimates either to incorporate new methodologies or, most commonly, to update recent historical data.

In each *Inventory* report, the results of all methodology changes and historical data updates are presented in the "Recalculations and Improvements" chapter; detailed descriptions of each recalculation appear within each source's description in the report, if applicable. In general, when methodological changes have been implemented, the entire time series (in the case of the most recent *Inventory* report, 1990 through 2007) has been recalculated to reflect the change, per *IPCC Good Practice Guidance*. Changes in historical data are generally the result of changes in statistical data supplied by other agencies. References for the data are provided for additional information. More information on the most recent changes is provided in the "Recalculations and Improvements" chapter of the *Inventory of U.S. Greenhouse Gas Emissions and Sinks: 1990–2007* (U.S. EPA/OAP 2009), and previous *Inventory* reports can further describe the changes in calculation methods and data since the 2006 *U.S. Climate Action Report*.

---

tropospheric and stratospheric ozone. These gases include carbon monoxide (CO), oxides of nitrogen (NO_x), and non-methane volatile organic compounds (NMVOCs). Aerosols, which are extremely small particles or liquid droplets, such as those produced by sulfur dioxide (SO_2) or elemental carbon emissions, can also affect the ability of the atmosphere to absorb radiation.

Although the direct GHGs CO_2, CH_4, and N_2O occur naturally in the atmosphere, human activities have changed their atmospheric concentrations. From the pre-industrial era (i.e., ending about 1750) to 2005, concentrations of these GHGs have increased globally by 36, 148, and 18 percent, respectively (IPCC 2007).

Beginning in the 1950s, the use of CFCs and other stratospheric ODS increased by nearly 10 percent per year until the mid-1980s, when international concern about stratospheric ozone depletion led to the entry into force of the *Montreal Protocol*. Since then, the production of ODS is being phased out. In recent years, use of ODS substitutes, such as HFCs and PFCs, has grown as they begin to be phased in as replacements for CFCs and HCFCs. Accordingly, atmospheric concentrations of these substitutes have been growing (IPCC 2007).

RECENT TRENDS IN U.S. GREENHOUSE GAS EMISSIONS AND SINKS

In 2007, total U.S. GHG emissions were 7,150.1 teragrams of CO_2 equivalents (Tg CO_2 Eq.). Overall, total U.S. emissions rose by 17 percent from 1990 to 2007. Emissions rose from 2006 to 2007, increasing by 1.4 percent (99.0 Tg CO_2 Eq.). The following factors were primary contributors to this increase:

[3] Article 4(1)(a) of the United Nations Framework Convention on Climate Change (also identified in Article 12). Subsequent decisions by the Conference of the Parties elaborated on the role of Annex I Parties in preparing national inventories. See http://unfccc.int/essential_background/convention/background/items/1362.php.

[4] See http://unfccc.int/resource/docs/cop8/08.pdf.

[5] Emission estimates of CFCs, HCFCs, halons, and other ODS are included in the annexes of the *Inventory of U.S. Greenhouse Gas Emissions and Sinks:1990–2007* for informational purposes (U.S. EPA/OAP 2009).

Box 3-2 Emissions Reporting Nomenclature

The global warming potential (GWP)-weighted emissions of all direct GHGs throughout this report are presented in terms of equivalent emissions of carbon dioxide (CO_2), using units of teragrams of CO_2 equivalents (Tg CO_2 Eq.). The GWP of a GHG is defined as the ratio of the time-integrated radiative forcing from the instantaneous release of 1 kilogram (kg) of a trace substance relative to that of 1 kg of a reference gas (IPCC 2001). The relationship between gigagrams (Gg) of a gas and Tg CO_2 Eq. can be expressed as follows:

$$\text{Tg } CO_2 \text{ Eq.} = (\text{Gg of gas}) \times (\text{GWP}) \times \left(\frac{\text{Tg}}{1,000 \text{ Gg}}\right)$$

The UNFCCC reporting guidelines for national inventories were updated in 2006 (UNFCCC 2006), but continue to require the use of GWPs from the IPCC Second Assessment Report (SAR) (IPCC 1996). The GWP values used in this report are listed below in Table 3-1, and are explained in more detail in Chapter 1 of the *Inventory of U.S. Greenhouse Gas Emissions and Sinks: 1990-2007* (U.S. EPA/OAP 2009).

Table 3-1 Global Warming Potentials (100-Year Time Horizon) Used in This Report

The concept of a GWP has been developed to compare the ability of each GHG to trap heat in the atmosphere relative to another gas. Carbon dioxide was chosen as the reference gas to be consistent with IPCC guidelines.

Gas	GWP	Gas	GWP
CO_2	1	HFC-227ea	2,900
CH_4*	21	HFC-236fa	6,300
N2O	310	HFC-4310mee	1,300
HFC-23	11,700	CF_4	6,500
HFC-32	650	C_2F_6	9,200
HFC-125	2,800	C_4F_{10}	7,000
HFC-134a	1,300	C_6F_{14}	7,400
HFC-143a	3,800	SF_6	23,900
HFC-152a	140		

* The methane GWP includes both the direct effects and those indirect effects of the production of tropospheric ozone and stratospheric water vapor. The indirect effect due to the production of CO_2 is not included.

GWP = global warming potential; CO_2 = carbon dioxide; CH_4 = methane; N_2O = nitrous oxide; HFC = hydrofluorocarbon; CF_4 = tetrafluoromethane; C_2F_6 = hexafluoroethane; C_4F_{10} = perfluorobutane; C_6F_{14} = perfluorohexane or tetradecafluorohexane; SF_6 = sulfur hexafluoride.

Source: IPCC 1996.

(1) cooler winter and warmer summer conditions in 2007 than in 2006, which increased the demand for heating fuels and contributed to the increase in the demand for electricity; (2) increased consumption of fossil fuels to generate electricity; and (3) a significant decrease (14.2 percent) in hydropower generation used to meet this demand.

Figures 3-1 through 3-3 illustrate the overall trends in total U.S. emissions by gas, annual changes, and absolute change since 1990. Table 3-2 provides a detailed summary of U.S. GHG emissions and sinks for 1990 through 2007.

Figure 3-4 illustrates the relative contribution of the direct GHGs to total U.S. emissions in 2007. The primary GHG emitted by human activities in the United States was CO_2, representing approximately 85.4 percent of total GHG emissions. The largest source of CO_2, and of overall U.S. anthropogenic GHG emissions, was fossil fuel combustion. CH_4 emissions,

---

[6] Global CO_2 emissions from fossil fuel combustion were taken from the U.S. Department of Energy, Energy Information Administration *International Energy Annual 2006* (U.S. DOE/EIA 2008b).

which have declined from 1990 levels, resulted primarily from enteric fermentation associated with domestic livestock, decomposition of wastes in landfills, and natural gas systems. Agricultural soil management and mobile source fuel combustion were the major sources of N_2O emissions. The emissions of substitutes for ODS and emissions of HFC-23 during the production of HCFC-22 were the primary contributors to aggregate HFC emissions. Electrical transmission and distribution systems accounted for most SF_6 emissions, while PFC emissions resulted as a by-product of primary aluminum production and from semiconductor manufacturing.

Overall, from 1990 to 2007, total emissions of CO_2 increased by 1,026.7 Tg CO_2 Eq. (20.2 percent), while CH_4 and N_2O emissions decreased by 31.2 Tg CO_2 Eq. (5.1 percent) and 3.1 Tg CO_2 Eq. (1.0 percent), respectively. During the same period, aggregate weighted emissions of HFCs, PFCs, and SF_6 rose by 59.0 Tg CO_2 Eq. (65.2 percent). From 1990 to 2007, HFCs increased by 88.6 Tg CO_2 Eq. (240.0 percent), PFCs decreased by 13.3 Tg CO_2 Eq. (64.0 percent), and SF_6 decreased by 16.3 Tg CO_2 Eq. (49.8 percent). Despite being emitted in smaller quantities relative to the other principal GHGs, emissions of HFCs, PFCs, and SF_6 are significant because many of them have extremely high GWPs. Conversely, U.S. GHG emissions were partly offset by carbon sequestration in forests, trees in urban areas, agricultural soils, and landfilled yard trimmings and food scraps, which, in aggregate, offset 14.9 percent of total emissions in 2007. The following sections describe each gas's contribution to total U.S. GHG emissions in more detail. Further information on source and sink categories, methods used to calculate emissions and fluxes, and trends across the entire time series may be found in the *Inventory of U.S. Greenhouse Gas Emissions and Sinks: 1990–2007* (U.S. EPA/OAP 2009).

Carbon Dioxide Emissions

The global carbon cycle is made up of large carbon flows and reservoirs. Billions of tons of carbon in the form of CO_2 are absorbed by oceans and living biomass (i.e., sinks) and are emitted to the atmosphere annually through natural processes (i.e., sources). When in equilibrium, carbon fluxes among these various reservoirs are roughly balanced. Since the Industrial Revolution, global atmospheric concentrations of CO_2 have risen about 36 percent (IPCC 2007), principally due to the combustion of fossil fuels. Within the United States, fuel combustion accounted for 94 percent of CO_2 emissions in 2007. Globally, approximately 29,195 Tg of CO_2 were added to the atmosphere through the combustion of fossil fuels in 2006, of which the United States accounted for about 20 percent.[6] Changes in land use and forestry practices can also emit CO_2 (e.g., through conversion of forest

Figure 3-1 Growth in U.S. Greenhouse Gas Emissions by Gas: 1990–2007

In 2007, total U.S. greenhouse gas emissions rose to 7,150.1 Tg CO_2 Eq., which was 17 percent above 1990 emissions, and 0.6 percent above 2005 emissions.

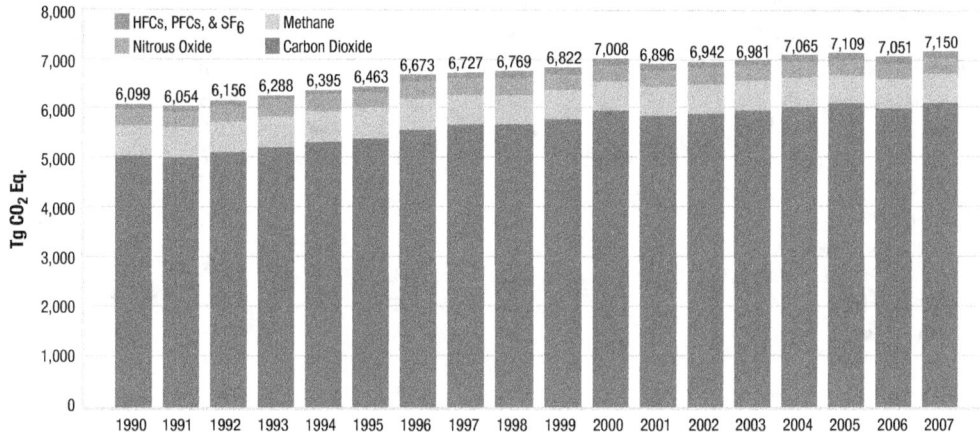

Figure 3-2 Annual Percentage Change in U.S. Greenhouse Gas Emissions: 1990–2007

Between 2005 and 2007, U.S. greenhouse gas emissions rose by 0.6 percent. The average annual rate of increase from 1990 through 2007 was 0.9 percent.

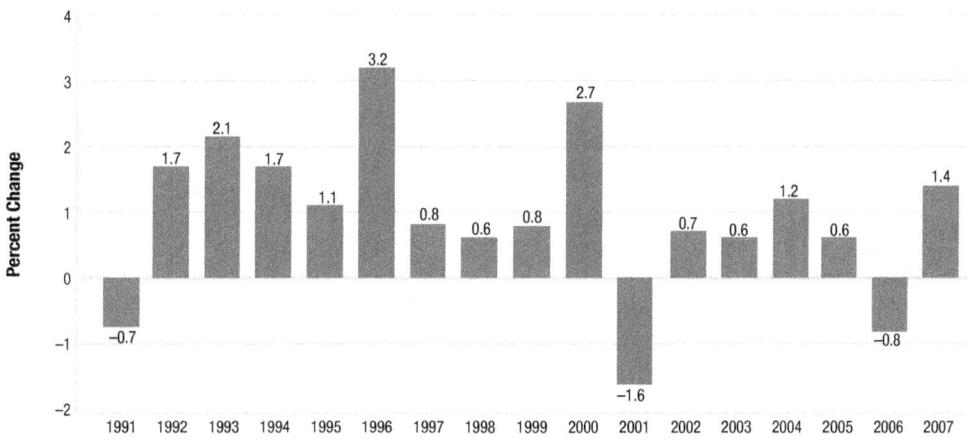

Figure 3-4 **2007 U.S. Greenhouse Gas Emissions by Gas** (presented on a CO_2 Eq. basis)

In 2007, CO_2 was the principal greenhouse gas emitted by human activities in the United States, driven primarily by emissions from fossil fuel combustion.

Figure 3-3 Cumulative Change in U.S. Greenhouse Gas Emissions: 1990–2007

From 1990 to 2007, total U.S. greenhouse gas emissions rose by 1,051 Tg CO_2 Eq., an increase of 17 percent. More recently, between 2005 and 2007, U.S. greenhouse gas emissions rose by 41 Tg CO_2 Eq., or 0.6 percent.

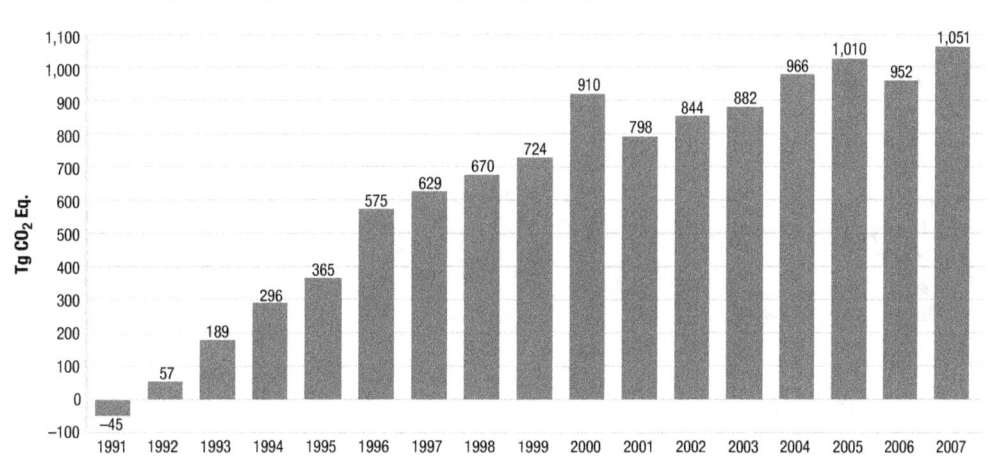

Table 3-2 **Recent Trends in U.S. Greenhouse Gas Emissions and Sinks** (Tg CO_2 Eq.)

In 2007, total U.S. greenhouse gas emissions were 7,150.1 Tg CO_2 Eq., representing a 17 percent rise since 1990.

Gas/Source	1990	1995	2000	2005	2006	2007
Carbon Dioxide (CO₂)	5,076.7	5,407.9	5,955.2	6,090.8	6,014.9	6,103.4
Fossil Fuel Combustion	4,708.9	5,013.9	5,561.5	5,723.5	5,635.4	5,735.8
Electricity Generation	1,809.7	1,938.9	2,283.2	2,381.0	2,327.3	2,397.2
Transportation	1,484.5	1,598.7	1,800.3	1,881.5	1,880.9	1,887.4
Industrial	834.2	862.6	844.6	828.0	844.5	845.4
Residential	337.7	354.4	370.4	358.0	321.9	340.6
Commercial	214.5	224.4	226.9	221.8	206.0	214.4
U.S. Territories	28.3	35.0	36.2	53.2	54.8	50.8
Non-Energy Use of Fuels	117.0	137.5	144.5	138.1	145.1	133.9
Iron and Steel Production & Metallurgical Coke Production	109.8	103.1	95.1	73.2	76.1	77.4
Cement Production	33.3	36.8	41.2	45.9	46.6	44.5
Natural Gas Systems	33.7	33.8	29.4	29.5	29.5	28.7
Incineration of Waste	10.9	15.7	17.5	19.5	19.8	20.8
Lime Production	11.5	13.3	14.1	14.4	15.1	14.6
Ammonia Production and Urea Consumption	16.8	17.8	16.4	12.8	12.3	13.8
Cropland Remaining Cropland	7.1	7.0	7.5	7.9	7.9	8.0
Limestone and Dolomite Use	5.1	6.7	5.1	6.8	8.0	6.2
Aluminum Production	6.8	5.7	6.1	4.1	3.8	4.3
Soda Ash Production and Consumption	4.1	4.3	4.2	4.2	4.2	4.1
Petrochemical Production	2.2	2.8	3.0	2.8	2.6	2.6
Titanium Dioxide Production	1.2	1.5	1.8	1.8	1.9	1.9
Carbon Dioxide Consumption	1.4	1.4	1.4	1.3	1.7	1.9
Ferroalloy Production	2.2	2.0	1.9	1.4	1.5	1.6
Phosphoric Acid Production	1.5	1.5	1.4	1.4	1.2	1.2
Wetlands Remaining Wetlands	1.0	1.0	1.2	1.1	0.9	1.0
Zinc Production	0.9	1.0	1.1	0.5	0.5	0.5
Petroleum Systems	0.4	0.3	0.3	0.3	0.3	0.3
Lead Production	0.3	0.3	0.3	0.3	0.3	0.3
Silicon Carbide Production and Consumption	0.4	0.3	0.2	0.2	0.2	0.2
Land Use, Land-Use Change, and Forestry (Sink)ᵃ	(841.4)	(851.0)	(717.5)	(1,122.7)	(1,050.5)	(1,062.6)
Biomass—Wood	215.2	229.1	218.1	208.9	209.9	209.8
International Bunker Fuels ᵇ	114.3	101.6	99.0	111.5	110.5	108.8
Biomass—Ethanol ᵇ	4.2	7.7	9.2	22.6	30.5	38.0
Methane (CH₄)	616.6	615.8	591.1	561.7	582.0	585.3
Enteric Fermentation	133.2	143.6	134.4	136.0	138.2	139.0
Landfills	149.2	144.3	122.3	127.8	130.4	132.9
Natural Gas Systems	129.6	132.6	130.8	106.3	104.8	104.7
Coal Mining	84.1	67.1	60.5	57.1	58.4	57.6
Manure Management	30.4	34.5	37.9	41.8	41.9	44.0
Forest Land Remaining Forest Land	4.6	6.1	20.6	14.2	31.3	29.0
Petroleum Systems	33.9	32.0	30.3	28.3	28.3	28.8
Wastewater Treatment	23.5	24.8	25.2	24.3	24.5	24.4
Stationary Combustion	7.4	7.1	6.6	6.7	6.3	6.6
Rice Cultivation	7.1	7.6	7.5	6.8	5.9	6.2
Abandoned Underground Coal Mines	6.0	8.2	7.4	5.6	5.5	5.7
Mobile Combustion	4.7	4.3	3.4	2.5	2.4	2.3
Composting	0.3	0.7	1.3	1.6	1.6	1.7
Petrochemical Production	0.9	1.1	1.2	1.1	1.0	1.0
Field Burning of Agricultural Residues	0.7	0.7	0.8	0.9	0.8	0.9
Iron and Steel Production & Metallurgical Coke Production	1.0	1.0	0.9	0.7	0.7	0.7
Ferroalloy Production	+	+	+	+	+	+
Silicon Carbide Production and Consumption	+	+	+	+	+	+
International Bunker Fuelsᵇ	0.2	0.1	0.1	0.1	0.1	0.1

Gas/Source	1990	1995	2000	2005	2006	2007
Nitrous Oxide (N₂O)	315.0	334.1	329.2	315.9	312.1	311.9
Agricultural Soil Management	200.3	202.3	204.5	210.6	208.4	207.9
Mobile Combustion	43.7	53.7	52.8	36.7	33.5	30.1
Nitric Acid Production	20.0	22.3	21.9	18.6	18.2	21.7
Manure Management	12.1	12.9	14.0	14.2	14.6	14.7
Stationary Combustion	12.8	13.3	14.5	14.8	14.5	14.7
Adipic Acid Production	15.3	17.3	6.2	5.9	5.9	5.9
Wastewater Treatment	3.7	4.0	4.5	4.8	4.8	4.9
N₂O from Product Uses	4.4	4.6	4.9	4.4	4.4	4.4
Forest Land Remaining Forest Land	0.5	0.8	2.4	1.8	3.5	3.3
Composting	0.4	0.8	1.4	1.7	1.8	1.8
Settlements Remaining Settlements	1.0	1.2	1.2	1.5	1.5	1.6
Field Burning of Agricultural Residues	0.4	0.4	0.5	0.5	0.5	0.5
Incineration of Waste	0.5	0.5	0.4	0.4	0.4	0.4
Wetlands Remaining Wetlands	+	+	+	+	+	+
International Bunker Fuels[b]	*1.1*	*0.9*	*0.9*	*1.0*	*1.0*	*1.0*
Hydrofluorocarbons (HFCs)	36.9	61.8	100.1	116.1	119.1	125.5
Substitution of Ozone-Depleting Substances[c]	0.3	28.5	71.2	100.0	105.0	108.3
HCFC-22 Production	36.4	33.0	28.6	15.8	13.8	17.0
Semiconductor Manufacture	0.2	0.3	0.3	0.2	0.3	0.3
Perfluorocarbons (PFCs)	20.8	15.6	13.5	6.2	6.0	7.5
Aluminum Production	18.5	11.8	8.6	3.0	2.5	3.8
Semiconductor Manufacture	2.2	3.8	4.9	3.2	3.5	3.6
Sulfur Hexafluoride (SF₆)	32.8	28.1	19.2	17.9	17.0	16.5
Electrical Transmission and Distribution	26.8	21.6	15.1	14.0	13.2	12.7
Magnesium Production and Processing	5.4	5.6	3.0	2.9	2.9	3.0
Semiconductor Manufacture	0.5	0.9	1.1	1.0	1.0	0.8
Total	6,098.7	6,463.3	7,008.2	7,108.6	7,051.1	7,150.1
Net Emissions (Sources and Sinks)	5,257.3	5,612.3	6,290.7	5,985.9	6,000.6	6,087.5

Tg CO₂ Eq. = teragrams of carbon dioxide equivalents; HCFC = hydrochlorofluorocarbon.

+ Does not exceed 0.05 Tg CO₂ Eq.

[a] Parentheses indicate negative values or sequestration. The net CO₂ flux total includes both emissions and sequestration, and constitutes a sink in the United States. Sinks are only included in the net emissions total.

[b] Emissions from International Bunker Fuels and Biomass Combustion are not included in the totals.

[c] Small amounts of PFC emissions also result from this source.

Note: Totals may not sum due to independent rounding. One teragram equals one million metric tons.

land to agricultural or urban use), or can act as a sink for CO₂ (e.g., through net additions to forest biomass) (Figure 3-5).

As the largest source of U.S. anthropogenic GHG emissions, CO₂ from fossil fuel combustion has accounted for approximately 79 percent of total GWP-weighted emissions since 1990, growing slowly from 77 percent of total GWP-weighted emissions in 1990 to 80 percent in 2007. Emissions of CO₂ from fossil fuel combustion increased at an average annual rate of 1.3 percent from 1990 to 2007. The fundamental factors influencing this trend include a generally growing domestic economy over the last 17 years, and the subsequent significant overall growth in emissions from electricity generation and transportation activities. Between 1990 and 2007, CO₂ emissions from fossil fuel combustion increased from 4,708.9 Tg CO₂

Eq. to 5,735.8 Tg CO₂ Eq.—a 21.8 percent total increase over the 18-year period. From 2006 to 2007, these emissions increased by 100.4 Tg CO₂ Eq. (1.8 percent).

Historically, changes in emissions from fossil fuel combustion have been the dominant factor affecting U.S. emission trends. Changes in CO₂ emissions from fossil fuel combustion are influenced by many long-term and short-term factors, including population and economic growth, energy price fluctuations, technological changes, and seasonal temperatures. On an annual basis, the overall consumption of fossil fuels in the United States generally fluctuates in response to changes in general economic conditions, energy prices, weather, and the availability of non-fossil alternatives. For example, in a year with increased consumption of goods and services, low fuel prices, severe summer and

Figure 3-5 2007 U.S. Emissions of Carbon Dioxide by Source

In 2007, CO_2 accounted for 85.4 percent of U.S. greenhouse gas emissions, with fossil fuel combustion accounting for 80.2 percent of emissions on a global warming potential (GWP)-weighted basis.

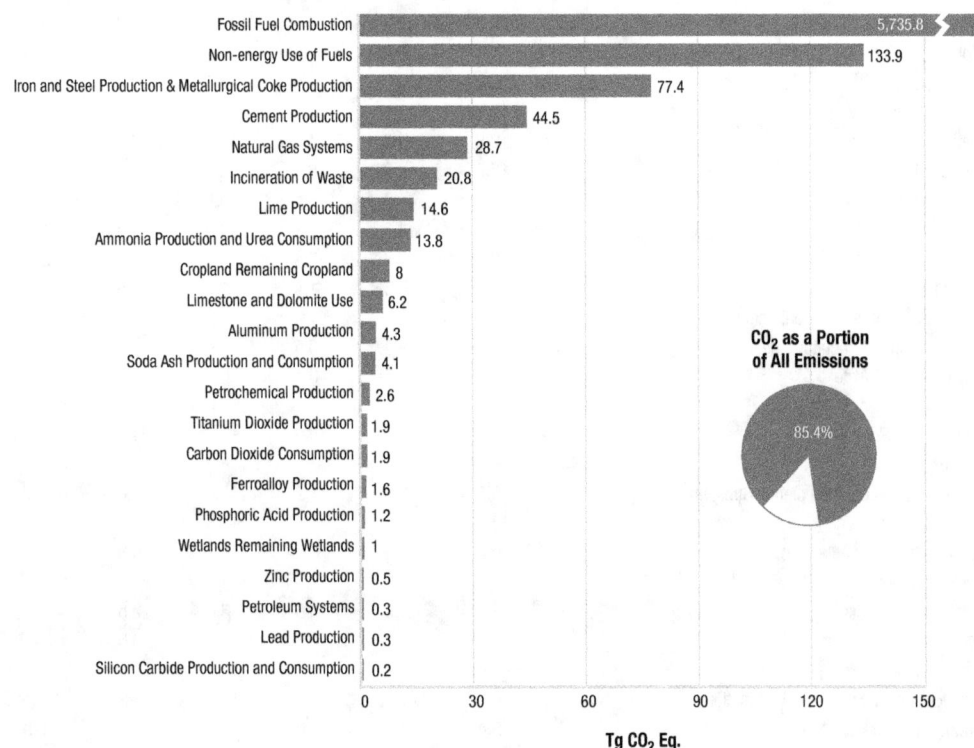

winter weather conditions, nuclear plant closures, and lower precipitation feeding hydroelectric dams, there would likely be proportionally greater fossil fuel consumption than in a year with poor economic performance, high fuel prices, mild temperatures, and increased output from nuclear and hydroelectric plants.

The five major fuel-consuming sectors contributing to CO_2 emissions from fossil fuel combustion are electricity generation, transportation, industrial, residential, and commercial. CO_2 emissions are produced by the electricity generation sector as it consumes fossil fuel to provide electricity to one of the other four sectors, or "end-use" sectors. In the following discussion, emissions from electricity generation have been distributed to each end-use sector on the basis of each sector's share of aggregate electricity consumption. The distribution of the electricity-related emissions assumes that each end-use sector consumes electricity that is generated from the national average mix of fuels according to their carbon intensity. Emissions from electricity generation are also addressed separately after the end-use sectors have been discussed. Note that emissions from U.S. territories are calculated separately due to a lack of specific consumption data for the individual end-use sectors.

Figures 3-6 and 3-7 summarize CO_2 emissions from fossil fuel combustion by sector and fuel type and by end-use sector.

Transportation End-Use Sector

Transportation activities (excluding international bunker fuels) accounted for 33 percent of CO_2 emissions from fossil fuel combustion in 2007.[7] Virtually all of the energy consumed in this end-use sector came from petroleum products. Nearly 60 percent of the emissions resulted from gasoline consumption for personal vehicle use. The remaining emissions came from other transportation activities, including the combustion of diesel fuel in heavy-duty vehicles and jet fuel in aircraft.

Industrial End-Use Sector

Industrial CO_2 emissions, resulting both directly from the combustion of fossil fuels and indirectly from the generation of electricity that is consumed by industry, accounted for 27 percent of CO_2 from fossil fuel combustion in 2007. Just over half of these emissions resulted from direct fossil fuel combustion to produce steam and/or heat for industrial processes. The remaining emissions resulted from consuming electricity for motors, electric furnaces, ovens, lighting, and other applications.

[7] If emissions from international bunker fuels are included, the transportation end-use sector accounted for 35 percent of U.S. emissions from fossil fuel combustion in 2007.

Residential and Commercial End-Use Sectors

The residential and commercial end-use sectors accounted for 21 and 18 percent, respectively, of CO_2 emissions from fossil fuel combustion in 2007. Both sectors relied heavily on electricity for meeting energy demands, with 72 and 79 percent, respectively, of their emissions attributable to electricity consumption for lighting, heating, cooling, and operating appliances. The remaining emissions were due to the consumption of natural gas and petroleum for heating and cooking.

Electricity Generation

The United States relies on electricity to meet a significant portion of its energy demands, especially for lighting, electric motors, heating, and air conditioning. Electricity generators consumed 36 percent of U.S. energy from fossil fuels and emitted 42 percent of the CO_2 from fossil fuel combustion in 2007.[8] The type of fuel combusted by electricity generators has a significant effect on their emissions. For example, some electricity is generated with low-CO_2-emitting energy technologies, particularly non-fossil options, such as nuclear, hydroelectric, wind, or geothermal energy. However, electricity generators rely on coal for over half of their total energy requirements and accounted for 94 percent of all coal consumed for energy in the United States in 2007. Consequently, changes in electricity demand have a significant impact on coal consumption and associated CO_2 emissions.

Other significant CO_2 trends included the following:

- CO_2 emissions from non-energy use of fossil fuels increased by 16.9 Tg CO_2 Eq. (14.5 percent) from 1990 through 2007. Emissions from non-energy uses of fossil fuels were 133.9 Tg CO_2 Eq. in 2007, which constituted 2.2 percent of total national CO_2 emissions, approximately the same proportion as in 1990.

- CO_2 emissions from iron and steel production and metallurgical coke production increased slightly from 2006 to 2007 (1.3 Tg CO_2 Eq.), but decreased by 29.5 percent to 77.4 Tg CO_2 Eq. from 1990 through 2007, due to restructuring of the industry, technological improvements, and increased scrap utilization.

- In 2007, CO_2 emissions from cement production decreased slightly by 2.0 Tg CO_2 Eq. (4.4 percent) from 2006 to 2007. This decrease occurred despite the overall increase over the time series. After falling in 1991 by 2 percent from 1990 levels, cement production emissions grew every year through 2006. Overall, from 1990 to 2007, emissions from cement production increased by 34 percent, an increase of 11.2 Tg CO_2 Eq.

- CO_2 emissions from incineration of waste (20.8 Tg CO_2 Eq. in 2007) increased by 9.8 Tg CO_2 Eq. (90 percent) from 1990 through 2007, as the volume of plastics and other fossil carbon-containing materials in the waste stream grew.

- Net CO_2 sequestration from land use, land-use change, and forestry increased by 221.1 Tg CO_2 Eq. (26 percent) from 1990 through 2007. This increase was primarily due to growth in the rate of net carbon accumulation in forest carbon stocks, particularly in above-ground and below-ground tree biomass. Annual carbon accumulation in landfilled yard trimmings and food scraps slowed over this period, while the rate of carbon accumulation in urban trees increased.

[8] The emissions of CO_2 attributable to electricity generation were allocated to the end-use sectors, as described in the end-use sector discussions.

Figure 3-6 **2007 U.S. CO_2 Emissions from Fossil Fuel Combustion by Sector and Fuel Type**

In 2007, U.S. transportation sector emissions were primarily from petroleum consumption, while electricity generation emissions were primarily from coal consumption.

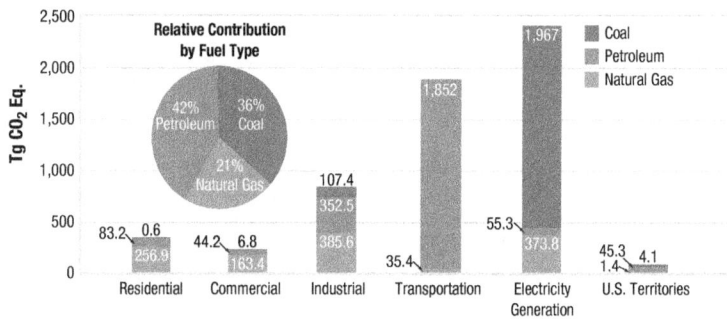

Note: Electricity generation also includes emissions of less than 1 Tg CO_2 Eq. from geothermal-based electricity generation.

Figure 3-7 **2007 U.S. End-Use Sector CO_2 Emissions from Fossil Fuel Combustion**

In 2007, direct fossil fuel combustion accounted for the vast majority of fossil fuel-related CO_2 emissions from the transportation sector (mostly petroleum combustion). Electricity consumption indirectly accounted for most of the fossil fuel-related CO_2 emissions from the commercial and residential sectors.

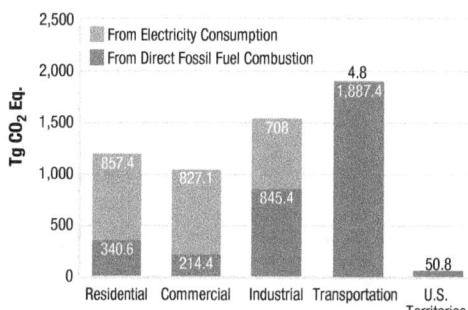

Methane Emissions

According to the IPCC, CH_4 is more than 20 times as effective as CO_2 at trapping heat in the atmosphere. Over the last 250 years, the concentration of CH_4 in the atmosphere increased by 148 percent (IPCC 2007). Anthropogenic sources of CH_4 include landfills, natural gas and petroleum systems, agricultural activities, coal mining, wastewater treatment, stationary and mobile combustion, and certain industrial processes (see Figure 3-8). Some significant trends in U.S. emissions of CH_4 include the following:

- Enteric fermentation is the largest anthropogenic source of CH_4 emissions in the United States. In 2007, enteric fermentation CH_4 emissions were 139.0 Tg CO_2 Eq. (approximately 24 percent of total CH_4 emissions), which represents an increase of 5.8 Tg CO_2 Eq., or 4.3 percent, since 1990.

- Landfills are the second-largest anthropogenic source of CH_4 emissions in the United States, accounting for approximately 23 percent of total CH_4 emissions (132.9 Tg CO_2 Eq.) in 2007. From 1990 to 2007, net CH_4 emissions from landfills decreased by 16.3 Tg CO_2 Eq. (11 percent), with small increases occurring in some interim years, including 2007. This downward trend in overall emissions is the result of increases in the amount of landfill gas collected and combusted,9 which has more than offset the additional CH_4 emissions resulting from an increase in the amount of municipal solid waste landfilled.

- CH_4 emissions from natural gas systems were 104.7 Tg CO_2 Eq. in 2007. Emissions have declined by 24.9 Tg CO_2 Eq. (19 percent) since 1990, due to improvements in technology and management practices, as well as some replacement of old equipment.

- In 2007, CH_4 emissions from coal mining were 57.6 Tg CO_2 Eq., a 0.8 Tg CO_2 Eq. (1.3 percent) decrease over 2006 emission levels. The overall decline of 26.4 Tg CO_2 Eq. (31 percent) from 1990 results from the mining of less gassy coal from underground mines and the increased use of CH_4 collected from degasification systems.

- CH_4 emissions from manure management increased by 44.7 percent, from 30.4 Tg CO_2 Eq. in 1990 to 44.0 Tg CO_2 Eq. in 2007. The majority of this increase was from swine and dairy cow manure, since the general trend in manure management is increasing use of liquid systems, which tends to produce higher CH_4 emissions. The increase in liquid systems is the combined result of a shift to larger facilities, and to facilities in the U.S. West and Southwest, all of which tend to use liquid systems. Also, new regulations limiting the application of manure nutrients have shifted manure management practices at smaller dairies from daily spread to manure managed and stored on site.

Figure 3-8 2007 U.S. Emissions of Methane by Source

In 2007, methane (CH_4) accounted for 8.2 percent of U.S. greenhouse gas emissions on a GWP-weighted basis. Enteric fermentation was the largest source of U.S. CH_4 emissions (24 percent), followed closely by emissions from landfills (23 percent) and natural gas systems (18 percent).

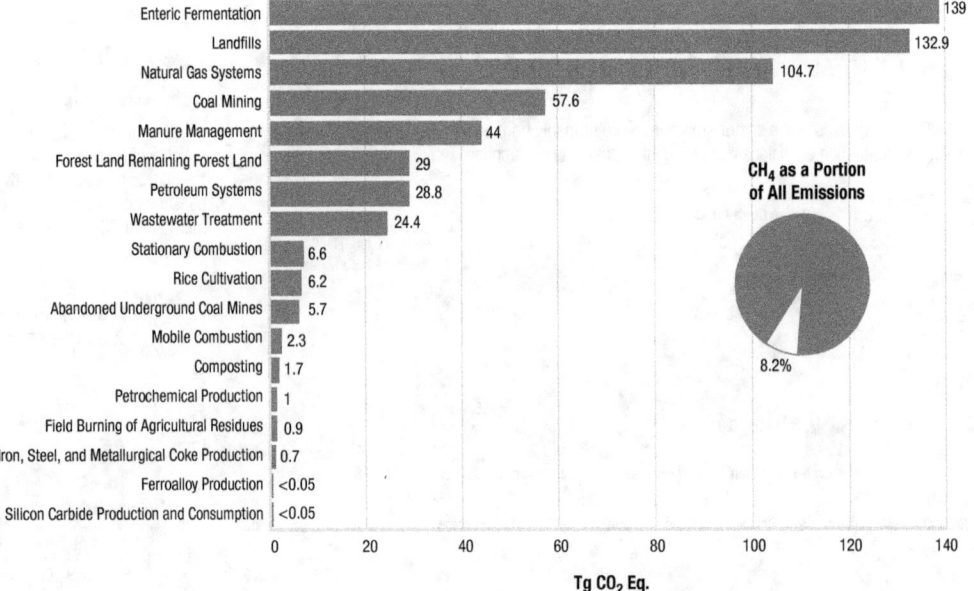

9 The CO_2 produced from combusted CH_4 at landfills is not counted in national inventories, as it is considered part of the natural carbon cycle of decomposition.

Nitrous Oxide Emissions

N_2O is produced by biological processes that occur in soil and water and by a variety of anthropogenic activities in the agricultural, energy-related, industrial, and waste management fields. While total N_2O emissions are much lower than CO_2 emissions, N_2O is approximately 300 times more powerful than CO_2 at trapping heat in the atmosphere. Since 1750, the global atmospheric concentration of N_2O has risen by approximately 18 percent (IPCC 2007). The main anthropogenic activities producing N_2O in the United States are agricultural soil management, fuel combustion in motor vehicles, nitric acid production, stationary fuel combustion, manure management, and adipic acid production (see Figure 3-9).

Some significant trends in U.S. emissions of N_2O include the following:

- Agricultural soils produced approximately 67 percent of N_2O emissions in the United States in 2007. Estimated emissions from this source in 2007 were 207.9 Tg CO_2 Eq. Annual N_2O emissions from agricultural soils fluctuated between 1990 and 2007, although overall emissions were 3.8 percent higher in 2007 than in 1990. N_2O emissions from this source have not shown any significant long-term trend, as they are highly sensitive to the amount of nitrogen applied to soils, which has not changed significantly over the time period, and to weather patterns and crop type.

- In 2007, N_2O emissions from mobile combustion were 30.1 Tg CO_2 Eq. (approximately 10 percent of U.S. N_2O emissions). From 1990 to 2007, N_2O emissions from mobile combustion decreased by 31 percent. However, from 1990 to 1998, emissions increased by 26 percent, due to control technologies that reduced NO_x emissions while increasing N_2O emissions. Since 1998, newer control technologies have led to a steady decline in N_2O from this source.

- N_2O emissions from adipic acid production were 5.9 Tg CO_2 Eq. in 2007, and have decreased significantly since 1996 from the widespread installation of pollution control measures. Emissions from adipic acid production have decreased by 61 percent since 1990, and emissions from adipic acid production have fluctuated by less than 1.2 Tg CO_2 Eq. annually since 1998.

HFC, PFC, and SF_6 Emissions

HFCs and PFCs are families of synthetic chemicals that are used as alternatives to the ODS, and along with SF_6, are potent GHGs. SF_6 and PFCs have extremely long atmospheric lifetimes, contributing to their high GWP values and resulting in their essentially irreversible accumulation in the atmosphere once

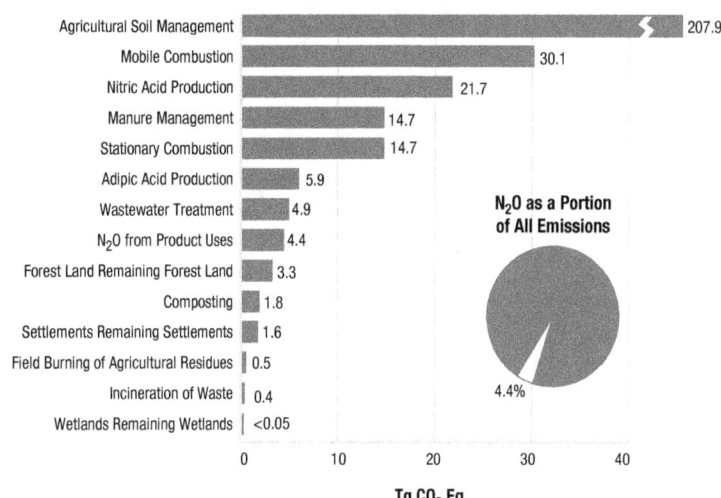

Figure 3-9 **2007 U.S. Emissions of Nitrous Oxide by Source**

In 2007, nitrous oxide (N_2O) accounted for 4.4 percent of U.S. greenhouse gas emissions on a GWP-weighted basis. Agricultural soil management was the largest U.S. source of N_2O, producing 67 percent of emissions.

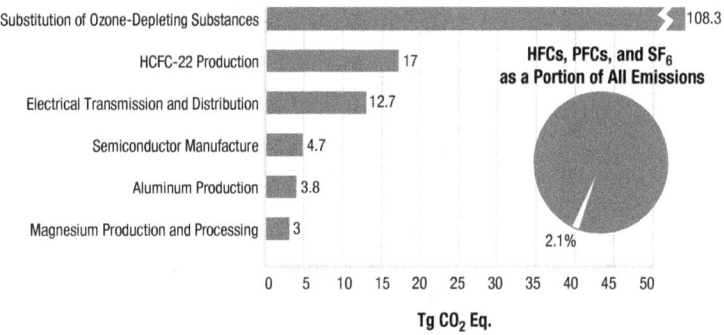

Figure 3-10 **2007 U.S. Emissions of HFCs, PFCs, and SF_6 by Source**

In 2007, HFCs, PFCs, and SF_6 accounted for 2.1 percent of U.S. greenhouse gas emissions on a GWP-weighted basis. Although the mass of these gases emitted is comparatively small, these emissions have high global warming potentials, and therefore have significant climate impacts.

emitted. SF_6 is the most potent GHG the IPCC has evaluated.

Other emissive sources of these gases include HCFC-22 production, electrical transmission and distribution systems, semiconductor manufacturing, aluminum production, and magnesium production and processing (see Figure 3-10).

Some significant trends in U.S. HFC, PFC, and SF_6 emissions include the following:

- Emissions resulting from the substitution of ODS (e.g., CFCs) have been increasing from small amounts in 1990 to 108.3 Tg CO_2 Eq. in 2007. Emissions from substitutes for ODS are both the largest and the fastest-growing source of HFC, PFC, and SF_6 emissions. These emissions have been increasing as phase-outs required under the Montreal Protocol come into effect, especially after 1994 when full market penetration was made for the first generation of new technologies featuring ODS substitutes.

- HFC emissions from the production of HCFC-22 decreased by 53 percent (19.4 Tg CO_2 Eq.) from 1990 through 2007, due to a steady decline in the emission rate of HFC-23 (i.e., the amount of HFC-23 emitted per kilogram of HCFC-22 manufactured) and the use of thermal oxidation at some plants to reduce HFC-23 emissions.

- SF_6 emissions from electric power transmission and distribution systems decreased by 53 percent (14.1 Tg CO_2 Eq.) from 1990 to 2007, primarily because of higher purchase prices for SF_6 and efforts by industry to reduce emissions.

- PFC emissions from aluminum production decreased by 79 percent (14.7 Tg CO_2 Eq.) from 1990 to 2007, due to both industry emission reduction efforts and lower domestic aluminum production.

OVERVIEW OF SECTOR EMISSIONS AND TRENDS

In accordance with the *Revised 1996 IPCC Guidelines for National Greenhouse Gas Inventories* (IPCC/UNEP/OECD/IEA 1997), and the 2003 *UNFCCC*

Guidelines on Reporting and Review (UNFCCC 2003), the *Inventory of U.S. Greenhouse Gas Emissions and Sinks: 1990–2007* (U.S. EPA/OAP 2009) is segregated into six sector-specific chapters. Figure 3-11 and Table 3-3 aggregate emissions and sinks by these chapters. Emissions of all gases can be summed from each source category using IPCC guidance. Over the 18-year period from 1990 to 2007, total emissions in the energy, industrial processes, and agriculture sectors climbed by 976.7 Tg CO_2 Eq. (19 percent), 28.5 Tg CO_2 Eq. (9 percent), and 28.9 Tg CO_2 Eq. (8 percent), respectively. Emissions decreased in the waste and the solvent and other product use sectors by 11.5 Tg CO_2 Eq. (6 percent) and less than 0.1 Tg CO_2 Eq. (0.4 percent), respectively. Over the same period, estimates of net carbon sequestration in the land use, land-use change, and forestry sector increased by 192.5 Tg CO_2 Eq. (23 percent).

Energy

The Energy chapter of the *Inventory of U.S. Greenhouse Gas Emissions and Sinks* contains emissions of all GHGs resulting from stationary and mobile energy activities, including fuel combustion and fugitive fuel emissions. Energy-related activities, primarily fossil fuel combustion, accounted for the vast majority of U.S. CO_2 emissions from 1990 through 2007. In 2007, approximately 85 percent of the energy consumed in the United States (on a British thermal unit basis) was pro-

Figure 3-11 **Recent Trends in U.S. Greenhouse Gas Emissions and Sinks by Chapter/IPCC Sector**

From 1990 to 2007, total emissions in the energy, industrial processes, and agriculture sectors climbed by 19 percent, 9 percent, and 8 percent, respectively. Over the same period, carbon uptake by the land use, land-use change, and forestry sector increased by 23 percent.

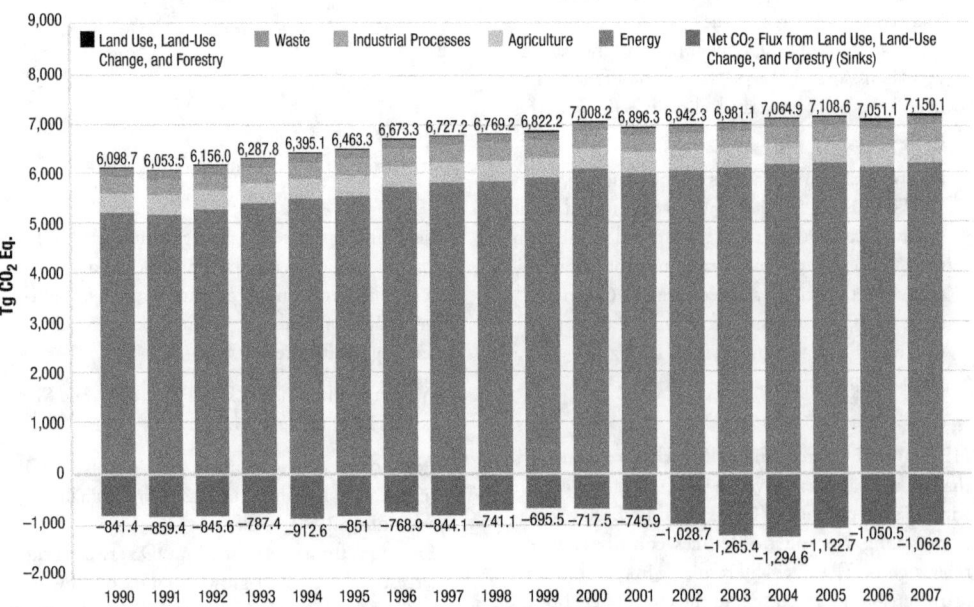

Note: Relatively smaller amounts of global warming potential-weighted emissions are also emitted from the Solvent and Other Product Use sectors.

Table 3-3 **Recent Trends in U.S. Greenhouse Gas Emissions and Sinks by Chapter/IPCC Sector** (Tg CO₂ Eq.)

From 1990 to 2007, total emissions in the energy, industrial processes, and agriculture sectors increased, emissions in the waste and the solvent and other product use sectors decreased, and net carbon sequestration in the land use, land-use change, and forestry sector rose by 23 percent.

Chapter/IPCC Sector	1990	1995	2000	2005	2006	2007
Energy	5,193.6	5,520.1	6,059.9	6,169.2	6,084.4	6,170.3
Industrial Processes	325.2	345.8	356.3	337.6	343.9	353.8
Solvent and Other Product Use	4.4	4.6	4.9	4.4	4.4	4.4
Agriculture	384.2	402.0	399.4	410.8	410.3	413.1
Land Use, Land-Use Change, and Forestry (Emissions)	14.2	16.2	33.0	26.4	45.1	42.9
Waste	177.1	174.7	154.6	160.2	163.0	165.6
Total Emissions	6,098.7	6,463.3	7,008.2	7,108.6	7,051.1	7,150.1
Net CO₂ Flux from Land Use, Land-Use Change, and Forestry (Sinks)*	(841.4)	(851.0)	(717.5)	(1,122.7)	(1,050.5)	(1,062.6)
Net Emissions (Sources and Sinks)	5,257.3	5,612.3	6,290.7	5,985.9	6,000.6	6,087.5

* The net CO₂ flux total includes both emissions and sequestration, and constitutes a sink in the United States. Sinks are only included in the net emissions total.

IPCC = Intergovernmental Panel on Climate Change; Tg CO₂ Eq. = teragrams of carbon dioxide equivalents.

Note: Totals may not sum due to independent rounding. Parentheses indicate negative values or sequestration.

duced through the combustion of fossil fuels. The remaining 15 percent came from other energy sources, such as hydropower, biomass, nuclear, wind, and solar energy (see Figure 3-12). Energy-related activities are also responsible for CH₄ and N₂O emissions (35 percent and 14 percent of total U.S. emissions of each gas, respectively). Overall, emission sources in the Energy chapter accounted for a combined 86.3 percent of total U.S. GHG emissions in 2007.

Industrial Processes

The Industrial Processes chapter of the *Inventory of U.S. Greenhouse Gas Emissions and Sinks* contains by-product or fugitive emissions of GHGs from industrial processes not directly related to energy activities, such as fossil fuel combustion. For example, industrial processes can chemically transform raw materials, which often release waste gases, such as CO₂, CH₄, and N₂O. These processes include iron and steel production and metallurgical coke production, cement production, ammonia production and urea consumption, lime manufacture, limestone and dolomite use (e.g., flux stone, flue gas desulfurization, and glass manufacturing), soda ash manufacture and use, titanium dioxide production, phosphoric acid production, ferroalloy production, CO₂ consumption, silicon carbide production and consumption, aluminum production, petrochemical production, nitric acid production, adipic acid production, lead production, and zinc production. Additionally, emissions from industrial processes release HFCs, PFCs, and SF₆. Overall, emission sources in the Industrial Process chapter accounted for 4.9 percent of U.S. GHG emissions in 2007.

Solvent and Other Product Use

The Solvent and Other Product Use chapter of the *Inventory of U.S. Greenhouse Gas Emissions and Sinks*

contains GHG emissions that are produced as a by-product of various solvent and other product uses. In the United States, emissions from N₂O from product uses—the only source of GHG emissions from this sector—accounted for less than 0.1 percent of total U.S. anthropogenic GHG emissions on a carbon-equivalent basis in 2007.

Agriculture

The Agriculture chapter of the *Inventory of U.S. Greenhouse Gas Emissions and Sinks* contains anthropogenic emissions from agricultural activities (except fuel combustion, which is addressed in the Energy chapter, and agricultural CO₂ fluxes, which are addressed in the Land Use, Land-Use Change, and Forestry chapter). Agricultural activities contribute directly to emissions of GHGs through a variety of processes, including the following source categories: enteric fermentation in domestic livestock, livestock manure management, rice cultivation, agricultural soil management, and field burning of agricultural residues. CH₄ and N₂O were the primary GHGs emitted by agricultural activities. In 2007, CH₄ emissions from enteric fermentation and manure management represented about 24 percent and 8 percent of total CH₄ emissions from anthropogenic activities, respectively. Agricultural soil management activities, such as fertilizer application and other cropping practices, were the largest source of U.S. N₂O emissions in 2007, accounting for 67 percent. In 2007, emission sources accounted for in the Agriculture chapter were responsible for 6 percent of total U.S. GHG emissions.

Land Use, Land-Use Change, and Forestry

The Land Use, Land-Use Change, and Forestry chapter of the *Inventory of U.S. Greenhouse Gas Emissions and Sinks* contains emissions of CH₄ and N₂O, and emissions and removals of CO₂ from forest manage-

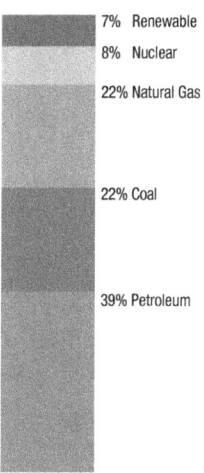

Figure 3-12 **2007 U.S. Energy Consumption by Energy Source**

In 2007, the combustion of fossil fuels accounted for approximately 85 percent of U.S. energy consumption, with the remaining 15 percent coming from other sources (nuclear, hydropower, wind, etc.).

7% Renewable

8% Nuclear

22% Natural Gas

22% Coal

39% Petroleum

ment, other land-use activities, and land-use change. Forest management practices, tree planting in urban areas, the management of agricultural soils, and the landfilling of yard trimmings and food scraps have resulted in a net uptake (sequestration) of carbon in the United States. Forests (including vegetation, soils, and harvested wood) accounted for approximately 86 percent of total 2007 net CO_2 flux, urban trees accounted for 9 percent, mineral and organic soil carbon stock changes accounted for 4 percent, and landfilled yard trimmings and food scraps accounted for 1 percent.

The net forest sequestration is a result of net forest growth and increasing forest area, as well as a net accumulation of carbon stocks in harvested wood pools. The net sequestration in urban forests is a result of net tree growth in these areas. In agricultural soils, mineral and organic soils sequester approximately 70 percent more carbon than is emitted through these soils, liming, and urea fertilization, combined. The mineral soil carbon sequestration is largely due to the conversion of cropland to permanent pastures and hay production, a reduction in summer fallow areas in semi-arid areas, an increase in the adoption of conservation tillage practices, and an increase in the amounts of organic fertilizers (i.e., manure and sewage sludge) applied to agriculture lands. The landfilled yard trimmings and food scraps net sequestration is due to the long-term accumulation of yard trimming carbon and food scraps in landfills.

Land use, land-use change, and forestry activities in 2007 resulted in a net carbon sequestration of 1,062.6 Tg CO_2 Eq. This represents an offset of approximately 17.4 percent of total U.S. CO_2 emissions, or 14.9 percent of total GHG emissions in 2007. Between 1990 and 2007, total land use, land-use change, and forestry net carbon flux resulted in a 26.3 percent increase in CO_2 sequestration, primarily due to an increase in the rate of net carbon accumulation in forest carbon stocks, particularly in above-ground and below-ground tree biomass. Annual carbon accumulation in landfilled yard trimmings and food scraps slowed over this period, while the rate of annual carbon accumulation increased in urban trees.

The application of crushed limestone and dolomite to managed land (i.e., soil liming) and urea fertilization resulted in CO_2 emissions of 8.0 Tg CO_2 Eq. in 2007, an increase of 13 percent relative to 1990. The application of synthetic fertilizers to forest and settlement soils in 2007 resulted in direct N_2O emissions of 1.6 Tg CO_2 Eq. Direct N_2O emissions from fertilizer application increased by approximately 61 percent between 1990 and 2007. Non-CO_2 emissions from forest fires in 2007 resulted in CH_4 emissions of 29.0 Tg CO_2 Eq., and in N_2O emissions of 2.9 Tg CO_2 Eq. CO_2 and N_2O emissions from peatlands in 2007

totaled 1.0 Tg CO_2 Eq. and less than 0.01 Tg CO_2 Eq., respectively.

Waste

The Waste chapter of the *Inventory of U.S. Greenhouse Gas Emissions and Sinks* contains emissions from waste management activities (except incineration of waste, which is addressed in the Energy chapter). Landfills were the largest source of anthropogenic CH_4 emissions in the Waste chapter, accounting for 23 percent of total U.S. CH_4 emissions.[10] Additionally, wastewater treatment accounted for 4 percent of U.S. CH_4 emissions. N_2O emissions from the discharge of wastewater treatment effluents into aquatic environments were estimated, as were N_2O emissions from the treatment process itself. Emissions of CH_4 and N_2O from composting grew from 1990 to 2007, and resulted in emissions of 1.7 Tg CO_2 Eq. and 1.8 Tg CO_2 Eq., respectively. Overall, in 2007, emission sources accounted for in the Waste chapter generated 2.3 percent of total U.S. GHG emissions.

EMISSIONS BY ECONOMIC SECTOR

Throughout the *Inventory of U.S. Greenhouse Gas Emissions and Sinks* report, emission estimates are grouped into six sectors (i.e., chapters) defined by the IPCC: Energy; Industrial Processes; Solvent Use; Agriculture; Land Use, Land-Use Change, and Forestry; and Waste (U.S. EPA/OAP 2009). While it is important to use this characterization for consistency with UNFCCC reporting guidelines, it is also useful to allocate emissions into more commonly used sectoral categories. This section reports emissions by the following economic sectors: residential, commercial, industry, transportation, electricity generation, agriculture, and U.S. territories. Table 3-4 summarizes emissions from each of these sectors, and Figure 3-13 shows the trend in emissions by sector from 1990 to 2007.

Using this categorization, emissions from electricity generation accounted for the largest portion (34 percent) of U.S. GHG emissions in 2007, while transportation activities, in aggregate, accounted for 28 percent, and emissions from industry accounted for 20 percent. In contrast to electricity generation and transportation, emissions from industry have in general declined over the past decade, due to structural changes in the U.S. economy (i.e., shifts from a manufacturing-based to a service-based economy), fuel switching, and energy efficiency improvements. The remaining 18 percent of U.S. GHG emissions were from the residential, agriculture, and commercial sectors, plus emissions from U.S. territories. The residential sector accounted for about 5 percent, and primarily consisted of CO_2 emissions from fossil fuel combustion. Activities related to agriculture accounted for roughly 7 percent of U.S. emissions; unlike

[10] Landfills also store carbon, due to incomplete degradation of organic materials, such as wood products and yard trimmings, as described in the Land-Use, Land-Use Change, and Forestry chapter of the *Inventory* report.

Table 3-4 **U.S. Greenhouse Gas Emissions Allocated to Economic Sectors** (Tg CO_2 Eq.)

In 2007, electricity generation accounted for 34 percent of total U.S. greenhouse gas emissions, transportation accounted for 28 percent, and Industry accounted for 30 percent.

Economic Sectors	1990	1995	2000	2005	2006	2007
Electric Power Industry	1,859.1	1,989.0	2,329.3	2,429.4	2,375.5	2,445.1
Transportation	1,543.6	1,685.2	1,919.7	1,998.9	1,994.4	1,995.2
Industry	1,496.0	1,524.5	1,467.5	1,364.9	1,388.4	1,386.3
Agriculture	428.5	453.7	470.2	482.6	502.9	502.8
Commercial	392.9	401.0	388.2	401.8	392.6	407.6
Residential	344.5	368.8	386.0	370.5	334.9	355.3
U.S. Territories	34.1	41.1	47.3	60.5	62.3	57.7
Total Emissions	6,098.7	6,463.3	7,008.2	7,108.6	7,051.1	7,150.1
Land Use, Land-Use Change, and Forestry (Sinks)	(841.4)	(851.0)	(717.5)	(1,122.7)	(1,050.5)	(1,062.6)
Net Emissions (Sources and Sinks)	5,257.3	5,612.3	6,290.7	5,985.9	6,000.6	6,087.5

Tg CO_2 Eq. = teragrams of carbon dioxide equivalents.

Note: Totals may not sum due to independent rounding. Emissions include carbon dioxide, methane, nitrous oxide, hydrofluorocarbons, perfluorocarbons, and sulfur hexafluoride.

Figure 3-13 **U.S. Greenhouse Gas Emissions Allocated to Economic Sectors: 1990–2007**

In 2007, electricity generation accounted for 34 percent of total U.S. greenhouse gas emissions, transportation accounted for 28 percent, and industry accounted for 20 percent.

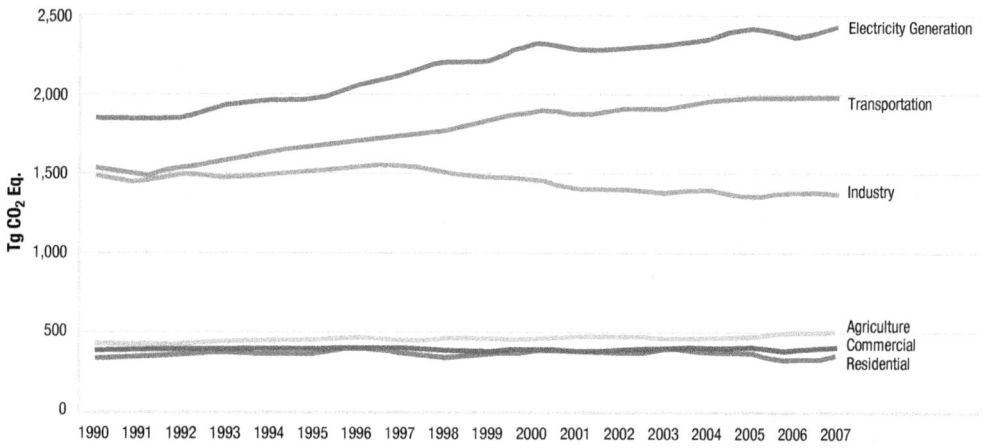

Table 3-5 **U.S. Greenhouse Gas Emissions by Economic Sector, with Electricity-Related Emissions Distributed Among Sectors** (Tg CO_2 Eq.)

In 2007, after distributing emissions from electricity generation to end-use sectors, industry accounted for 30 percent of total U.S. greenhouse gas emissions, and the transportation sector accounted for 28 percent.

Implied Sectors	1990	1995	2000	2005	2006	2007
Industry	2,166.5	2,219.8	2,235.5	2,081.2	2,082.3	2,081.2
Transportation	1,546.7	1,688.3	1,923.2	2,003.6	1,999.0	2,000.1
Commercial	942.2	1,000.2	1,140.0	1,214.6	1,201.5	1,251.2
Residential	950.0	1,024.2	1,159.2	1,237.0	1,176.1	1,229.8
Agriculture	459.2	489.7	503.2	511.7	530.0	530.1
U.S. Territories	34.1	41.1	47.3	60.5	62.3	57.7
Total Emissions	6,098.7	6,463.3	7,008.2	7,108.6	7,051.1	7,150.1
Land Use, Land-Use Change, and Forestry (Sinks)	(841.4)	(851.0)	(717.5)	(1,122.7)	(1,050.5)	(1,062.6)
Net Emissions (Sources and Sinks)	5,257.3	5,612.3	6,290.7	5,985.9	6,000.6	6,087.5

Tg CO_2 Eq. = teragrams of carbon dioxide equivalents.

other economic sectors, agricultural sector emissions were dominated by N_2O emissions from agricultural soil management and CH_4 emissions from enteric fermentation, rather than CO_2 from fossil fuel combustion. The commercial sector accounted for about 6 percent of emissions, while U.S. territories accounted for approximately 1 percent.

CO_2 was also emitted and sequestered by a variety of activities related to forest management practices, tree planting in urban areas, the management of agricultural soils, and landfilling of yard trimmings.

Electricity is ultimately consumed in the economic sectors described above. Table 3-5 presents GHG emissions from economic sectors with emissions related to electricity generation distributed into end-use categories (i.e., emissions from electricity generation are allocated to the economic sectors in which the electricity is consumed). To distribute electricity emissions among end-use sectors, emissions from the source categories assigned to electricity generation were allocated to the residential, commercial, industry, transportation, and agriculture economic sectors according to retail sales of electricity.[11] These source categories include CO_2 from fossil fuel combustion and the use of limestone and dolomite for flue gas desulfurization, CO_2 and N_2O from incineration of waste, CH_4 and N_2O from stationary sources, and SF_6 from electrical transmission and distribution systems.

When emissions from electricity are distributed among these sectors, industry accounts for the largest share of U.S. GHG emissions (30 percent) in 2007. Emissions from the residential and commercial sectors also increase substantially when emissions from elec-

tricity are included, due to their relatively large share of electricity consumption (lighting, appliances, etc.). Transportation activities remain the second-largest contributor to total U.S. emissions (28 percent). In all sectors except agriculture, CO_2 accounts for more than 80 percent of GHG emissions, primarily from the combustion of fossil fuels. Figure 3-14 shows the trend in these emissions by sector from 1990 to 2007, while Box 3-3 shows recent trends in various U.S. GHG emissions-related data.

INDIRECT GREENHOUSE GASES

The reporting requirements of the UNFCCC request that information be provided on indirect GHGs, which include CO, NO_x, NMVOCs, and SO_2 (UNFCCC 2003). These gases do not have a direct global warming effect, but indirectly affect terrestrial radiation absorption by influencing the formation and destruction of tropospheric and stratospheric ozone and methane, or, in the case of SO_2, by affecting the absorptive characteristics of the atmosphere. Additionally, some of these gases may react with other chemical compounds in the atmosphere to form compounds that are GHGs.

Since 1970, the United States has published estimates of annual emissions of CO, NO_x, NMVOCs, and SO_2 (U.S. EPA/OAQPS 2008), which are regulated under the Clean Air Act.[12] Table 3-7 shows that fuel combustion accounts for the majority of emissions of these indirect GHGs. Industrial processes—such as the manufacture of chemical and allied products, metals processing, and industrial uses of solvents—are also significant sources of CO, NO_x, and NMVOCs.

[11] Emissions were not distributed to U.S. territories, since the electricity generation sector only includes emissions related to the generation of electricity in the 50 states and the District of Columbia.

[12] NO_x and CO emissions from field burning of agricultural residues were estimated separately, and therefore not taken from U.S. EPA/OAQPS 2008.

Figure 3-14 **U.S. Electricity-Related Greenhouse Gas Emissions Distributed to Economic Sectors: 1990–2007**

In 2007, when electricity generation-related emissions are distributed to economic sectors, the transportation sector accounted for 28 percent of total U.S. greenhouse gas emissions, and the industrial sector accounted for 30 percent (emissions from industrial processes and from electricity use at industrial facilities).

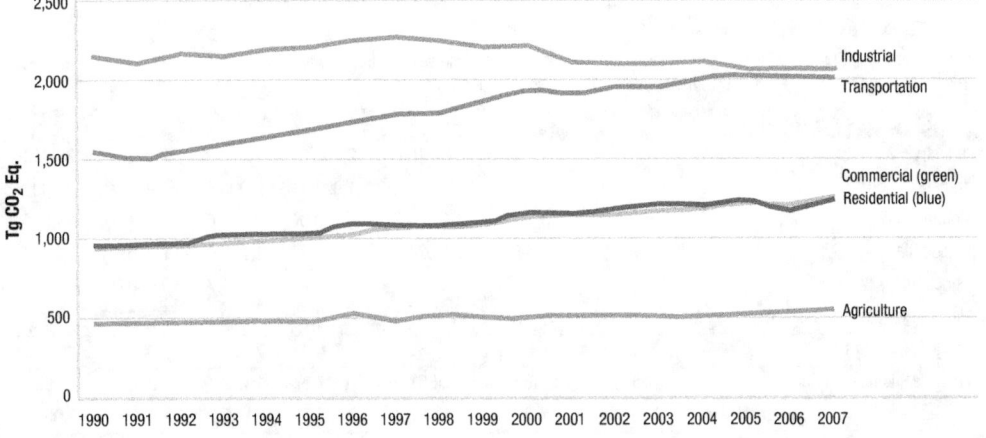

Box 3-3 Recent Trends in Various U.S. Greenhouse Gas Emissions-Related Data

Total emissions can be compared to other economic and social indices to highlight changes over time. These comparisons include: (1) emissions per unit of aggregate energy consumption, because energy-related activities are the largest sources of emissions; (2) emissions per unit of fossil fuel consumption, because almost all energy-related emissions involve the combustion of fossil fuels; (3) emissions per unit of electricity consumption, because the electric power industry—utilities and non-utilities combined—was the largest source of U.S. GHG emissions in 2007; (4) emissions per unit of total gross domestic product as a measure of national economic activity; or (5) emissions per capita.

Table 3-6 provides data on various statistics related to U.S. GHG emissions normalized to 1990 as a baseline year. U.S. GHG emissions have grown at an average annual rate of 0.9 percent since 1990. This rate of growth is slightly slower than that for total energy consumption or fossil fuel consumption and much slower than that for either electricity consumption or overall gross domestic product. Total U.S. GHG emissions have also grown slightly slower than the U.S. population since 1990 (see Figure 3-15).

Table 3-6 Recent Trends in Various U.S. Data (Index 1990 = 100) (Tg CO_2 Eq.)

U.S. GHG emissions have grown at an average annual rate of 0.9 percent since 1990. This rate of growth is slightly slower than that for total energy consumption or fossil fuel consumption and much slower than that for either electricity consumption or overall gross domestic product.

Variable	1990	1995	2000	2005	2006	2007	Growth Rate[a]
Gross Domestic Product[b]	100	113	138	155	159	162	2.9%
Electricity Consumption[c]	100	112	127	134	135	137	1.9%
Fossil Fuel Consumption[c]	100	107	117	119	117	119	1.1%
Energy Consumption[c]	100	108	117	119	118	120	1.1%
Population[d]	100	107	113	118	119	120	1.1%
Greenhouse Gas Emissions[e]	100	106	115	117	115	117	0.9%

[a] Average annual growth rate.

[b] Gross domestic product in chained 2000 dollars (U.S. DOC/BEA 2008).

[c] Energy content-weighted values (U.S. DOE/EIA 2008a).

[d] U.S. DOC/Census 2008a.

[e] Global warming potential-weighted values.

Figure 3-15 Recent Trends in U.S. Greenhouse Gas Emissions per Capita and per Dollar of Gross Domestic Product: 1990–2007

Since 1990, U.S. greenhouse gas emissions have grown at an average annual rate of 0.9 percent. This is significantly slower than the average annual 2.9 percent growth rate in the U.S. gross domestic product (GDP).

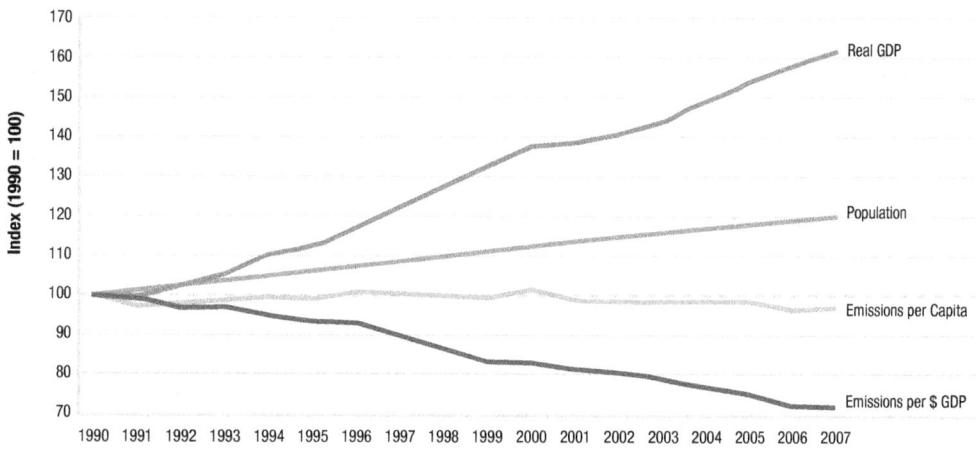

Sources: U.S. DOC/BEA 2008, U.S. DOC/Census 2008a, and emission estimates in this report.

Table 3-7 **Emissions of NO$_x$, CO, NMVOCs, and SO$_2$** (Gg)

Fuel combustion accounts for the majority of emissions of indirect greenhouse gases. Industrial processes—such as the manufacture of chemical and allied products, metals processing, and industrial uses of solvents—are also significant sources of CO, NO$_x$, and NMVOCs.

Gas/Activity	1990	1995	2000	2005	2006	2007
Nitrogen Oxides (NO$_x$)	21,450	21,070	19,004	15,612	14,701	14,250
Mobile Fossil Fuel Combustion	10,920	10,622	10,310	8,757	8,271	7,831
Stationary Fossil Fuel Combustion	9,689	9,619	7,802	5,857	5,445	5,445
Industrial Processes	591	607	626	534	527	520
Oil and Gas Activities	139	100	111	321	316	314
Incineration of Waste	82	88	114	98	98	97
Agricultural Burning	28	29	35	39	38	37
Solvent Use	1	3	3	5	5	5
Waste	0	1	2	2	2	2
Carbon Monoxide (CO)	130,461	109,032	92,776	71,672	67,453	63,875
Mobile Fossil Fuel Combustion	119,360	97,630	83,559	62,519	58,322	54,678
Stationary Fossil Fuel Combustion	5,000	5,383	4,340	4,778	4,792	4,792
Industrial Processes	4,125	3,959	2,216	1,744	1,743	1,743
Incineration of Waste	978	1,073	1,670	1,439	1,438	1,438
Agricultural Burning	691	663	792	860	825	892
Oil and Gas Activities	302	316	146	324	323	323
Waste	1	2	8	7	7	7
Solvent Use	5	5	45	2	2	2
Non-Methane Volatile Organic Compounds (NMVOCs)	20,930	19,520	15,227	14,562	14,129	13,747
Mobile Fossil Fuel Combustion	10,932	8,745	7,229	6,292	5,954	5,672
Solvent Use	5,216	5,609	4,384	3,881	3,867	3,855
Industrial Processes	2,422	2,642	1,773	2,035	1,950	1,878
Stationary Fossil Fuel Combustion	912	973	1,077	1,450	1,470	1,470
Oil and Gas Activities	554	582	388	545	535	526
Incineration of Waste	222	237	257	243	239	234
Waste	673	731	119	115	113	111
Agricultural Burning	N/A	N/A	N/A	N/A	N/A	N/A
Sulfur Dioxide (SO$_2$)	20,935	16,891	14,830	13,348	12,259	11,725
Stationary Fossil Fuel Combustion	18,407	14,724	12,849	11,641	10,650	10,211
Industrial Processes	1,307	1,117	1,031	852	845	839
Mobile Fossil Fuel Combustion	793	672	632	600	520	442
Oil and Gas Activities	390	335	287	233	221	210
Incineration of Waste	38	42	29	22	22	22
Waste	0	1	1	1	1	1
Solvent Use	0	1	1	0	0	0
Agricultural Burning	N/A	N/A	N/A	N/A	N/A	N/A

Gg = gigagrams; N/A = not available.

Note: Totals may not sum due to independent rounding.

Source: U.S. EPA 2008, disaggregated based on U.S. EPA 2003, except for estimates from field burning of agricultural residues.

4
Policies and Measures

At no time in its history has the United States been more engaged—both at home and abroad, at the federal, state, and local levels—in enhancing its efforts to reduce climate change. Since assuming office, President Obama has moved quickly to establish new federal policies and measures designed to reassert American leadership in solving the global climate challenge. President Obama has outlined a comprehensive plan to address global climate change through investments that will save or create many jobs. The plan will:

- Help transform the economy through investments in research, development, demonstration, and deployment of new forms of clean energy and through improvements in energy efficiency.

- Ensure the United States is on a path to reduce its dependence on oil, in part by promoting the next generation of cars and trucks and the alternative fuels on which they will run.

- Reduce the pollution that causes global warming. By stemming carbon pollution through a market-based cap, the United States will protect the national heritage for generations to come, and address many of the energy challenges the nation faces.

This chapter presents key policies and measures undertaken in these past months to fulfill this vision, including the American Recovery and Reinvestment Act of 2009 (ARRA), a new program to simultaneously promote fuel economy and limit tailpipe emissions of greenhouse gases (GHGs), and other significant actions (Box 4-1). The chapter also outlines significant actions since the most recent *2006 U.S. Climate Action Report* (2006 CAR), as well as ongoing policies and measures at the federal level. In addition, for many

years, U.S. states and localities have led the way in advancing clean energy climate policies. This chapter does not attempt to cover the broad landscape of U.S. state and local efforts on climate change. Instead, it provides a brief overview and sampling of the many programs that are occurring at the sub-national level.

[1] See http://www.whitehouse.gov/the_press_office/Statement-By-The-President-On-House-Passage-Of-The-American-Clean-Energy-And-Security-Act/.

A cornerstone of the President's platform involves comprehensive new energy and climate legislation, which requires approval by Congress and the executive branch. In February 2009, before a joint session of Congress, President Obama announced his intent to work with Congress to achieve legislation that would reduce GHG emissions through a market-based cap, drive the production of more renewable energy, and invest billions in low-emission technologies, to put the United States on a path to reduce its GHG emissions by more than 80 percent below 2005 levels by 2050.

In June 2009, the U.S. House of Representatives passed the landmark American Clean Energy and Security Act, which includes economy-wide GHG reduction goals of 3 percent below 2005 levels in 2012, 17 percent below 2005 levels in 2020, and 83 percent below 2005 levels in 2050. President Obama praised the passage of the House bill as a "bold and necessary step" in reducing GHGs.[1] Through a cap-and-trade program and other complementary measures, the bill would promote the development and deployment of new clean energy technologies that would fundamentally change the way we produce, deliver, and use energy. The bill would (1) advance energy efficiency and reduce reliance on oil; (2) stimulate innovation in clean coal technology to reduce GHG emissions before they enter the atmosphere; (3) accelerate the use of renewable sources of energy, including biomass, wind, solar, and geothermal; (4) create strong demand for a domestic manufacturing market for these next-generation technologies that will enable American workers to serve in a central role in U.S. clean energy transformation; and (5) play a critical role in the American economic recovery and job growth—from retooling shuttered manufacturing plants to make wind turbines, to using equipment and expertise from oil drilling to develop clean energy from underground geothermal sources, to tapping into American ingenuity to engineer coal-fired power plants that do not contribute to climate change.

This is an exciting and critical time for global climate change in U.S. history. The U.S. Senate is currently considering its own legislation to promote clean energy and reduce GHG emissions. Under the U.S. system, legislation passed by the House and Senate is reconciled, and the resulting bill is passed to the President for signature. As new policies and measures are enacted, existing climate change strategies and investments will continue to be evaluated and adjusted as necessary. Though much remains to be done, this fifth national communication demonstrates new momentum in U.S. efforts to achieve the objective of the United Nations Framework Convention on Climate Change.

FEDERAL POLICIES AND MEASURES

The United States recognizes that a strong set of national policies and measures is critical to achieving the

President's emissions reduction goal. To reach this goal, a combination of near- and long-term, voluntary and regulatory activities will be needed across the economy, including in the residential, commercial, industrial, transportation, waste, and agricultural sectors. While significant GHG reductions have been made through existing initiatives, the administration recognizes the need to expand upon successful initiatives and introduce new policies and measures. Through the programs outlined sector by sector in this chapter, the federal government will build on its partnerships, invest in the research, development, demonstration, and deployment of emission-reducing technologies, and advance the implementation of critical regulatory policies.

New Initiatives Since the 2006 CAR

In addition to the significant funding for energy efficiency and alternative forms of energy provided through ARRA, the United States has advanced legislation, regulations, and initiatives to reduce GHG emissions since the nation last reported to the United Nations in the 2006 CAR. These efforts constitute a "bottom-up" approach to addressing climate change. They are both distinct from and complementary to the top-down targets described in Chapter 5. These activities are only a sample of U.S. efforts; they do not comprehend all that the United States does to mitigate and adapt to climate change. Most of all, such bottom-up efforts are necessary, tangible evidence of America's commitment to reducing dangerous GHG emissions. This section and Table 4-1 introduce these initiatives, and the following broader section discusses progress resulting from both recent and ongoing initiatives.

American Recovery and Reinvestment Act

On February 17, 2009, President Obama took an important step toward reaching the administration's climate change goals with his signing of ARRA. This law provides unprecedented investments in clean energy improvements for the U.S. economy, allocating more than $90 billion for clean energy programs, such as weatherization assistance for low-income homes, and billions more for science and infrastructure, including the efficient modernization of mass transit systems (EOP/CEA 2010). Key components of the legislation include:

- Appropriating funding for numerous grant programs and tax incentives for clean energy technologies, including solar, wind, biomass, geothermal, marine, hydropower, fuel cells, plug-in electric vehicles, and other technologies that have the potential to reduce U.S. GHG emissions.

- Emphasizing energy-efficient technologies, practices, and policies, including a 30 percent tax credit for residential energy efficiency investments, as well as mandates for improved energy efficiency standards for electric heat pumps, central air conditioning...

Table 4-1 **Recent Initiatives Not Featured in CAR 2006**

In addition to the significant funding for energy efficiency and alternative forms of energy provided through the American Recovery and Reinvestment Act of 2009, the United States has advanced legislation, regulations, and initiatives to reduce GHG emissions since the nation last reported to the United Nations in the 2006 *U.S. Climate Action Report*.

Policy/Measure	Agency
Energy: Residential and Commercial	
Energy Efficiency and Conservation Block Grants Net-Zero Energy Commercial Building Initiative	DOE
Energy: Industrial	
Energy-Intensive Industries Program	DOE
Energy: Supply	
Biorefinery Assistance	USDA
Energy Smart Parks Indian Education Renewable Energy Challenge University-National Park Energy Partnership Program	DOI/DOE
Energy Transmission Infrastructure Geothermal Energy Deployment Program Solar Energy Deployment Program Wind Energy Development Program	DOI
Transportation	
Alternative Transport Systems and Use of Clean Vehicles	DOI
Loan Programs (Advanced Technology Vehicles Manufacturing Incentive)	DOE
Transit Investments for Greenhouse Gas and Energy Reduction (TIGGER)	DOT
Industry (Non-Carbon Dioxide)	
Responsible Appliance Disposal Program	EPA
Forestry	
Enhancing Ecosystems Services on Forest, Grasslands, Parks, and Wildlife Reserves	DOI
Cross-Sectoral	
Carbon Monitoring and Sequestration Climate Friendly Parks	DOI
Climate Showcase Communities Grant Program	EPA
National Action Plan for Energy Efficiency	EPA/DOE
Interagency Partnership for Sustainable Communities	DOT/HUD/EPA

DOE = U.S. Department of Energy; DOI = U.S. Department of the Interior; DOT = U.S. Department of Transportation; EPA = U.S. Environmental Protection Agency; HUD = U.S. Department of Housing and Urban Development; USDA = U.S. Department of Agriculture.

ers, water heaters, wood stoves, oil furnaces, and hot-water boilers.

- Increasing the investments allocated to new clean renewable energy bonds and qualified energy conservation bonds.

- Investing in critical energy infrastructure by providing loan guarantees for new or upgraded electric power transmission projects, and by providing funding for the Smart Grid and new Smart Grid technologies.

- Asserting an energy efficiency leadership role for the federal government, investing in the "green" conversion of federal facilities, and purchasing vehicles for government use with higher fuel economy, including hybrid and electric vehicles.

Executive Order 13514: Federal Leadership in Environmental, Energy, and Economic Performance[2]

On October 5, 2009, President Obama signed an Executive Order[3] that sets sustainability goals for federal agencies and focuses on improving their environmental, energy, and economic performance. The Executive Order requires federal agencies to set a 2020 GHG emission reduction target within 90 days, increase energy efficiency, reduce fleet petroleum consumption, conserve water, reduce waste, support sustainable communities, and leverage federal purchasing power to promote environmentally responsible products and technologies.

The new Executive Order requires agencies to measure, manage, and reduce GHG emissions toward agency-defined targets. It describes a process by which agency goals will be set and reported to the President by the Chair of the Council on Environmental Quality (CEQ). The Executive Order also requires agencies to meet a number of energy, water, and waste reduction targets, including:

- 30 percent reduction in vehicle fleet petroleum use by 2020, relative to 2005;

- 26 percent improvement in water efficiency by 2020;

- 50 percent recycling and waste diversion by 2015;

- 95 percent of all applicable contracts in compliance with sustainability requirements;

- implementation of the 2030 net-zero-energy building requirement;

- implementation of the stormwater provisions of the Energy Independence and Security Act of 2007 (EISA), section 438; and

- development of guidance for sustainable federal building locations in alignment with the Livability Principles put forward by the U.S. Department of Housing and Urban Development (HUD), the U.S. Department of Transportation (DOT), and the U.S. Environmental Protection Agency (EPA).

Implementation of the Executive Order will focus on integrating achievement of sustainability goals with agency mission and strategic planning to optimize performance and minimize implementation costs. Each agency will develop and carry out an integrated Strategic Sustainability Performance Plan that prioritizes the agency's actions toward the goals of the Executive Order based on life-cycle return on investments. Implementation will be managed through the previously established Office of the Federal Environmental Executive, working in close partnership with the Office of Management and Budget, CEQ, and the federal agencies.

Energy Independence and Security Act of 2007[4]

In December 2007, EISA was signed into law. This major energy policy bill enacted numerous key provisions designed to increase energy efficiency and the availability of renewable energy, including:

- *Corporate Average Fuel Economy (CAFE)*—The law set a minimum target of 35 miles per gallon for the combined fleet of cars and light trucks by model year (MY) 2020. In March 2009, DOT issued a final rule increasing fuel economy standards for MY 2011 passenger cars and light trucks. In May 2010, EPA and DOT published a final regulation (FR 2010). This rule is expected to save some 960 million metric tons of carbon dioxide ($MMTCO_2$) over the life of the regulated vehicles (U.S. EPA/OTAQ 2010).

- *Renewable Fuels Standard (RFS)[5]*—The law established a modified standard that taps the potential of renewable fuels to reduce life-cycle GHG emissions and provide economic growth. The RFS starts at 9 billion gallons in 2008 and increases to 36 billion gallons by 2022, including 21 billion gallons from cellulosic ethanol and other advanced biofuels.

- *Energy Efficiency Equipment Standards*—EISA contains a variety of new standards for reducing energy use in lighting and for residential and commercial appliance equipment, including incandescent and fluorescent lamps, residential refrigerators, freezers, electric motors, and residential boilers.

EISA also established a national goal to achieve zero-net-energy use for new commercial buildings built after 2025, and a goal to retrofit all pre-2025 commercial buildings to zero-net-energy use by 2050. For federal buildings, the law requires total energy use to be reduced by 30 percent by 2015 compared to 2005.

EISA authorized the Energy Efficiency and Conservation Block Grant Program to develop and implement projects that help reduce energy use and emissions at the local and regional levels. Over $2.7 billion in grants are available through the program, which is funded by ARRA. EISA also established loans, grants, and debentures to help small businesses develop, invest in, and purchase energy-efficient buildings, fixtures, equipment, and technology.

EISA directs U.S. government agencies to support accelerated research, development, and demonstration (RD&D) and commercial application of clean technologies, such as solar, geothermal, marine and hydrokinetic, and energy storage. It also calls for expanded research and development (R&D) for carbon capture and sequestration, including large-scale demonstration projects, and assessment of ecosystem capacities to sequester carbon. It establishes a federal policy to modernize the electric utility transmission and distribution system, including investments in smart grid technologies.

[2] See http://www.whitehouse.gov/the_press_office/President-Obama-signs-an-Executive-Order-Focused-on-Federal-Leadership-in-Environmental-Energy-and-Economic-Performance/.

[3] See http://www.whitehouse.gov/assets/documents/2009fedleader_eo_rel.pdf.

[4] See http://frwebgate.access.gpo.gov/cgi-bin/getdoc.cgi?dbname=110_cong_bills&docid=f:h6enr.txt.pdf.

[5] See http://www.epa.gov/otaq/renewablefuels/index.htm.

Energy Improvement and Extension Act of 2008

Signed into law in October 2008, the Energy Improvement and Extension Act of 2008 offers an array of incentives for U.S. energy production and conservation, including provisions for renewable energy production, clean coal and carbon sequestration, and efficient transportation and end-use standards and incentives (Box 4-2).

Measures to encourage investment in capital-intensive projects with otherwise high financial risk are emphasized, such as incentives for wind, biomass, solar, geothermal, landfill gas, trash combustion, combined heat and power systems, marine and hydropower facilities, fuel cells, microturbines, and advanced coal-based generation technology projects. The legislation provides a new tax credit for investment in qualified energy conservation bonds to reduce energy consumption in public buildings, implement green community programs, and promote mass commuting facilities. Qualifying commercial and residential improvements and provisions for energy-efficient household appliances also are included.

The act also accelerates the deployment of the next generation of vehicles by supporting renewable and alternative fuels and alternative-fuel vehicles. Other low-carbon transportation measures include incentives for bicycle commuting and idling-reduction devices in heavy trucks.

National Policy to Establish Vehicle GHG Emissions and CAFE Standards[7, 8]

EPA and DOT's National Highway Traffic Safety Administration (NHTSA) signed a joint proposal on September 15, 2009, to establish a national program consisting of new standards for light-duty vehicles that will reduce GHG emissions and improve fuel economy. In May 2010, EPA and DOT published a final regulation (FR 2010). This joint rulemaking implements the National Autos Policy announced by President Obama on May 19, 2009, responding to the country's critical need to address global climate change and reduce oil consumption. For the first time, EPA promulgated federal emission standards for GHGs, using its authority under the Clean Air Act, and NHTSA proposed CAFE standards under the Energy Policy and Conservation Act. These standards apply to passenger cars, light-duty trucks, and medium-duty passenger vehicles, covering MY 2012–2016, and represent a harmonized and consistent national program. Under the program, automobile manufacturers could build a single light-duty national fleet that satisfies all requirements under both standards, while ensuring that consumers still have a full range of vehicle choices.

The new standards, covering MY 2012–2016, require an average fuel economy of 35.5 miles per gallon and an average 250 grams per mile of carbon dioxide

Box 4-2 Energy Improvement and Extension Act of 2008[6]

The Emergency Economic Stabilization Act of 2008 (Public Law 110-343) [4], which was signed into law on October 3, 2008, incorporates the Energy Improvement and Extension Act of 2008 in Division B as follows:

- Extension of the residential and business tax credits for renewable energy as well as for the purchase and production of certain energy-efficient appliances, many of which were originally enacted in the Energy Policy Act of 2005.

- Removal of the cap on the tax credit for purchases of residential solar photovoltaic installations and an increase in the tax credit for residential ground-source heat pumps.

- Addition of a business investment tax credit for combined heat and power, small wind systems, and commercial ground-source heat pumps.

- Provision of a tax credit for the purchase of new, qualified, plug-in electric drive motor vehicles.

- Extension of the income and excise tax credits for biodiesel and renewable diesel to the end of 2009, and an increase in the amount of the tax credit for biodiesel and renewable diesel produced from recycled feedstock.

- Provision of tax credits for the production of liquid petroleum gas, compressed natural gas, and aviation fuels from biomass.

- Provision of an additional tax credit for the elimination of CO_2 that would otherwise be emitted into the atmosphere in enhanced oil recovery and non-enhanced oil recovery operations.

- Extension and modification of key renewable energy tax provisions that were scheduled to expire at the end of 2008, including production tax credits (PTCs) for wind, geothermal, landfill gas, and certain biomass and hydroelectric facilities.

- Expansion of the PTC-eligible technologies to include plants that use energy from offshore, tidal, or river currents (in-stream turbines), ocean waves, or ocean thermal gradients.

(CO_2) in 2016. In turn, they are projected to save 1.8 billion barrels of oil over the life of the program, with a fuel economy gain averaging more than 5 percent per year, and a cumulative reduction of approximately 960 teragrams of carbon dioxide equivalents ($Tg\ CO_2$ Eq.) in GHG emissions over the lifetime of the vehicles sold in MY 2012–2016. This substantially accelerates increases in average fuel economy mandated under the CAFE law passed by Congress in 2007.

Mandatory Greenhouse Gas Reporting Rule[9]

In 2009, EPA issued the Mandatory Reporting of Greenhouse Gas Emissions Final Rule. The rule requires reporting of GHG emissions from large U.S. sources, and is intended to collect accurate and timely emissions data to inform future policy decisions. Under the rule, suppliers of fossil fuels or industrial GHGs, manufacturers of vehicles and engines, and facilities that emit 25,000 metric tons or more per year of GHGs are required to submit annual reports to EPA. The gases covered by the proposed rule are CO_2, methane (CH_4), nitrous oxide (N_2O), hydrofluorocarbons (HFCs), perfluorocarbons (PFCs), sulfur hexafluoride (SF_6), and other fluorinated gases, including nitrogen trifluoride (NF_3) and hydrofluorinated ethers (HFEs).

This comprehensive national reporting system's accurate and timely GHG emissions data will serve as a cornerstone of the domestic U.S. effort to combat climate change. The reporting program covers about 85 percent of total U.S. emissions from roughly 10,000

[6] See http://frwebgate.access.gpo.gov/cgibin/getdoc.cgi?dbname=110_cong_public_laws&docid=f:publ343.110.pdf.

[7] See http://www.epa.gov/otaq/climate/regulations.htm.

[8] See http://www.nhtsa.dot.gov/portal/fueleconomy.jsp.

[9] See www.epa.gov/climatechange/emissions/ghgrulemaking.html.

facilities. The reporting threshold of 25,000 metric tons per year is roughly equivalent to the annual GHG emissions from just over 4,500 passenger vehicles. Annual reporting will begin in 2011 for calendar year 2010 emissions. Once the system is in place, EPA will make the GHG emissions data freely available to the public.

Proposed Regulation Facilitating Geologic Sequestration of CO_2[10]

EPA proposed new federal requirements under the Safe Drinking Water Act for the underground injection of CO_2 for the purpose of long-term underground storage, or geologic sequestration. The proposed rule would establish a new class of injection well—Class VI—and technical criteria for geologic site characterization, area of review and corrective action, well construction and operation, mechanical integrity testing and monitoring, well plugging, post-injection site care, and site closure for protecting underground sources of drinking water. Elements of the proposal are based on the existing regulatory framework of EPA's Underground Injection Control Program, while modifications address the unique nature of CO_2 injection for geologic sequestration. EPA is currently evaluating public comments on the proposal, and expects to issue final regulations by 2011.

Department of the Interior Secretarial Order 3289

Under President Obama, the U.S. Department of the Interior (DOI) has taken a key leadership role in the federal government's plans to adapt to the impacts of climate change. To these efforts, DOI brings considerable expertise in climate change and adaptation science, management of U.S. natural landscapes and resources, fish and wildlife, ecosystem services, and biological carbon sequestration.

DOI initiatives will help to address climate change, adapt to a warming planet, and create the new jobs of a clean energy economy. In response to the need for fully integrated information to achieve these objectives, DOI issued Secretarial Order 3285 on March 11, 2009. The order made the production and transmission of renewable energy on public lands a priority and created a new DOI Energy and Climate Change Task Force co-chaired by the Deputy Secretary and Counselor to the Secretary. Building upon this effort, DOI issued Secretarial Order 3289 on September 14, 2009, establishing a Department-wide approach for applying scientific tools to increase understanding of climate change and to coordinate an effective response to its impacts on tribes and on the land, water, ocean, fish and wildlife, and cultural heritage resources that DOI manages.

To fulfill President Obama's vision for a clean energy economy, DOI manages America's public lands and oceans, not just for balanced oil, natural gas, and coal

development, but also to promote environmentally responsible renewable energy development. Sun, wind, biomass, and geothermal energy from U.S. public lands is creating new jobs and will soon power millions of American homes.

Progress and Projections for Reducing U.S. GHG Emissions

The U.S. government is continuing to make important progress toward reducing GHG emissions through energy-related policies and measures that promote increased investment in end-use efficiency, clean energy development, and reductions in GHG emissions in agriculture and through efforts focused on the most potent GHGs. The policies and measures in this chapter highlight the successful U.S. government initiatives focused on reducing GHG emissions. While many of the policies and measures include projections for reducing GHGs, several do not for a variety of reasons, such as double benefit counting and stage of implementation. As a result, the projections presented in this chapter should not be compared to the information presented in Chapter 5. Table 4-3 at the end of this chapter summarizes the current representative U.S. programs and their estimated GHG mitigation impacts through 2020.

In addition, any mitigation levels or projections in this chapter are estimates generated using a range of methods and assumptions. Amounts are subject to change in the future and may have changed relative to amounts presented in past reports due to improvements in calculation methodologies. As noted in Table 4-3, estimates of mitigation impacts for individual policies or measures should not be aggregated to the sectoral level and may not be directly comparable, due to differences in calculation methodologies and possible synergies and interactions among policies and measures that may result in double counting.

Energy: Residential and Commercial Sectors[11]

The residential and commercial sectors represent approximately 35 percent of U.S. GHG emissions, making them an integral focus of U.S. climate change policies and measures. The use of electricity for such services as lighting, heating, cooling, and running electronic equipment and appliances accounts for the majority of CO_2 emissions in these sectors. Much progress has been made in achieving GHG emission reductions in these sectors, but significant additional potential can be realized through both regulatory and voluntary programs that set standards, provide information, develop measurement tools, and build partnerships. With the use of commercially available energy-efficient products, technologies, and best practices, many commercial buildings and homes could save significantly on energy bills and substantially reduce GHG emissions. These savings are hindered due to a range of pervasive and persistent market barriers.

[10] See http://www.epa.gov/OGWDW/uic/wells_sequestration.html.

[11] See http://www.eere.energy.gov/buildings/.

Following are descriptions of key policies and measures aimed at addressing these barriers, saving energy, and avoiding GHG emissions in the residential and commercial sectors.

Appliances and Commercial Equipment Standards Program, Appliance Energy Efficiency Standards,[12] Lighting Energy Efficiency Standards

The U.S. Department of Energy's (DOE's) Appliances and Commercial Equipment Standards Program develops test procedures and minimum efficiency standards for residential appliances and commercial equipment. The rules and regulations that are developed apply to products manufactured for sale in, as well as those imported into, the United States. By law, DOE is required to set efficiency standards at levels that achieve the maximum improvement in energy efficiency that is technologically feasible and economically justified. Standards benefit consumers by requiring that appliance manufacturers reduce the energy and water use of their products—and thus the costs to operate them. They are a cost-effective means of saving energy, reducing consumer utility bills, and lowering CO_2 emissions.

Less than three weeks into his term, on February 5, 2009, President Obama issued a memorandum requiring DOE to set more stringent efficiency standards for appliances, consistent with EISA and the Energy Policy Act of 2005 (EPAct). EPAct requires DOE to issue semi-annual reports describing its rulemaking schedule and plan for implementing the schedule. DOE published the first report on January 31, 2006. Efficiency and environmental advocates, states, utilities, manufacturers, retailers, and consumers are encouraged to participate in all stages of the rulemaking process. DOE is currently developing standards and test procedures for the following products:

- *Commercial Equipment*—clothes washers, distribution transformers, electric motors, furnaces and boilers, high-intensity discharge lamps, metal halide lamp fixtures, small electric motors, and walk-in coolers and freezers.

- *Residential Products*—battery chargers, external power supplies, central air conditioners and heat pumps, clothes washers and dryers, direct heating equipment, furnaces and boilers, fluorescent lamp ballasts, high-intensity discharge lamps, pool heaters, microwave ovens, refrigerators and freezers, room air conditioners, and water heaters.

In addition, the program shares responsibility with the Federal Trade Commission for labeling commercial equipment.

Building Energy Codes Program[13]

DOE's Building Energy Codes Program is an advocate for and information resource on national model energy codes, working with other federal government agencies, state and local jurisdictions, national code organizations, and industry to promote stronger model building energy codes and help states adopt, implement, and enforce those codes. The program is involved in three major areas to help improve the energy efficiency of residential and commercial buildings: propose and advocate improvements to national model energy codes; lead DOE's development and promulgation of improved federal energy codes for manufactured housing (mobile homes); and provide financial and technical assistance to help states adopt, implement, and enforce building energy codes. The technical assistance includes development and distribution of easy-to-use compliance tools and materials, and collaboration with stakeholders to address industry needs and provide information on compliance products, Web-based training, and energy code-related news.

ENERGY STAR Labeled Products

EPA and DOE have continued to expand the energy-efficient products available for homes and businesses through the ENERGY STAR program. The label is now available on more than 60 product categories. In 2008, the level of public awareness of ENERGY STAR increased to more than 75 percent. In addition to maintaining the integrity of the brand, DOE and EPA continue to identify new product categories for ENERGY STAR as well as revise existing product specifications to more stringent levels. Between 2000 and 2008, more than 2.5 billion ENERGY STAR-qualified products were sold. EPA estimates that consumption of these products helped to reduce emissions by about 64.5 Tg CO_2 Eq. in 2007 and could help to reduce emissions by 141.2 Tg. CO_2 Eq. by 2020.

ENERGY STAR for the Commercial Market

Commercial buildings use nearly 20 percent of the total U.S. energy consumed and contribute nearly the same proportion of GHGs to the atmosphere. Addressing these emissions through energy efficiency is critical to reducing GHG emissions.

EPA has continued to expand the ENERGY STAR[14] program in the commercial market, offering thousands of businesses and other organizations a strategy for superior energy management and standardized measurement tools. Since 2006, EPA has expanded and improved its national performance rating system, Portfolio Manager. Introduced in 1999, Portfolio Manager evaluates building energy efficiency and helps identify cost-effective opportunities for improvements for a wide range of building types, including hospitals, schools, grocery stores, office buildings, warehouses, retail spaces, residence halls, and hotels. By 2008, about 16 percent of U.S. floor space had been rated using this building rating system. Utilities can down-

[12] See http://www1.eere.energy.gov/buildings/appliance_standards/index.html.

[13] See http://www.energycodes.gov/.

[14] See www.energystar.gov.

load customers' data directly into Portfolio Manager, and a number of states and municipalities are requiring the disclosure of this information as part of the key building transactions.

In addition, more than 6,000 buildings have earned the ENERGY STAR label for top performance, and are using 35–40 percent less energy than average buildings. Further, through ENERGY STAR Leaders, EPA is recognizing organizations that reduce the energy use in their buildings by as much as 30 percent or by achieving top-performing portfolios in Portfolio Manager. EPA estimates that in 2007, ENERGY STAR in the commercial sector helped avoid 66 Tg CO_2 Eq., and that continued efforts could result in reductions of about 93 Tg CO_2 Eq. by 2020.

ENERGY STAR for the Residential Market

The ENERGY STAR programs in the residential sector—in both new and existing housing markets—have continued to expand since the 2006 CAR. Despite the recent downturn in the new housing market, ENERGY STAR for New Homes has gained momentum, with nearly 17 percent of all new homes being built to ENERGY STAR specifications in 2008. The program is working with more than 6,000 builders and, in November 2009, passed the 1 million homes threshold across the country, even as EPA proceeds to increase the stringency of program requirements.[15]

EPA has also worked with partners across the country to expand ENERGY STAR into the existing homes market with "Home Performance with ENERGY STAR," a whole-house retrofit program. As of 2008, more than 50,000 homeowners relied on trained and certified contractors to conduct whole-house energy audits and implement improvements. To help homeowners assess the current efficiency of their homes, EPA released interactive Web-based tools. In addition, EPA launched the ENERGY STAR HVAC Quality Installation Program to increase the number of properly installed heating, ventilation, and air conditioning systems, helping homeowners save 25 percent or more on energy.

EPA is also making ENERGY STAR tools and resources available to improve the energy efficiency of housing for lower-income families by providing housing finance agencies with recommendations and cost/benefit analysis for considerations as criteria in competitively allocating low-income housing credits. Finally, through the ENERGY STAR Mortgage program, EPA is helping to finance new energy-efficient homes, as well as energy efficiency improvements in existing homes. EPA estimates these programs helped to reduce emissions by nearly 2 Tg CO_2 Eq. in 2007 and could help to reduce emissions by 44 Tg CO_2 Eq. in 2020.

Net-Zero Energy Commercial Building Initiative[16]

DOE's Net-Zero Energy Commercial Building Initiative (CBI, formerly known as Commercial Building Integration) aims to achieve marketable net-zero-energy commercial buildings by 2025. Net-zero-energy buildings generate as much energy as they consume through efficiency technologies and on-site power generation. CBI encompasses all activities that support this goal, including industry partnerships, research, and tool development. The initiative provides key design and evaluation steps for developing energy-efficient and net-zero-energy buildings, including energy simulation software for evaluating building performance. DOE estimates that this initiative could generate 15.5 Tg CO_2 Eq. of emission reductions in 2020.

Under the authorization of EISA in 2007, CBI has advanced net-zero-energy commercial buildings by launching three Commercial Building Energy Alliances: the Retailer Energy Alliance, the Commercial Real Estate Energy Alliance, and the Hospital Energy Alliance. These alliances link commercial building owners and operators, by sector, who want to reduce the energy consumption, GHG emissions, and operating expenses of their buildings with the advanced technologies, analytical tools, and capabilities emerging from DOE and the national laboratories. CBI has also established the National Accounts program, which partners companies from the private sector and representatives from the national laboratories in order to construct and operate a new building and retrofit an existing building to achieve energy savings of 50 percent and 30 percent, respectively, over baselines. As of August 2009, 23 companies were signed up for the National Accounts program.

Building America[17]

DOE's Building America program helps design, build, and evaluate energy-efficient homes that use 30–40 percent less energy than comparable traditional homes with little or no increase in construction costs, and helps industry to adopt these practices for new home construction. The program optimizes building energy performance and savings through the integration of new technologies with innovative residential building practices. Ongoing research also focuses on integrating on-site power systems, including renewable energy technologies.

Hundreds of industry partners have followed the Building America approach in constructing more than 41,000 homes in 41 states (U.S. DOE/EERE 2010). The energy technologies and solutions being advanced by the program will contribute to a 70 percent reduction in energy use of new prototype residential buildings that, when combined with on-site energy technologies, will result in "zero-energy homes" by 2020, and an additional 20 percent reduction in energy use of exist-

[15] See http://www.energystar.gov/index.cfm?fuseaction=mil_homes.showSplash.

[16] See http://www1.eere.energy.gov/buildings/commercial_initiative/goals.html.

[17] See http://www1.eere.energy.gov/buildings/building_america/.

ing homes. DOE estimates these efforts could help to reduce GHG emissions by 19.8 Tg CO_2 Eq. in 2020.

Energy Efficiency and Conservation Block Grants[18]

DOE's Energy Efficiency and Conservation Block Grants (EECBG) program represents a presidential priority to invest in the cheapest, cleanest, and most reliable energy technologies that can be deployed immediately. Through formula and competitive grants to U.S. cities, counties, states, territories, and Native American nations, the program empowers local communities to make strategic investments to meet the nation's long-term goals for energy independence and leadership on climate change. Over $2.7 billion in formula grants are now available to U.S. states, territories, local governments, and Native American nations under this program, funded for the first time under ARRA. Authorized in EISA Title V, Subtitle E and signed into Public Law (PL 110-140) on December 19, 2007, the EECBG program provides funds to units of local and state governments, Native American nations, and territories to develop and implement projects to improve energy efficiency and reduce energy use and fossil fuel emissions.

Weatherization Assistance Program[19]

DOE's Weatherization Assistance Program (WAP) increases residential energy efficiency and reduces energy costs for low-income families. The program provides technical and financial assistance in support of state and local weatherization agencies throughout the United States. These providers manage one of the largest residential energy retrofit programs in the country.

Since the WAP's inception in 1976, over 6.2 million homes have been weatherized with DOE funds, with an estimated 100,000 homes weatherized in 2009. An average of 30.5 million British thermal units (Btus) of energy per household is saved as a result of weatherization, approximately a 16 percent reduction in primary heating fuel use. At 2009 prices, low-income families will save an average of $350 in reduced first-year energy costs. In addition, weatherization projects create both direct and indirect job opportunities. ARRA dedicated $5 billion in additional funding to support the administration's goal of weatherizing 1 million homes each year. DOE estimates that 8.9 Tg CO_2 Eq. of GHGs will be reduced in 2020.

Energy: Industrial Sector

The industrial sector contributes approximately 29 percent of U.S. GHG emissions, largely from fossil fuel combustion on site or at the power generation source.[20] Many U.S. industries are energy-intensive, with energy use contributing to a significant portion of operating costs. It is estimated that energy efficiency alone can reduce industry emissions by nearly 10 percent (Creyts et al. 2007). The policies and measures highlighted in this section are focused on opportuni-

ties that improve energy efficiency and reduce GHG emissions by helping the industrial sector adopt cost-effective, efficient technologies that improve productivity, while reducing energy costs, energy consumption, and waste.

ENERGY STAR for Industry

EPA's ENERGY STAR for Industry program has continued to grow since the 2006 CAR. EPA's ENERGY STAR Industrial Focuses, which directly address barriers to energy efficiency by providing industry-specific energy management tools and resources, have grown to include 16 industrial sectors with the launch of the Steelmaking Focus in 2008. In 2006 EPA began recognizing energy-efficient industrial plants with the ENERGY STAR label, and by the end of 2008 45 plants had earned this label. EPA has further developed its existing relationships with industrial partners. EPA estimates that the industrial sector, with the help of the ENERGY STAR program, prevented about 23 Tg CO_2 Eq. in 2007, and could avoid 36.6 Tg CO_2 Eq. in 2020.

Save Energy Now[21]

DOE's Industrial Technologies Program (ITP) works with industry to identify plant-wide opportunities for energy savings and process efficiency. By implementing new technologies and system improvements, many companies are realizing the benefits of applying DOE's software tools and resources. In fiscal year (FY) 2006, ITP introduced Save Energy Now to address high U.S. natural gas prices. The goal of this initiative is to achieve a 25 percent reduction in U.S. industrial energy intensity over 10 years.

Since 2006, Save Energy Now has completed 2,324 energy assessments in small, medium, and large U.S. industrial plants, with resulting annual energy cost savings of $218 million and related CO_2 reductions of 2.3 Tg CO_2 Eq.[22] The assessments have ranged from plant-wide to system-specific to process-specific. On average, energy assessments identify $1.5 million in energy cost savings per plant, or 8 percent of total plant energy costs.[23] Save Energy Now partnership development and outreach further extends the program's impact by increasing technology deployment and efficiency implementation and leveraging financial and technical resources. DOE estimates this program will reduce GHG emissions by 28.9 Tg CO_2 Eq. by 2020.

Industrial Assessment Centers[24]

DOE funds 26 Industrial Assessment Centers (IACs) housed at universities across the nation where DOE Energy Experts and IAC faculty and students conduct no-cost energy assessments for small- and medium-sized manufacturers, identifying an average of 1,300 metric tons of potential CO_2 savings per assessment per year.[25] The IACs also serve as a training ground for engineers who conduct energy audits or industrial

[18] See http://www.eecbg.energy. gov/.

[19] See http://apps1.eere.energy. gov/weatherization/.

[20] See http://www.eia.doe.gov/ oiaf/archive/aeo08/index.html.

[21] See http://www.eere.energy.gov/ industry/saveenergynow.

[22] See http://apps1.eere.energy. gov/industry/saveenergynow/ partners/results.cfm.

[23] As of September 1, 2009. The 8 percent is a conservative estimate based on the *Results from the U.S. DOE 2007 Save Energy Now Assessment Initiative* in 2006 and 2007, which is 10 percent (ORNL 2009). See http://apps1. eere.energy.gov/industry/ saveenergynow/partners/results. cfm.

[24] See http://www1.eere.energy. gov/industry/bestpractices/iacs. html.

[25] Prymak, Bill. "Energy Assessments: What Are the Benefits to Small and Medium Facilities?" February 19, 2009. Available at: http://www1.eere. energy.gov/industry/pdfs/ webcast_2009-0219_small_ medium_assessment_benefits.pdf

assessments, and provide recommendations to manufacturers to help them identify opportunities to improve productivity, reduce waste, and save energy.

Recommendations from the IACs have averaged $55,000 in potential annual savings for each manufacturer.[26] The savings identified for IACs between 2006 and 2008 were 28.8 trillion Btus per year (TBtus/yr) and $277.2 million per year. The continuing efforts of this program may help to reduce an estimated 5.1 Tg CO_2 Eq. in 2020 (U.S. DOE/EERE 2010).

Industry-Specific and Cross-Cutting RDD&D[27]

DOE funds both industry-specific and cross-cutting research, development, demonstration, and deployment (RDD&D) of emerging industrial energy efficiency technologies and best practices. Industry-specific RDD&D focuses on energy-intensive industries, including aluminum, chemicals, forest products, glass, metal casting, mining, petroleum refining, and steel. Cross-cutting RDD&D focuses on key technology areas common to most energy-intensive industries, such as combustion, distributed energy, energy-intensive processes, fuel and feedstock flexibility, Industrial Materials for the Future, nanomanufacturing, and sensors and automation. From the program's inception in 1975 through 2006, and over the useful life of the newly developed technologies, industry-specific and cross-cutting technologies have saved 5,650 TBtus of energy and about 375 Tg CO_2 Eq.[28] The program has supported more than 600 RDD&D projects, producing 220 commercialized technologies, with more than 141 technologies expected to emerge within the next one to two years.[29]

Energy: Supply

Electricity generation from fossil fuels is a major contributor to U.S. CO_2 emissions. Federal policies and measures aimed at the American energy supply promote CO_2 reductions through a variety of means, including energy efficiency for power generation and transmission, cleaner fuels, and the use of nuclear power and renewable energy resources. Solar, wind, and geothermal energy, and hydroelectric and biomass are some of the renewable energy resources consumed and generated by the United States. The nation's programs include the provision of tax credits and R&D, which help increase domestic investments in renewable energy and continue to accelerate the cost-competitiveness of these emerging technologies.

Solar Energy Development Program

While no commercial-scale solar energy facilities currently exist on public lands managed by DOI's Bureau of Land Management (BLM), federal tax incentives, loan guarantees, grants, and state renewable energy portfolio standards are driving an interest in utility-scale solar energy development projects on public lands. BLM has identified approximately 12 million

hectares (ha) (29.5 million acres [ac]) of public lands in six southwestern states with solar energy potential, and is evaluating a number of alternatives to determine best management practices for environmentally responsible utility-scale solar energy development on public lands.[30]

Wind Energy Development Program

Wind energy is the fastest-growing renewable energy resource in the nation, with an annual growth rate of over 30 percent.[31] As of November 2009, the total installed capacity of U.S. wind energy was over 31,000 megawatts (MW).[32] As of July 2009, BLM had approved 28 wind energy projects on BLM-managed public lands with installed capacity of 327 MW, and another 249 MW under construction. BLM completed a Programmatic Environmental Impact Statement (EIS) and Record of Decision (ROD) in December 2005 that established best management practices for the development of wind energy resources on the public lands. The EIS identified 8.3 million ha (20.6 million ac) of BLM-managed lands with wind energy resource potential.[33]

Geothermal Energy Development Program

In 2009, BLM managed 612 geothermal leases. Of these, 58 leases were in producing status and generated approximately 1,275 MW of capacity—about 50 percent of total U.S. geothermal energy capacity. BLM has also completed a Programmatic EIS and ROD, which allocated about 45 million ha (111 million ac) of BLM-managed lands as open to geothermal leasing. In 2007–2009, BLM sold 232 geothermal lease parcels.

Energy Transmission Infrastructure

One of President Obama's top energy priorities is to speed the development of a 21st-century network to move American electricity more cleanly, efficiently, and securely around the nation. To create this network, the United States is exploring ways to develop a unified, forward-looking strategy for siting, allocating the cost of, and coordinating the permitting for proposed transmission projects.

On October 23, 2009, nine agencies signed a Memorandum of Understanding (MOU) to expedite the permitting, siting, and construction of electric transmission infrastructure on federal lands. This will be accomplished through improved coordination among project applicants, federal agencies, and other stakeholders involved in the siting process. By reducing the expense and uncertainty associated with siting new lines, the MOU will speed approval of new lines and cut costs passed on to consumers. In addition, BLM has identified and designated more than 5,000 miles of energy transport corridors on public lands in the western United States.[34]

Energy SmartPARKS[35]

In November 2008, DOI's National Park Service (NPS) joined DOE to establish an innovative partner-

[26] See http://www1.eere.energy.gov/industry/bestpractices/about_iac.html.

[27] See http://www1.eere.energy.gov/industry/technologies/emerging_tech.html.

[28] Prymak, Bill. " Energy Assessments: What Are the Benefits to Small and Medium Facilities?" February 19, 2009. Available at: http://www1.eere.energy.gov/industry/pdfs/webcast_2009-0219_small_medium_assessment_benefits.pdf.

[29] See http://www1.eere.energy.gov/industry/technologies/emerging_tech.html.

[30] See http://www.blm.gov/pgdata/etc/medialib/blm/wo/MINERALS__REALTY__AND_RESOURCE_PROTECTION_/energy.Par.28512.File.dat/09factsheet_Solar.pdf.

[31] See http://www.blm.gov/pgdata/etc/medialib/blm/wo/MINERALS__REALTY__AND_RESOURCE_PROTECTION_/energy.Par.82982.File.dat/09factsheet_Wind.pdf.

[32] See http://www.awea.org/publications/reports/3Q09.pdf

[33] See http://www.blm.gov/wo/st/en/prog/energy/renewable_energy.html.

[34] Record of Decision, January 14, 2009. See http://www.blm.gov/pgdata/etc/medialib/blm/wo/MINERALS__REALTY__AND_RESOURCE_PROTECTION_/lands_and_realty.Par.27853.File.dat/Energy_Corridors_final_signed_ROD_1_14_2009.pdf.

[35] See http://www.nps.gov/energy/.

ship called Energy SmartPARKS. This partnership will showcase sustainable energy practices in national parks and inspire a green energy future for America. Energy SmartPARKS will enable NPS to showcase sustainable energy best practices and further its leadership mission. With combined federal government and private-sector support, Energy SmartPARKS will spark a green energy movement in America.

Clean Energy Initiative

Launched in 2001, EPA's Clean Energy Initiative consists of two partnership programs that promote cost-effective technologies that offer improved efficiencies and lower emissions than traditional energy supply options. In 2007 alone, these programs helped to reduce GHG emissions by about 17.6 Tg CO_2 Eq. and could result in reductions of 73 Tg CO_2 Eq. in 2020.

Green Power Partnership[36]

EPA's Green Power Partnership facilitates the purchase of environmentally friendly electricity from renewable energy sources by addressing the market barriers that stifle demand. The program now includes more than 1,000 partners who have committed to purchasing 16 billion kilowatt-hours (kWh) of green power. There has been increasing interest in on-site renewable generation, and the partnership has developed new resources and forms of recognition to encourage this trend further. Another innovation was the creation of the Green Power Communities designation, which eligible municipalities can achieve through minimum purchases of green power and community-wide campaigns.

Combined Heat and Power Partnership[37]

EPA's Combined Heat and Power (CHP) Partnership provides technical assistance to organizations across multiple sectors who invest in CHP projects, and assists state governments in designing regulations that encourage investment in CHP. As a result, the program now includes almost 270 partners who have installed over 4,700 MW of operational CHP. The CHP Partnership has targeted key sectors for action, including dry mill ethanol production, wastewater treatment facilities, and utilities. Technical guidance documents and project assistance have been provided to partners in these sectors.

Indian Education Renewable Energy Challenge[38]

In September 2009, DOE's Argonne National Laboratory (ANL) and DOI's Bureau of Indian Education (BIE) and Indian Affairs Office of Indian Energy and Economic Development announced a competition for students attending tribal and BIE high schools and tribal colleges to promote careers in the fields of green and renewable energy and build sustainable tribal economies. During Phase I of the Indian Education Renewable Energy Challenge, competing teams of students designed a small wind turbine that will harness wind energy, store it mechanically or electrically,

and use it to power an array of light-emitting diodes. At the end of Phase I, five high school and five college design teams with the best submissions received $1,300 each to construct prototypes of their inventions. During Phase II, the 10 teams conducted performance data collections to submit to ANL, along with detailed reports and videos of their prototypes in operation, for evaluation by a team of judges. In April 2010, two college teams and one high school team were announced as Energy Challenge winners.

University-National Park Energy Partnership Program[39]

In 1997, NPS partnered with the Rochester Institute of Technology and the Federal Energy Management Program to launch the University-National Park Energy Partnership Program (UNPEPP). This nationwide program partners university students and faculty with NPS energy management personnel to address universities' energy concerns and provide students real-world problem-solving experience in the energy field. Since 1997, UNPEPP has established nearly 50 partnerships nationwide, creating nearly 70 separate projects ranging from energy audits to solar power implementation to public education.[40]

Biorefinery Assistance[41]

The Food, Conservation, and Energy Act of 2008 (Farm Bill) includes three programs designed to provide targeted assistance to biorefineries and producers of advanced biofuels. The Biorefinery Assistance Program (Section 9003) provides loan guarantees for the development, construction, and retrofitting of commercial-scale biorefineries, and grants to help pay for the development and construction costs of demonstration-scale biorefineries. The Repowering Assistance Program (Section 9004) provides for payments to biorefineries (that were in existence at the time the 2008 Farm Bill was passed) to replace fossil fuels used to produce heat or power to operate the biorefineries with renewable biomass. The Bioenergy Program for Advanced Biofuels (Section 9005) provides for payments to eligible biofuel producers to support and ensure expanded production of advanced biofuels. The three programs are administered by the U.S. Department of Agriculture's (USDA's) Rural Development mission area.

Nuclear Technologies[42, 43]

DOE's Office of Nuclear Energy funds a diverse portfolio of programs to research and develop nuclear energy technologies. In FY 2009, major efforts included R&D on nuclear waste management techniques and advanced reactor designs through the Generation IV Nuclear Energy Systems program (Gen IV). Another effort funded in 2009 was the Nuclear Power 2010 program. Nuclear Power 2010 is an industry cost-shared effort to demonstrate revised Nuclear Regulatory Commission licensing processes; the program will be brought to closure in FY 2010. Gen IV worked to

[36] See http://www.epa.gov/greenpower/index.htm.

[37] See http://www.epa.gov/chp/index.htm.

[38] See http://www.bie.edu/downloads/641F7A3E9A724657B79FFDFF764986E7/Indian%20Education%20Renewable%20Energy%20Challenge.doc.

[39] See http://www.energypartnerships.org/.

[40] See http://www.rurdev.usda.gov/rbs/busp/baplg9003.htm.

[41] See http://www.energypartnerships.org/docs/UNPEPP_10Year_Report.pdf.

[42] See http://www.ne.doe.gov/np2010/overview.html.

[43] See http://climatetechnology.gov/Strategy-Intensity-Reducing-Technologies.pdf.

address critical unanswered questions about advanced nuclear reactor technologies through R&D and included support for the Next Generation Nuclear Plant project. The fuel-cycle R&D program focused on the development of fuel-cycle technologies that minimize waste and improve proliferation resistance. Many Gen IV and fuel-cycle R&D activities will be continued in FY 2010.

Renewable Energy Deployment Grants[44]

Community renewable energy deployment grants, totaling nearly $22 million, provide financial assistance for the implementation of integrated renewable energy deployment plans for communities, and construction of renewable energy systems. The grants support utility-scale renewable energy projects in up to four communities nationwide. The projects funded by these grants are expected to create jobs and avoid 50,000 tons of CO_2 annually.

Renewable Energy Production Incentive[45]

The Renewable Energy Production Incentive (REPI) program was created by the Energy Policy Act of 1992, and amended in 2005, to provide financial incentives for renewable energy electricity produced and sold by qualified renewable energy generation facilities. REPI provides financial incentive payments for electricity generated and sold by new qualifying renewable energy generation facilities. Qualifying facilities are eligible for annual incentive payments of 1.5 cents/kWh (1993 dollars and indexed for inflation) for the first 10-year period of their operation, subject to the availability of annual appropriations in each federal fiscal year of operation.

Rural Energy for America Program[46]

Formerly known as the Renewable Energy Systems and Energy Efficiency Improvements Program, USDA's Rural Energy for America Program provides loan guarantees and grants to agricultural producers and rural small businesses to purchase renewable energy systems and improve energy efficiency. Between 2002 and 2008, the program helped finance 694 renewable energy systems (including 271 biomass, 245 wind, 108 solar, 52 geothermal, and 18 hybrid projects) and 1,329 energy efficiency improvements. USDA estimates that these projects have achieved energy savings amounting to 46.6 million barrels of oil and an estimated reduction in GHG emissions of 8.2 Tg CO_2 Eq. Reduced GHG emissions for 2020 are projected to be 21 Tg CO_2 Eq.

Solar Energy Technologies Program[47]

DOE's Solar Energy Technologies Program is improving the performance of energy systems and reducing development, production, and installation costs to competitive levels, thereby accelerating large-scale use across the nation. When federal solar energy research began in the 1970s, the cost of electricity from solar resources was about $2.00/kWh. Technological ad-

vances over the last two decades have significantly reduced solar electricity costs, which range from as low as $0.12/kWh for concentrating solar power to $0.18/kWh for certain photovoltaic applications. DOE estimates that realizing the program's R&D goals could result in solar energy displacing 2.5 Tg CO_2 Eq. in 2020.

Wind Energy Program[48]

Wind energy is the world's fastest-growing energy supply technology. Today, the United States has more than 31,000 MW of wind-generating capacity. DOE's Wind Energy Program has helped lower the cost of wind energy to between 5 and 8 cents/kWh, making large wind farms in certain areas of the United States cost-competitive with fossil fuel and nuclear power plants.

Since 2002, the program has focused most of its efforts on utility-scale technologies and, through its public–private partnerships, has improved the cost of energy for large systems in Class 4 onshore winds.[49] The Wind Energy Program also supports U.S. offshore wind power development prospects. DOE estimates that realizing the program's R&D goals could result in wind energy displacing more than 67.6 Tg CO_2 Eq. in 2020.

Biomass Program[50]

DOE has contributed to the advancement of biomass technology by working with industry, academia, and national laboratory partners on a balanced portfolio of RD&D efforts geared toward biomass feedstocks and conversion technologies. A major effort includes deployment and further development of infrastructure and opportunities for market penetration of bio-based fuels and products. DOE seeks to develop advanced technologies for producing biofuels—including ethanol—from wood chips, crop residues, and dedicated energy crops, such as switchgrass, while bridging the gap from technology validation to deployment and promoting terrestrial carbon sequestration. The DOE research portfolio is supporting national goals related to reducing GHG emissions associated with bioenergy production and use when compared to conventional petroleum-based fuels.

DOE has contributed to the advancement of biomass technology by testing and demonstrating biomass co-firing with coal, developing advanced technologies for biomass gasification, developing and demonstrating small modular systems, and developing and testing high-yield, low-cost biomass feedstocks. This research has helped biomass become a proven commercial electricity generation option in the United States. DOE estimates that these efforts could help reduce emissions by about 55.2 Tg CO_2 Eq. in 2020.

Coal Technologies[51]

The mission of DOE's Office of Fossil Energy Coal Program is to ensure the availability of near-zero atmospheric emissions, and abundant, affordable, do-

[44] See http://apps1.eere.energy.gov/news/daily.cfm/hp_news_id=185.

[45] See http://apps1.eere.energy.gov/repi/.

[46] See http://www.rurdev.usda.gov/rbs/farmbill/index.html.

[47] See http://www1.eere.energy.gov/solar/.

[48] See http://www1.eere.energy.gov/windandhydro/.

[49] Onshore sites with a Class 4 rating (7.0–7.5 meters per second at 50 ft) or higher (on a scale of 7 classes) are preferred for large-scale wind plants. See American Wind Energy Association: http://www.awea.org/faq/basicwr.html.

[50] See http://www1.eere.energy.gov/biomass/.

[51] See http://www.fossil.energy.gov/programs/powersystems/cleancoal/.

mestic energy to fuel economic prosperity, strengthen energy security, and enhance environmental quality. In the late 1980s and early 1990s, DOE conducted a joint program with industry and state agencies to demonstrate technologies that addressed the environmental challenges of the time—primarily concerns about the impact of acid rain on forests and watersheds. In the 21st century, the focus of environmental concern has shifted to the global climate-altering impact of GHGs.

With coal likely to remain one of the nation's lowest-cost energy resources for the foreseeable future, the United States is actively funding applied R&D of advanced coal technologies that improve efficiency and reduce the intensity of CO_2 emissions. In addition, the Clean Coal Power Initiative is a cost-shared partnership between the government and industry to develop and demonstrate advanced coal-based power generation technologies. By 2020, these initiatives could prevent the emission of 23.1 Tg CO_2 Eq.

Geothermal Technologies Program[52]

Geothermal energy is poised for widespread expansion, if enhanced geothermal systems (EGS) technology can be proven commercially viable. DOE is working to overcome a variety of technical, market, and institutional barriers. Studies indicate that these challenges can be overcome with further investment in research, changes in policy, and market conditioning. The Massachusetts Institute of Technology (MIT) and the U.S. Geological Survey (USGS) have conducted studies that point to the potential contribution of EGS to domestic electricity supply ranging from 100,000 megawatts electric (MWe) to 517,800 MWe, respectively.[53, 54] As these studies suggest, geothermal energy, once restricted to naturally occurring hydrothermal fields in remote areas, could someday be operating in more locations and in greater proximity to large end-use markets.

DOE's Geothermal Technologies Program was relaunched in 2008. ARRA has dedicated $400 million in investments across the spectrum of geothermal technologies, from co-produced geothermal energy from oil and gas fields to EGS technologies and more widespread deployment of geothermal heat pumps. DOE estimates that U.S. electricity generated from geothermal power could displace 1.8 Tg CO_2 Eq. in 2020.

Loan Guarantee Program[55]

DOE's Loan Guarantee Program provides financial support for projects that employ new or significantly improved technologies that avoid, reduce, or sequester air pollutants or anthropogenic GHG emissions. As authorized by Title XVII of EPAct, the program promotes early commercial use of advanced technologies for projects with a reasonable prospect of repayment of the principal and interest. In addition to $4 billion in loan guarantee authority provided in FY 2007, DOE has the authority to issue $18.5 billion for new nuclear plants, $2 billion for uranium enrichment, $8 billion for advanced coal, and $18.5 billion for renewable or energy-efficient systems and manufacturing projects. An additional $4 billion in credit subsidy was provided in ARRA to support up to $40 billion in loans for renewable energy and transmission projects, including commercial and advanced technologies.

Water Power Program[56]

Today, power from water resources represents the largest U.S. renewable energy source and serves as a foundation for other renewable forms of energy being developed. DOE's Water Power Program seeks to identify and undertake RDD&D to assess the potential extractable energy from water resources and to facilitate the development and deployment of renewable, environmentally sound, and cost-effective energy from U.S. rivers, estuaries, and marine waters. In 2008, the program began supporting a comprehensive suite of R&D projects geared toward advances in marine and hydrokinetic technology that has the potential to provide 43 gigawatts of additional capacity from waves, tides, and currents in U.S. waters. DOE is also supporting capacity additions for both small and large conventional hydropower, as well as advances in water power grid services to make better use of existing hydroelectric capacity and its highly valuable ancillary services.

Transportation
Renewable Fuel Standard[57]

EISA made several changes to the RFS, as originally implemented under EPAct, including a significant increase in the volume of renewable fuel that must be used in transportation fuel each year. By 2022, 36 billion gallons of renewable fuel are required—a fivefold increase over the volumes included in EPAct. The statute also includes volume requirements for biomass-based diesel and other advanced biofuels, including 16 billion gallons of cellulosic biofuel by 2022.

The revised requirements also include new definitions and criteria for both renewable fuels and the feedstocks used to produce them, including new life-cycle GHG emission thresholds for renewable fuels. EPA, which issued a final rule in February 2010, is currently working to implement these changes to the RFS program, which is anticipated to achieve significant reductions in both petroleum use and GHGs.

National Clean Diesel Campaign[58]

EPA's National Clean Diesel Campaign (NCDC) works aggressively to reduce diesel emissions across the country through the implementation of proven emission control technologies and innovative strategies with the involvement of national, state, and local partners. Many of the clean diesel strategies that NCDC

[52] See http://www1.eere.energy.gov/geothermal/.

[53] Tester et al. *The Future of Geothermal Energy: Impact of Enhanced Geothermal Systems (EGS) on the United States in the 21st Century.* 2006. Refers to 100,000 MWe.

[54] Williams et al. "Assessment of Moderate- and High-Temperature Geothermal Resources of the United States." 2008. Refers to 517,800 MWe. Note that this assessment is only applicable to the western states, and does not include an eastern assessment.

[55] See http://www.lgprogram.energy.gov/index.html.

[56] See http://www1.eere.energy.gov/windandhydro/.

[57] See http://www.epa.gov/otaq/renewablefuels/index.htm.

[58] See http://www.epa.gov/diesel/.

promotes to mitigate nitrogen oxides (NO_x) and particulate matter (PM)—such as retrofits, engine repair, engine replacement, engine repower, idle reduction, and cleaner fuels—can also reduce CO_2 emissions through diesel fuel savings and help mitigate black carbon emissions. Black carbon, a component of PM, has been found to both increase atmospheric warming and speed Arctic melting. Removing PM may have a significant effect on slowing global warming due to the short-lived nature of black carbon.

The Diesel Emissions Reduction Act (DERA) provisions in EPAct are a significant funding source for NCDC. In the first year of the DERA program (FY 2008), EPA awarded $49.2 million for projects across the country—including projects funded through the SmartWay Clean Diesel Finance Program, State Clean Diesel Grants, National Clean Diesel Funding Assistance Program, and Emerging Technologies Program—which will lead to emission reductions of approximately 41,700 metric tons of NO_x and 2,000 metric tons of fine particulate matter ($PM_{2.5}$). EPA estimates that the idle-reduction technologies funded through FY 2008 grants alone will save more than 3.2 million gallons of fuel and 32,300 metric tons of CO_2 per year. Engine replacement, repowering, and vehicle replacement projects are contributing additional fuel savings and CO_2 reductions. With an additional $300 million of ARRA funding to be awarded for clean diesel projects and $60 million appropriated for FY 2009 and FY 2010, EPA anticipates that the level of emissions, including CO_2 and black carbon, from the existing fleet of diesel engines will continue to decrease significantly.

SmartWay℠ Transport Partnership[59]

The SmartWay℠ Transport Partnership is an innovative collaboration with the freight industry to increase energy efficiency while significantly reducing GHGs and air pollution. EPA provides tools and models to help SmartWay Transport partners—including producers and the trucking, rail, and marine shipping companies that deliver their products—adopt cost-effective strategies to save fuel and reduce GHG emissions. To date, more than 2,000 companies and organizations have joined the partnership. Freight shippers meet their goals by using participating carriers, while trucking and rail companies meet their goals by improving freight transport efficiency.

The SmartWay Clean Diesel Finance Program is a complementary initiative that aims to accelerate the deployment of energy-efficient and emission-control technologies by helping truck owners overcome financial obstacles. In 2008, EPA awarded $3.4 million to support three loan programs to help small trucking companies reduce fuel costs and emissions. In 2009, EPA will award the program $30 million from ARRA funding to support the development of new financing programs.

Other SmartWay initiatives include identification of clean and efficient SmartWay-certified vehicles, including heavy-duty trucks, upgrade kits, and components. SmartWay-designated tractor-trailers can save 10–20 percent annually in fuel and CO_2 emissions compared to a typical long-haul truck. SmartWay also promotes a national idle-reduction program for trucks and locomotives and has developed guidance on idle-reduction policies and programs for states. SmartWay's Supply Chain initiative is developing new tools to help companies quantify and track freight transport environmental performance across all modes, including truck, marine, rail, and aviation.

The SmartWay program is also working with other governments and organizations around the world to establish international benchmarks for cleaner, efficient freight transportation. EPA held the first international SmartWay workshop in December 2008, and several countries have initiated projects and programs modeled after the U.S. SmartWay program. EPA estimates that SmartWay could help the industry reduce up to 43 Tg CO_2 Eq. in 2020.

Aviation Fuel Efficiency[60]

In the United States, aviation makes up about 3 percent of the national GHG inventory and about 12 percent of transportation emissions. Currently, measuring and tracking fuel efficiency from aircraft operations provide the data for assessing the improvements in aircraft and engine technology, operational procedures, and the airspace transportation system that reduce aviation's contribution to CO_2 emissions.

Although there are no mandatory U.S. or international goals or requirements with respect to aviation fuel efficiency, the Federal Aviation Administration (FAA) has a number of initiatives to improve aviation fuel efficiency. For example, FAA has aviation GHG reduction goals as part of the Next Generation Air Transportation System (NextGen). In the near term, new technologies to improve air traffic management will help reduce fuel consumption and, thus, emissions. In the long term, new engines and aircraft will feature more efficient components and aircraft aerodynamics, enhanced engine cycles, and reduced weight, thereby improving fuel efficiency and fuel economy.

Commercial Aviation Alternative Fuels Initiative[61]

FAA's Commercial Aviation Alternative Fuels Initiative (CAAFI) seeks to enhance energy security and environmental sustainability for aviation through alternative jet fuels. CAAFI is a government and private-sector coalition that focuses the efforts of commercial aviation to engage the emerging alternative fuels industry. It enables its diverse participants—representing all the leading stakeholders in the field of aviation—to build relationships, share and collect data, identify resources, and direct RD&D of alternative jet fuels.

[59] See http://www.epa.gov/smartway/.

[60] See http://ntl.bts.gov/lib/32000/32700/32779/DOT_Climate_Change_Report_-_April_2010_-_Volume_1_and_2.pdf.

[61] See http://www.caafi.org.

Clean Automotive Technology[62]

EPA's Clean Automotive Technology program searches for cost-effective advanced automotive technologies that greatly cut GHG emissions, increase fuel efficiency, reduce emissions, and are affordable for mainstream consumer and commercial vehicles. The program has developed several historic engine and drivetrain technology breakthroughs, and currently holds 60 powertrain patents with 28 more in process.

EPA has been instrumental in moving advanced vehicle technologies from the lab to the road by partnering with industry companies, such as UPS, FedEx, Navistar, Freightliner, Eaton, and Parker, to get the first series of hydraulic hybrid package-delivery vehicles on the road. The first generation of this advanced technology has improved real-world fuel efficiency by 50 percent. EPA is working to incorporate the next generation of advanced engine and fuel technologies into series hybrids to boost these gains for commercial trucks to near 100 percent.

Fuel Cell Technologies Program[63]

DOE's Fuel Cells Technologies Program is implementing RD&D efforts needed for the widespread use of fuel cells and hydrogen in the stationary, portable, and transportation sectors. Through partnerships with the public and private sectors, national laboratories, and academia, the program accelerates the pace of R&D to reduce U.S. dependence on oil, GHG emissions, and criteria pollutants. Technological advances through R&D over the last seven years have successfully reduced fuel cell costs from $275 per kilowatt (kW) to approximately $61/kW in 2009, based on a production volume of 500,000 units per year, doubled fuel cell durability (to 60,000 miles), and reduced high-volume hydrogen costs to $3 per gallon gasoline equivalent.

Federal Transit Program[64]

DOT's Federal Transit Administration (FTA) grows and sustains public transportation as a low-emission alternative to automobiles by providing grants, technical assistance, research, and policy leadership to communities throughout the United States. FTA's main area of support to communities is providing more than $10 billion per year in grants for the construction and operation of transit services ranging from local buses and paratransit services to heavy-rail subways and commuter rail. Over 1,500 transit providers in communities throughout the United States benefit from FTA grant programs as a way to assist their own efforts to provide public transportation options to their customers. FTA also encourages adoption of clean technologies by funding a significant percentage of the cost of purchasing lower-emission vehicles.

Through its technical assistance efforts focused on transportation planning and transit-oriented de-velopment, FTA provides communities with the tools to effectively coordinate land-use and transportation decisions. Combining investment in public transportation with compact, mixed-use development around transit stations has a synergistic effect that amplifies the GHG reductions of each activity. FTA also provides environmental management systems training to transit agencies to help them continually assess and reduce the environmental impact of their operations.

FTA research on alternative fuels and high-fuel-efficiency vehicles has yielded the introduction of low-emission technologies, such as hybrid electric, compressed natural gas, and biodiesel. Current research is also supporting the development of a commercially viable fuel cell bus through a multi-year, multi-organization research effort. Finally, FTA policy research is aimed at providing the analysis to give decision makers the information to make informed decisions about the role of public transportation in reducing GHG emissions.

Transit Investments for Greenhouse Gas and Energy Reduction[65]

Administered by FTA, DOT's Transit Investments for Greenhouse Gas and Energy Reduction (TIGGER) grant program aims to position the public transportation industry as a leader in the effort to reduce America's dependence on foreign oil and help address global climate change. Created by ARRA, the program provides $100 million in discretionary grants to U.S. transit agencies for capital investments that will help reduce energy consumption or GHG emissions. In particular, TIGGER seeks to fund projects that will yield long-term energy savings and emission reductions from targeted investments in public transit facilities and vehicle operations.

Vehicle Technologies Program[66]

DOE's Vehicle Technologies Program develops energy-efficient and environmentally friendly highway transportation technologies that will reduce use of petroleum in the United States. The long-term aim is to develop "leapfrog technologies" that will provide Americans with greater freedom of mobility and energy security, while lowering costs and reducing impacts on the environment. Program areas include hybrid and vehicle systems, energy storage, power electronics and electrical machine technologies, advanced combustion engines, fuel and lubricant technologies, materials, EPAct support, and educational activities. Clean Cities, the main deployment arm of the program, is a public–private partnership designed to reduce petroleum consumption in the transportation sector by advancing the use of alternative fuels and vehicles, idle-reduction technologies, hybrid electric vehicles, fuel blends, and fuel economy measures. Industry partnerships include the FreedomCAR and Fuel Partnership and the 21st Century Truck Partnership.

[62] See http://www.epa.gov/otaq/technology/.

[63] See http://www1.eere.energy.gov/hydrogenandfuelcells/.

[64] See http://www.fta.dot.gov/.

[65] See http://www.fta.dot.gov/index_9440_9920.html.

[66] See http://www1.eere.energy.gov/vehiclesandfuels/.

Congestion Mitigation and Air Quality Improvement Program[67]

Administered by DOT's Federal Highway Administration (FHA) and FTA in consultation with EPA, the Congestion Mitigation and Air Quality Improvement (CMAQ) Program provides states with funding to reduce congestion and improve air quality through transportation control measures and other transportation strategies that will contribute to attainment or maintenance of the national ambient air quality standards for ozone, carbon monoxide, and PM. Funds are apportioned to states by statutory formula, and the amount of funding is primarily based on the severity of the air quality problem and the population of the area. State and local governments select CMAQ projects and coordinate them through metropolitan planning organizations (MPOs), where appropriate. The projects typically include transit improvements, alternative fuel programs, shared-ride services, traffic flow improvements, demand management strategies, freight and intermodal facilities, diesel engine retrofits, pedestrian and bicycle programs, and inspection and maintenance programs.

The Safe, Accountable, Flexible, Efficient, Transportation Equity Act: A Legacy for Users (SAFETEA-LU)[68] directed states and MPOs to give priority to two categories of funding: (1) diesel retrofits and other cost-effective emission reduction activities, taking into consideration air quality and health effects; and (2) cost-effective congestion mitigation activities that provide air quality benefits. States and local governments, however, retain their project selection authority through SAFETEA-LU, maintaining the full range of eligible CMAQ projects, such as idle-reduction efforts and public education and outreach programs.

Alternative Transport Systems and Use of Clean Vehicles

Alternative Transportation Systems[69]

Alternative Transportation Systems integrate all modes of travel within a park managed by NPS, including public transit, bicycle and pedestrian linkages, automobiles, and a whole range of technologies, facilities, and transportation management strategies. There are 98 National Park Units supporting 110 Alternative Transportation Systems: 17 systems are owned and operated by NPS; 71 systems are operated by a concessionaire; and 22 systems are run by NPS, in partnership with local public transit service.

Alternative-Fuel Vehicles

BLM has reduced the size of its vehicles and fleet, has purchased more than 500 alternative-fuel vehicles, and has established a transportation policy to reduce vehicle use, which saves money, reduces fossil fuel use, and lowers GHG emissions. Current BLM targets include reducing the fleet's total consumption of petroleum products by 2 percent annually through the end of FY 2015, increasing the total fuel consumption that is non-petroleum-based by 10 percent annually, and using plug-in hybrid electric vehicles (PHEVs) when they are commercially available at a cost reasonably comparable to non-PHEVs, on the basis of life-cycle cost.

Industry: Non-CO$_2$

Methane Programs[70]

U.S. industries and state and local governments collaborate with EPA to implement several voluntary programs that promote profitable opportunities for reducing emissions of methane, an important GHG. These programs are designed to overcome a wide range of informational, technical, and institutional barriers to reducing methane emissions, while creating profitable activities for the coal, natural gas, and petroleum industries. The collective results of EPA's voluntary methane partnership programs have been substantial. Total U.S. methane emissions in 2004 were 10 percent lower than emissions in 1990, despite robust economic growth over that period. EPA expects that these programs will maintain emissions below 1990 levels through and beyond 2020 due to expanded industry participation and the continuing commitment of the participating companies to identify and implement cost-effective technologies and practices.

Coalbed Methane Outreach Program[71]

The fraction of coal mine methane captured and used by degasification systems grew from 25 percent in 1990 to more than 80 percent in 2006. Initiated in 1994, EPA's Coalbed Methane Outreach Program (CMOP) is working to demonstrate technologies that can eliminate the remaining emissions from active mine degasification systems, including mitigating methane emissions in mine ventilation air. The program also addresses opportunities to recover and use methane emitted from abandoned (closed) underground mines. EPA estimates that CMOP reduced 7 Tg CO$_2$ Eq. in 2007. Based on a number of anticipated conditions, including enhanced market opportunities for natural gas and power, further refinement of technical options for the capture and utilization of mine methane, a growing reliance on methane degasification in the U.S. West, and CMOP's anticipated success in reducing ventilation air methane over the next few years, EPA projects that the program could help reduce emissions by 12 Tg CO$_2$ Eq. in 2020.

Natural Gas STAR[72]

Through this partnership program, EPA works with oil and natural gas companies to promote proven, cost-effective technologies and practices that improve operational efficiency and reduce methane (i.e., natural gas) emissions. Methane is emitted by oil production and all sectors of the natural gas industry, from drilling and production, through processing and storage, to transmission and distribution. Since its launch

---

[67] See http://www.fhwa.dot.gov/environment/cmaqpgs/.

[68] Public Law 109-59.

[69] See http://www.nps.gov/transportation/tmp/shuttles.htm.

[70] See http://www.epa.gov/methane/index.html.

[71] See http://www.epa.gov/cmop/index.html.

[72] See http://www.epa.gov/gasstar/index.htm.

in 1993, Natural Gas STAR has been successful in working with U.S. oil and natural gas companies to reduce methane emissions and bring more energy to markets. As of 2007, Natural Gas STAR partner companies represented almost 60 percent of the U.S. natural gas industry. For calendar year 2007, Natural Gas STAR domestic partners reported emission reductions of approximately 37 Tg CO_2 Eq. EPA projects that the program could help reduce methane emissions by 46.9 Tg CO_2 Eq. in 2020.

High Global Warming Potential Programs[73]

The United States is one of the first nations to develop and implement a national strategy to control emissions of high global warming potential (GWP) gases. The strategy is a combination of industry partnerships and regulatory mechanisms to minimize atmospheric releases of hydrofluorocarbons (HFCs), perfluorocarbons (PFCs), and sulfur hexafluoride (SF_6), which are potent GHGs that contribute to global warming, while ensuring a safe, rapid, and cost-effective transition away from chlorofluorocarbons (CFCs), hydrochlorofluorocarbons (HCFCs), halons, and other ozone-depleting substances (ODS) across multiple industry sectors.

Environmental Stewardship Initiative

EPA's Environmental Stewardship Initiative aims to limit emissions of HFCs, PFCs, and SF_6 in three industrial applications: semiconductor production,[74] electric power distribution,[75] and magnesium production.[76] Since 2002, the SF_6 emission reduction partnership for magnesium has worked toward its goal of eliminating emissions of SF_6 by the end of 2010. Additional sectors are being assessed for the availability of cost-effective emission reduction opportunities and are being added to this initiative. EPA estimates that partnerships in this initiative reduced emissions by 17.3 Tg CO_2 Eq. in 2007 and projects that the programs could help reduce emissions by 44.7 Tg CO_2 Eq. in 2020.

Voluntary Code of Practice for the Reduction of Emissions of HFC & PFC Fire Protection Agents[77]

In 2002, EPA and several hundred equipment and chemical manufacturers and distributors representing the U.S. fire protection industry launched the Voluntary Code of Practice for the Reduction of Emissions of HFC & PFC Fire Protection Agents (VCOP). Successful implementation of VCOP achieves the dual goals of minimizing nonfire emissions of HFCs and PFCs (predominantly HFCs), which are used as fire-suppression alternatives to ozone-depleting halons, and can effectively protect people and property from the threat of fire. In addition, approximately 22 manufacturers annually report to the HFC Emissions Estimating Program, tracking industry-wide emissions of HFCs and progress under VCOP.

HFC-23 Emission Reduction Partnership[78]

EPA's HFC-23 Emission Reduction Partnership continued to encourage companies to develop and implement technically feasible, cost-effective processing practices or technologies to reduce HFC-23 emissions from the manufacture of HCFC-22. Despite a 4 percent increase in the production of HCFC-22 compared to 1990, EPA estimates that total HFC emissions in 2007 were significantly below 1990 levels. Compared to business as usual, EPA estimates the partnership reduced emissions by 17.8 Tg CO_2 Eq. in 2007, and projects that this partnership could help reduce GHG emissions by 20.9 Tg CO_2 Eq. for 2020, due to manufacturers switching away from the production of this chemical.

Mobile Air Conditioning Climate Protection Partnership[79]

EPA's Mobile Air Conditioning Climate Protection Partnership has been reducing GHG emissions from vehicle air conditioning fuel use and refrigerant emissions. In 2007, the partnership demonstrated new, commercially available technology that would reduce vehicle air conditioner fuel use by over 30 percent, cut refrigerant emissions by 50 percent, reduce cooling loads, and improve refrigerant recovery and recycling. These technologies are currently being integrated into vehicles and refrigerant recovery and recycling equipment sold in the United States and worldwide. In addition, partnership members have announced plans to switch to a low-GWP refrigerant. This transition, which will begin in 2010, could help displace 24.6 Tg CO_2 Eq. of GHG in 2020.[80]

GreenChill Advanced Refrigeration Partnership[81]

Formerly known as Green Grocer, the GreenChill Advanced Refrigeration Partnership is an EPA cooperative alliance with the supermarket industry to promote advanced technologies, strategies, and practices that reduce supermarkets' impact on the ozone layer and climate system. Through GreenChill, EPA works with supermarkets to reduce the amount of refrigerants they use in their stores. Refrigerants are responsible for high-GWP and ozone-depleting gases, so their minimization is especially beneficial for the environment.

EPA launched GreenChill in November 2007 with 10 founding partners. GreenChill now has 45 partners with more than 6,500 supermarkets (18 percent of all U.S. supermarkets) in 47 states. Upon joining, partners measure their refrigerant emissions annually and set goals to reduce those emissions. On average, more than 20 percent of the refrigerant used each year in the supermarket industry is released into the atmosphere in the form of dangerous GHGs. In 2008—GreenChill's first year of measuring supermarket emission reductions—partners reduced their aggregate total corporate emission rate from 13 percent to 11.9 per-

[73] See http://www.epa.gov/highgwp/.

[74] See http://www.epa.gov/semiconductor-pfc/.

[75] See http://www.epa.gov/highgwp/electricpower-sf6/index.html.

[76] See http://www.epa.gov/magnesium-sf6/.

[77] See http://epa.gov/ozone/snap/fire/vcopdocument.pdf.

[78] See http://www.epa.gov/highgwp/voluntary.html.

[79] See http://www.epa.gov/cppd/mac/.

[80] According to the latest U.S. GHG Inventory Report (U.S. EPA/OAP 2009), HFC-134a emissions from U.S. vehicle air conditioners account for 50 MMTCO$_2$ Eq. per year. Transition to a low-GWP refrigerant (either CO_2, GWP 1, or HFO-1234yf, GWP 4) will reduce CO_2 Eq. emissions by well over 99 percent).

[81] See http://www.epa.gov/greenchill/.

cent per year. As the partnership continues to grow, EPA estimates it could help reduce annual GHG emissions in 2020 by 1.9 Tg CO_2 Eq.

Responsible Appliance Disposal Program[82]

Announced in 2006, the Responsible Appliance Disposal Program is reducing GHG emissions from refrigerant-containing home appliances that have reached their end of life. Through this voluntary program, partners ensure the disposal of refrigerant-containing appliances using the best environmental practices available. Through such responsible disposal practices, partners are able to recover and recycle refrigerants and foam, thereby reducing emissions of high-GWP gases. They also prevent the release of hazardous materials (e.g., used oil, polychlorinated biphenyls, and mercury), and they save landfill space and energy by recycling durable materials. EPA estimates that the annual emission reduction in 2010 from this effort will be 0.1 Tg CO_2 Eq. In 2020, the program could help to save 0.4 Tg CO_2 Eq. (not including possible CO_2 reductions from energy savings).

Significant New Alternatives Policy Program[83]

Since the 2006 CAR, EPA's Significant New Alternatives Policy Program has continued to identify substitutes for ODS, such as CFCs and HCFCs. EPA has worked closely with industry to research, identify, and implement climate- and ozone-friendly alternatives, supporting a smooth transition to these new technologies. In addition, EPA has initiated programs with different industry sectors to monitor and minimize emissions of global-warming gases, such as HFCs and PFCs, used as substitutes for ozone-depleting chemicals. By limiting use of these gases in specific applications where safe alternatives are available, EPA reduced annual emissions by an estimated 115 Tg CO_2 Eq. in 2007, and the program could help to reduce GHG emissions by an estimated 240 Tg CO_2 Eq. in 2020.

Voluntary Aluminum Industry Partnership[84]

EPA's Voluntary Aluminum Industry Partnership continued to reduce PFCs, tetrafluoromethane, and hexafluoroethane where cost-effective technologies and practices are technically feasible. Since 2006, the partnership has intensified its efforts to further reduce PFC emissions and direct carbon emissions from anode consumption. EPA estimates that the partnership reduced PFC emissions by 10.0 Tg CO_2 Eq. in 2007, and EPA projects reductions of 8.2 Tg CO_2 Eq. in 2020.

Agriculture

Through a portfolio of conservation, renewable energy, and energy efficiency programs, USDA provides incentives and other support for voluntary actions by private landowners to reduce GHG emissions and increase carbon sequestration. Depending on the program and activity, USDA support can include financial incentives, technical assistance, demonstrations, pilot programs, education and capacity building, and

frameworks and tools for assessing success in achieving GHG benefits. Major elements of the USDA actions to reduce GHGs are described in the sections below.

AgSTAR[85]

AgSTAR is a voluntary effort jointly sponsored by EPA, USDA, and DOE. The program encourages the use of methane recovery (biogas) technologies at confined animal feeding operations that manage manure as liquids or slurries. These technologies reduce methane emissions while achieving other environmental benefits. Although the overall impact of AgSTAR on GHG emissions has been comparatively small on a national scale, livestock producers in the dairy and swine sector have demonstrated that the practices can reduce GHG emissions and achieve other pollution control benefits while increasing farm profitability. The practices recommended under AgSTAR have been incorporated into USDA's broader technical, conservation, and cost-share programs.

Conservation Reserve Program[86]

The Conservation Reserve Program (CRP) encourages farmers to convert environmentally sensitive acreage to native grasses, wildlife plantings, trees, restored wetlands, filter strips, or riparian buffers. Administered by USDA's Farm Service Agency (FSA), the CRP sequesters more carbon on private lands than any other federally administered program. CRP contracts last 10–15 years, and landowners retain the right to put land back into production once contracts end. Hence, the benefits of many contracts are not permanent. FSA allows the private sale of carbon credits for lands enrolled in the CRP. FSA has also included carbon sequestration potential in its ranking process by which offers are selected for enrollment. In addition to increasing carbon sequestration, CRP lands produce GHG benefits in the form of reduced CO_2 emissions from fewer field operations and reduced N_2O emissions from avoided fertilizer applications. For 2008, FSA estimates the net GHG benefits of the CRP were emission reductions of 56 Tg CO_2 Eq. This value is expected to decrease to 53 Tg CO_2 Eq. in 2012 and remain at that level through 2020.

Environmental Quality Incentives Program[87]

The Environmental Quality Incentives Program (EQIP) provides financial assistance for conservation practices on working farm and ranch lands. NRCS has provided guidance to its state offices to recognize actions that provide GHG benefits within the EQIP ranking systems. A wide array of conservation practices can reduce GHG emissions, including residue management, irrigation and water management, nutrient management, crop rotations, cover crops, restoring wetlands, and grazing land management. However, these benefits are not permanent, as EQIP contracts last for 10 years, and producers retain the right to put land back into production after the contract ends. In

[82] See http://www.epa.gov/Ozone/partnerships/rad/.

[83] See http://www.epa.gov/ozone/snap/.

[84] See http://www.epa.gov/highgwp/aluminum-pfc/index.html.

[85] See http://www.epa.gov/agstar/ and http://www.rurdev.usda.gov/rbs/farmbill/index.html.

[86] See http://www.fsa.usda.gov/dafp/cepd/crp.htm.

[87] See http://www.nrcs.usda.gov/programs/eqip/.

2009, NRCS estimated the GHG mitigation benefits associated with 17 conservation practices that it identified as sequestering carbon and/or reducing emissions. For 2007, total GHG mitigation attributable to these practices is estimated at 3.9 Tg CO_2 Eq. This value is projected to increase to 14.2 Tg CO_2 in 2020.

Conservation Stewardship Program[88]

Formerly known as the Conservation Security Program, USDA's Conservation Stewardship Program (CSP) is a voluntary nationwide program that provides financial and technical assistance to promote the conservation and improvement of soil, water, air, energy, plant and animal life, and other conservation purposes on tribal and private working lands. Working lands include cropland, grassland, prairie land, improved pasture, and range land, as well as forested land that is an incidental part of an agriculture operation. CSP contracts last 5 years, and landowners retain the right to put land back into production once contracts end. Hence, the benefits of many contracts are not permanent. CSP has the potential to support activities and actions that increase carbon sequestration and reduce GHG emissions. The program began in FY 2009. GHG emission reductions will depend on the contracts enrolled and practices those contracts put into place.

Wetlands Reserve Program[89]

The Wetlands Reserve Program (WRP) is a voluntary program offering landowners the opportunity to protect, restore, and enhance wetlands on their property. NRCS provides technical and financial support to help landowners with their wetland restoration efforts, toward the goal of achieving the greatest wetland functions and values, along with optimum wildlife habitat, on every acre enrolled in the program. This program offers landowners an opportunity to establish long-term conservation and wildlife practices and protection. Activities associated with wetland conservation often increase carbon sequestration and reduce GHG emissions. For 2007, NRCS estimates the WRP reduced GHG emissions by 0.18 Tg CO_2 Eq. These reductions are projected to increase to 0.25 Tg CO_2 Eq. in 2020.

Grassland Reserve Program[90]

NRCS's Grassland Reserve Program (GRP) is a voluntary conservation program that emphasizes support for working grazing operations, enhancement of plant and animal biodiversity, and protection of grassland under threat of conversion to other uses. Participants voluntarily limit future development and cropping uses of the land, while retaining the right to conduct common grazing practices and operations related to the production of forage and seeding, subject to certain restrictions during nesting seasons of bird species that are in significant decline or are protected under federal or state law. A grazing management plan is required for participants. Many of the conservation practices encouraged under the GRP increase the quantity of carbon sequestered in the affected soils. However, only part of these benefits is permanent, as the program includes both permanent easements and 10- to 20-year rental contracts. Once rental contracts end, farmers retain the right to put land back into production. For 2007, NRCS estimates the GRP reduced GHG emissions by 0.007 Tg CO_2 Eq. These reductions are projected to increase to 0.027 Tg CO_2 Eq. in 2020.

Wildlife Habitat Incentives Program[91]

The 2008 Farm Bill reauthorized the Wildlife Habitat Incentives Program (WHIP) as a voluntary approach to improving the nation's wildlife habitat. NRCS administers WHIP to provide both technical assistance and up to 75 percent cost-share assistance to establish and improve fish and wildlife habitat. WHIP cost-share agreements between NRCS and the participant generally begin one year after the last conservation practice is implemented and end not more than 10 years from the date the agreement is signed. Establishing and enhancing wildlife habitats often increase carbon sequestration and reduce GHG emissions. However, these benefits are not permanent, as landowners retain the right to put land back into production once the contracts end. For 2007, NRCS estimates WHIP reduced GHG emissions by 0.25 Tg CO_2 Eq. These reductions are projected to increase to 0.50 Tg CO_2 Eq. in 2020.

Forestry

The U.S. government supports efforts to sequester carbon in both forests and harvested wood products to minimize unintended carbon emissions from forests by reducing the catastrophic risk of wildfires.

Enhancing Ecosystem Services on Forests, Grasslands, Parks, and Wildlife Reserves

To address the effects of climate change on federal lands administered by DOI and other agencies, DOI, in conjunction with USDA, EPA, other federal agencies, academic institutions, and the private sector, is conducting thorough ecoregional analyses. These analyses will help DOI conserve, enhance, restore, and adapt ecosystems, find opportunities for carbon sequestration, and provide opportunities for renewable energy development. As a result of extreme fire events, other natural disasters, and human activities, public lands present a substantial opportunity to optimize the potential benefits of a carbon sequestration program. Those benefits include long-term capture and storage of CO_2, improved biodiversity and wildlife habitat condition and connectivity, improved water quality and quantity, reduced soil erosion, decreased invasive species, improved environmental esthetics, and enhanced recreational experiences.[92]

---

[88] See http://www.nrcs.usda.gov/programs/CSP/.

[89] See http://www.nrcs.usda.gov/Programs/WRP/.

[90] See http://www.nrcs.usda.gov/programs/grp/.

[91] See http://www.nrcs.usda.gov/programs/whip/.

[92] U.S. Forest Service, Forest Inventory and National Analysis Program. Forest Inventory Data Online. See http://fiatools.fs.fed.us/fido/. The estimate was prepared by DOI. The footnote refers generally to the USDA Forest Inventory data.

Healthy Forest Initiative

Today, up to 81 million ha (200 million ac) of federal lands are currently at risk for catastrophic wildfires, in large part due to significant changes in forest structure and density during the last 60–70 years, prolonged drought, and other environmental changes. The need for innovative, large-scale management to restore the health and productivity of at-risk ecosystems prompted the development of the Healthy Forest Initiative, which now includes the National Fire Plan[93] and the joint federal–state 10-Year Strategy Implementation Plan.[94] One goal of these efforts is to increase biomass and wood fiber utilization as an integral component of restoring the nation's forests, woodlands, and rangelands. Addressing hazardous fuels on federal lands is a key element of the National Fire Plan and related efforts, with almost 7.3 million ha (18 million ac) being treated by USDA since 2001.

Woody Biomass Utilization Grant Program[95]

The Woody Biomass Utilization Grant Program focuses on creating markets for small-diameter material and low-valued trees removed from forest restoration activities, such as reducing hazardous fuels, handling insect and disease conditions, or treating forestlands impacted by catastrophic weather events. Most of this woody biomass would have been piled and burned in the open, so the program reduces GHG emissions when that material is used for energy or substitutes for fossil fuel-intensive products instead. For 2008, the U.S. Forest Service estimates this grant program reduced GHG emissions by 0.43 Tg CO_2 Eq. These reductions are projected to increase to 0.77 Tg CO_2 Eq. in 2020.

Waste Management

The U.S. government's waste management programs reduce municipal solid waste and GHG emissions through energy savings, increased carbon sequestration, and avoided methane emissions from landfill gas—the largest contributor to U.S. anthropogenic methane emissions.

Stringent Landfill Rule[96]

Promulgated under the Clean Air Act in March 1996, the New Source Performance Standards and Emissions Guidelines (Landfill Rule) require large landfills to capture and combust their landfill gas emissions. The implementation of the rule began at the state level in 1998. Recent data on the rule's impact indicate that increasing its stringency has significantly increased the number of landfills that must collect and combust their landfill gas. The current EPA projection is that reductions will be about 10 Tg CO_2 Eq. in 2020.

Landfill Methane Outreach Program[97]

EPA's Landfill Methane Outreach Program (LMOP) reduces GHG emissions at landfills by supporting the recovery and use of landfill gas for energy. Capturing and using landfill gas reduces methane emissions directly and reduces CO_2 emissions by displacing the use of fossil fuels through the utilization of landfill gas as a source of energy.

Since the 2006 CAR, the LMOP continues to partner with landfill owners and operators, state energy and environmental agencies, utilities and other energy suppliers, corporations, industry, and other stakeholders to lower the barriers to installing cost-effective landfill gas energy projects. LMOP focuses its efforts on smaller landfills not required to collect and combust their landfill gas, as well as larger, regulated operations that are combusting their gas but not utilizing it as a clean energy source. LMOP has developed a range of technical resources and tools to help the landfill gas industry overcome barriers to energy project development, including feasibility analyses, project evaluation software, a database of more than 500 candidate landfills across the country, a project development handbook, commercial and industrial sector analyses, and economic analyses. Due to these efforts, the number of landfill gas energy projects has grown from approximately 100 in 1990 to 500 projects today. EPA estimates that LMOP reduced GHG emissions from landfills by 19 Tg CO_2 Eq. in 2007, and projects reductions of 30.8 Tg CO_2 Eq. in 2020.

WasteWise[98]

WasteWise encourages waste reduction through preventing and recycling waste and purchasing recycled-content products. EPA is implementing a number of targeted efforts within this program and is working with organizations to reduce solid waste through voluntary waste reduction activities. New efforts since the 2006 CAR include WasteWise Communities, a WasteWise campaign in support of local governments to reduce residential municipal solid waste and its impact on climate change; WasteWise Re-TRAC, a valuable new Web-based tool to assist organizations with tracking and analyzing their waste reduction activities; and the Office Carbon Footprint Tool, which assists office-based organizations in making decisions to reduce the GHG emissions associated with their activities. In addition to program implementation, EPA's climate and waste programs support outreach, technical assistance, and research efforts on the linkages between climate change and waste management. EPA estimates GHG emission reductions in 2007 were 20 Tg CO_2 Eq. EPA projects reductions will increase to 38 Tg CO_2 Eq. in 2020.

Cross-Sectoral

Carbon Monitoring and Sequestration

Under EISA, DOI's USGS is charged with assessing the status of U.S. carbon stores.[99] This includes the potential for global, active, long-term containment of carbon in subsurface geologic areas and the natural capacity of ecosystems—including forests, soils, wet-

---

[93] See http://www.forestsandrange lands.gov/reports/documents/ 2001/8-20-en.pdf.

[94] See http://www.forestsandrange lands.gov/plan/documents/10-YearStrategyFinal_Dec2006.pdf.

[95] See http://www.fs.fed.us/ woodybiomass/opportunities. shtml.

[96] See http://www.epa.gov/ttn/ atw/landfill/landflpg.html.

[97] See http://www.epa.gov/lmop/.

[98] See http://www.epa.gov/waste/ partnerships/wastewise/index.htm.

[99] See http://frwebgate.access.gpo. gov/cgi-bin/getdoc.cgi?dbname= 110_cong_public_laws&docid= f:publ140.110.pdf.

lands, and coastal areas—to store carbon. DOI, in consultation with DOE, USDA, and others, conducts national assessments of biologic carbon sequestration, ecosystem GHG fluxes, and potential effects of management practices and policies on ecosystem carbon sequestration and GHG emissions. This work is essential to developing science-based best management practices for GHG mitigation within the United States and globally.

USGS scientists are helping to assess ways to limit human-caused CO_2 emissions and remove GHGs from the atmosphere through both geologic and biological carbon sequestration. They are also closely evaluating the potential environmental risks and economic costs of capturing and storing CO_2, as well as other carbon management strategies, and are studying the global carbon cycle by observing both natural and human sources for CO_2 and how it moves and interacts in the environment. These data are crucial to establishing a baseline of carbon and GHG emissions essential to source attribution and reduced uncertainty in global carbon and climate models.

USGS also operates remote-sensing and satellite monitoring systems, such as Landsat, which help monitor afforestation and prevention of deforestation efforts globally. The Landsat data set contains over 2.4 million scenes of the Earth's surface spanning over 37 years and is the only global, radiometrically accurate, terrestrial database available today. As such, it is critical to studying land-use trends and ecosystem performance that occurs as a consequence of climate change. Landsat is Web-enabled and delivered over 1 million scenes to 166 countries during FY 2009.

Interagency Partnership for Sustainable Communities[100]

In June 2009, DOT, HUD, and EPA announced the Interagency Partnership for Sustainable Communities.[101] This partnership has identified six principles to guide the alignment of federal transportation, environmental protection, and housing policies:

- Provide more transportation choices.
- Promote equitable, affordable housing.
- Enhance economic competitiveness.
- Support existing communities.
- Coordinate polices and leverage investment.
- Value communities and neighborhoods.

The interagency partnership will pursue these principles by establishing incentives for integrated regional design, identifying and removing federal barriers to sustainable design strategies, and providing information and training to federal employees to incorporate these principles at the local level. Key benefits of this partnership include reduced vehicle miles traveled, lower per-capita GHG emissions, and reduced dependence on fossil fuels.

Voluntary Greenhouse Gas Reporting in Agriculture and Forestry[102]

In 2006, USDA completed the first phases of its development of comprehensive accounting rules and guidelines for forest and agriculture GHG emissions and carbon sequestration. These technical guidelines enable farmers, ranchers, and forest landowners to construct entity-level GHG inventories that account for emissions and removals from virtually all agriculture and forestry sources and sinks. By preparing annual inventories, farmers and forest landowners can quantify and track changes in GHG emissions and terrestrial carbon sequestration associated with changes in production activities and land-use practices. DOE has adopted USDA's technical guidelines for use in this voluntary GHG reporting program, which was originally established by Section 1605(b) of the Energy Policy Act of 1992. USDA will continue to develop technical guidelines and science-based methods for energy efficiency and quantifying GHG emissions and removals from agriculture and forestry sources and sinks, as directed by EISA.

Climate Leaders[103]

EPA launched Climate Leaders in 2002. In recent years, the program has initiated work in several new areas—providing lower-emitting companies with more streamlined tools and guidance, advancing measurement of indirect emissions associated with firms' supply chains, and allowing partners to meet their voluntary reduction goals with offsets. Companies that join the partnership receive a number of benefits, such as understanding and managing their emissions, increased identification of cost-effective reduction opportunities, and strategic preparation for the future as the climate change policy discussion evolves. Climate Leader partners set aggressive, corporate-wide GHG reduction goals and conduct annual inventories of their emissions to measure progress. The program has expanded from its original 12 Charter Partners to over 250 partners across a number of industrial sectors from heavy manufacturing to banking and retail. The GHG emissions from these partners comprise more than 8 percent of total U.S. emissions.

Climate Showcase Communities Grant Program[104]

In 2009, EPA issued $10 million in grants through the Climate Showcase Communities program, an initiative to help local and tribal governments take steps to reduce GHG emissions while achieving additional environmental, economic, and social benefits.[105] The goal of these grants was to create models of community action that generate cost-effective and persistent GHG reductions and can be replicated across the country. Local and tribal activities funded through the grants included energy performance in municipal, residential, commercial, and industrial operations; land use and transportation; waste management; renewable

[100] See http://www.epa.gov/dced/2009-0616-epahuddot.htm.

[101] See http://www.dot.gov/affairs/2009/dot8009.htm.

[102] See http://www.usda.gov/oce/global_change/gg_reporting.htm.

[103] See http://www.epa.gov/climateleaders/.

[104] See http://www.epa.gov/RDEE/energy-programs/state-and-local/showcase.html.

[105] See http://www.epa.gov/RDEE/energy-programs/state-and-local/showcase.html.

energy; heat island management; removal of barriers to GHG management; and other innovative activities that generate measurable reductions of GHGs. EPA is offering peer exchange, training, and technical support to grant recipients, and showcasing the recipients' successes to spur additional action.

State Climate and Energy Partner Network[106]

The State Climate and Energy Partner Network is the next generation of EPA's Clean Energy-Environment State Partnership, which operated from 2005 to 2009 and included 16 partner states (California, Colorado, Connecticut, Georgia, Hawaii, Massachusetts, Minnesota, New Jersey, New Mexico, New York, North Carolina, Ohio, Pennsylvania, Texas, Utah, and Virginia). Through the partnership, each state took important steps toward developing clean energy action plans and integrating energy and environmental strategies to achieve multiple benefits. Lessons learned from the partnership will be shared broadly through the new partner network.

Climate Friendly Parks[107]

Climate Friendly Parks (CFP) was launched in 2003 as a collaborative partnership between NPS and EPA. Run independently by NPS since July 2009, the CFP program is dedicated to helping NPS and the general public understand the interaction between climate change and national parks. The program now involves more than 70 parks and focuses on providing the necessary tools and resources so that parks can (1) measure their GHG emissions, (2) plan ways to reduce their impact on the global climate, (3) adapt to a changing climate, and (4) effectively promote sound science by educating park staff and the public about climate change. CFP's Climate Leadership in Parks tool, an Excel-based calculator designed for parks to assess their own GHG emissions, focuses on in-park operational activities, such as electricity use, transportation, waste and wastewater treatment, and other GHG-emitting activities inside parks.

National Action Plan for Energy Efficiency[108]

Since 2005, DOE and EPA have facilitated the National Action Plan for Energy Efficiency, which engages more than 60 leading electric and gas utilities, state utility regulators and energy agencies, energy consumers, and others. The Action Plan focuses on the critical state-level policies that have a profound impact on the overall level of investment in energy efficiency across the country. Through its "Vision for 2025," this leadership group offers a complete policy framework to implement substantial cost-effective energy efficiency measures by 2025. A number of best-practice-based guides, reports, and tools are available to help organizations expand and meet their commitments to energy efficiency.

State Energy Program[109]

The State Energy Program (SEP) provides grants and technical assistance to states and U.S. territories to promote energy conservation and reduce the growth of energy demand in ways that are consistent with national energy goals. State energy offices use SEP funds to develop state plans that identify new opportunities for states to adopt renewable energy and energy efficiency technologies. SEP funds are also used to implement programs to improve energy sustainability.

SEP is the only program in the DOE Office of Energy Efficiency and Renewable Energy that supports outreach for energy technologies in every sector of the economy: industry; businesses; residences; public facilities, schools, and hospitals; and transportation. SEP effectively leverages investment in renewable energy and energy efficiency. Based on the nationally peer-reviewed 2005 Oak Ridge National Laboratory methodology, for each $1 of federal investment in SEP, states report $10 of nonfederal investment in energy projects and $7.22 savings in energy costs. State energy offices propagate this financial leverage of SEP funds by co-sponsoring energy projects with local stakeholders and private-sector partners. Those partners, in turn, provide feedback and help DOE direct requests for technical assistance.

SEP is playing a central role in implementing ARRA. Under ARRA, states have received $3.1 billion for energy projects through SEP. These funds are allocated among the states according to the following formula: one-third equally among states and territories, one-third according to population, and one-third according to energy consumption. States will use this funding to upgrade the efficiency of state and local facilities, expand utility energy efficiency programs that help families save money on their energy bills, promote consumer products that carry the ENERGY STAR® label for energy efficiency, and invest in alternative fuel infrastructure.

Federal Government Programs
Federal Energy Management Program[110]

The federal government is the largest single user of energy in the nation. DOE's Federal Energy Management Program (FEMP) works to reduce the cost and environmental impact of the federal government by advancing energy efficiency and water conservation, promoting the use of distributed and renewable energy, and improving utility management decisions at federal sites. FEMP accomplishes its mission by leveraging both federal and private resources to provide federal agencies with the technical and financial assistance they need to achieve their energy goals and statutory requirements. FEMP provides project transaction services (assistance with alternative financing mechanisms), applied technology services, and decision support services. FEMP also provides guidance

[106] See http://www.epa.gov/cleanenergy/energy-programs/state-and-local/index.html.

[107] See http://www.nps.gov/climatefriendlyparks/.

[108] See http://www.epa.gov/eeactionplan/.

[109] See http://apps1.eere.energy.gov/state_energy_program/.

[110] See http://www1.eere.energy.gov/femp/.

and assistance for implementing and managing energy-efficient and alternative-fuel vehicles within federal fleets.

As of 2007, FEMP had assisted federal agencies in reducing the energy intensity of their buildings by 11 percent compared to 2003. DOE estimates that realizing FEMP's goal of providing financing and technical assistance to federal agencies to further the use of cost-effective energy efficiency and renewable energy could result in energy savings of about 3.4 Tg CO_2 Eq. in 2020.

NONFEDERAL POLICIES AND MEASURES

Within the United States, several regional, state, and local initiatives supplement the federal effort to reduce GHG emissions. These policies either directly regulate GHG emissions or encourage investments in energy efficiency and renewable energy, thereby leading to GHG reductions. Through the U.S. Department of State and its Office of Global Intergovernmental Affairs, the results of and feedback from state and local climate protection efforts will play an integral role in the development of the federal actions to address climate change. In addition, some of these actions serve as a model for countries that are beginning to formulate their response to climate change because they can be tailored to local and regional conditions, are often scalable, and can create economic opportunities and job growth through the promotion of clean energy.

Direct Greenhouse Gas Policies and Measures
Regional Initiatives

Many states have joined regional initiatives to reduce GHG emissions and promote clean energy. States participating in regional GHG cap-and-trade initiatives are shown in Figure 4-1.

Regional Greenhouse Gas Initiative

Launched on January 1, 2009, the Regional Greenhouse Gas Initiative (RGGI) is the first mandatory market-based U.S. cap-and-trade program to reduce GHG emissions. Emissions from large electricity generators in 10 Northeast and Mid-Atlantic states are capped at approximately 188 million short tons (approximately 171 million metric tons) of CO_2 per year from 2009 to 2014. Then the cap will be reduced by 2.5 percent in each of the four years from 2015 through 2018, for a total reduction of 10 percent. As of August 2009, the RGGI states auctioned more than 110 million allowances and raised $367 million. Most states have decided to auction the majority of allowances for public benefit and use the proceeds to invest in energy efficiency improvements and renewable energy programs at the state and local levels.[111]

Western Climate Initiative

The Western Climate Initiative (WCI) was launched in February 2007 by the governors of five western states, with the goal of aggregate GHG emission re-

ductions of 15 percent below 2005 levels by 2020. Beginning in 2012, WCI aims to cap emissions from the electricity sector (including imported electricity) and large industrial sources (including combustion and process emissions). The second phase is set to begin in 2015, when the program expands to include transportation fuels and residential, commercial, and industrial fuels not otherwise covered.[112] The WCI also includes commitments to "complementary policies" promoting clean energy.

Midwestern Greenhouse Gas Reduction Accord

The Midwestern Greenhouse Gas Reduction Accord includes six midwestern states that have agreed to develop a market-based and multisector cap-and-trade mechanism, along with complementary policies to help achieve GHG reduction targets. Details and recommendations are under development, but the group is considering mid-term targets that would reduce emissions by 16–20 percent from 2005 levels by 2020 and long-term targets that would reduce emissions by 80 percent[113] from 2005 levels by 2050.[114]

State Policies and Measures

Many state governments are reducing GHG emissions through a wide range of policies and measures, from voluntary measures, such as tax credits, to legislative mandates, such as state GHG caps. Appreciating the value of collaboration, states are working with public- and private-sector stakeholders to develop the most cost-effective mitigation strategies. Table 4-2 illustrates the range of actions that states are taking on climate change.

State Emission Targets and Caps

As of November 2009, 23 of the 50 states had adopted a state GHG reduction target, although these vary in stringency, timing, and enforceability. In 2006, Cali-

Figure 4-1 **U.S. Regional Climate Initiatives**

Many states across the nation have joined regional initiatives to reduce greenhouse gas emissions and promote clean energy.

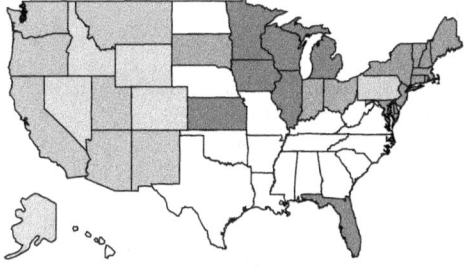

■ Regional Greenhouse Gas Initiative (RGGI)
▨ RGGI Observer
■ Midwest Greenhouse Gas Reduction Accord (MGGRA)
▨ MGGRA Observer
▨ Western Climate Initiative (WCI)
▨ WCI Observer
■ Individual State Cap-and-Trade Program

Source: Pew Center on Global Climate Change.

[111] See http://www.rggi.org/.

[112] See http://www.westernclimateinitiative.org/.

[113] See http://www.pewclimate.org/node/6572.

[114] See http://www.midwesternaccord.org/.

fornia became the first state to adopt legislation specifying mandatory GHG reductions. It was soon joined by Hawaii, New Jersey, Washington, Connecticut, Massachusetts, Minnesota, and Maryland. These laws cap state GHG emissions by a certain percentage relative to a baseline year, such as 1990. For example, California's Global Warming Solution Act has capped the state's GHG emissions at 1990 levels by 2020.[115] In May 2009, Maryland passed a law requiring that the state achieve a 25 percent reduction in GHG emissions from 2005 levels by 2020. Maryland expects that decreasing emissions by this amount will have a positive net economic benefit of $2 billion by 2020.[116]

In another example of state-based action on climate change, Florida passed legislation in 2008 that authorizes its Department of Environmental Protection to develop a cap-and-trade program for the electric utility sector. The state is currently evaluating whether to implement its own program or join one of the other existing regional cap-and-trade programs.

Performance Standards for Electric Power

As of July 2009, five states had enacted legislation requiring entities that sell electricity to customers to adhere to a standard of maximum allowable emissions per megawatt-hour (MWh) of electricity produced. In California, for example, the maximum emission level is 1,100 pounds (500 kg) of CO_2 per MWh of electricity produced.[117]

Local Policies and Measures
County Climate Protection Program

The County Climate Protection Program, under the aegis of the National Association of Counties (NACo), works to encourage and support counties in their actions to combat climate change. Several counties have made significant strides in implementing and achieving the objectives of emissions reduction and air quality improvement. The Fresh AIRE Program of Arlington, Virginia, and the Climate Protection Campaign of Sonoma County, California, are two examples of work being done under the County Climate Protection Program.

Local Governments for Sustainability

Several counties and municipalities are also making commitments to Local Governments for Sustainability (ICLEI), such as California's Alameda and San Francisco counties, and Florida's Miami-Dade County. These partnerships encourage the establishment and implementation of goals to reduce GHG emissions. Special attention is paid to producing identifiable and measurable results from these goals.

National League of Cities

The National League of Cities (NLC) is a driving force in strengthening and promoting local government in the United States. NLC is actively coordinating and encouraging climate change initiatives across the na-

tion. NLC's commitment to sustainability is illustrated by the climate protection agreements and pledges made at the 2009 Congress of Cities, and also by NLC's determination to urge Congress and the Obama administration to study and address climate change.

Examples of initiatives undertaken by NLC members include the Clean and Green Program in Riverside, California; the Cambridge Energy Alliance in Cambridge, Massachusetts; and the Energy Efficiency and Climate Change Rally for residents and local businesses in Waterbury, Vermont.

U.S. Conference of Mayors Climate Protection Agreement

To date, 965 mayors have signed the U.S. Conference of Mayors Climate Protection Agreement to reduce GHG emissions in their cities to 7 percent below 1990 levels by 2012.[118] While signatories are embracing a wide range of measures to meet this ambitious goal, recent best practices award winners include Denver, which has reduced emissions by over 60,000 metric tons of CO_2 through investment in rapid transit, and San Francisco, whose restaurant grease recycling program is reducing emissions by over 6,000 metric tons of CO_2 annually.[119]

Cities for Climate Protection Campaign

As of August 2009, 569 U.S. cities and counties are participating in the International Council for Local Environmental Initiatives' Cities for Climate Protection Campaign.[120] Participants pledge to reduce GHG emissions from local government operations and throughout their communities, while the program provides training and technical assistance. Participants are developing plans for their respective communities, such as the Arlington (Virginia) Initiative to Reduce Emissions, which aims to achieve a 10 percent reduction in GHG emissions from county operations by 2012 through building efficiency, employee engagement, and fleet efficiency.[121]

Clean Energy and Energy Efficiency Policies and Measures
State Policies and Measures

A number of state governments have made energy efficiency and renewable energy a high priority, recognizing the significant economic and environmental benefits and widespread public support. Table 4-2 illustrates the range of actions that states are taking on clean energy and energy efficiency.

Lead by Example

Many state and local governments lead by example by establishing programs that achieve substantial energy cost savings within their own operations and buildings (owned or leased). These "lead-by-example" programs include energy standards for new buildings, binding usage reduction targets for existing buildings, and innovations in financing efficiency projects. In addition

[115] See http://www.pewclimate.org/what_s_being_done/in_the_states/emissionstargets_map.cfm.

[116] See http://www.climatestrategies.us/What_New.cfm.

[117] Rubin, E.S. "A Performance Standards Approach to Reducing CO_2 Emissions from Electric Power Plants." Coal Initiative Series. Pew Center on Global Climate Change. June 2009.

[118] See http://www.usmayors.org/climateprotection/revised/.

[119] See http://usmayors.org/pressreleases/uploads/ClimateBestPractices061209.pdf.

[120] See http://www.iclei.org/index.php?id=1484&region=NA.

[121] See http://www.arlingtonva.us/portals/Topics/Climate.aspx.

Table 4-2 **State Actions on Climate Change** (as of August 2009)

A number of state governments have made energy efficiency and renewable energy a high priority, recognizing the significant economic and environmental benefits of and widespread public support for taking action.

Type of Action	Participating States	Number of States
Greenhouse Gas Policies and Measures		
Mandatory Greenhouse Gas Reductions[1]	California, Connecticut, Hawaii, Maryland, Massachusetts, Minnesota, New Jersey, Washington	8
Emission Targets and Goals[2]	Arizona, Colorado, Florida, Illinois, Maine, Michigan, Montana, New Hampshire, New Mexico, New York, Oregon, Rhode Island, Utah, Vermont, Virginia	15
Greenhouse Gas Performance Standards for Power Plant Emissions[3]	California, Illinois, Minnesota, Montana, Oregon, Washington	5
Offsets and Limits on Power Plant Emissions[4]	California, Massachusetts, Montana, Oregon, New Hampshire, Washington	6
Greenhouse Gas Emission Standards for Vehicles[5]	Arizona, California, Connecticut, Florida, Maine, Maryland, Massachusetts, New Jersey, New Mexico, New York, Oregon, Pennsylvania, Rhode Island, Vermont, Washington	15
Low-Carbon Fuel Standard[6]	California	1
Renewable Fuels Standards[7]	California, Florida, Hawaii, Iowa, Louisiana, Massachusetts, Minnesota, Missouri, Montana, New Mexico, Oregon, Washington	12
Energy Efficiency and Renewable Energy Policies		
Green Building Standards for State Buildings[8]	Arizona, California, Colorado, Connecticut, Florida, Illinois, Indiana, Kentucky, Maine, Maryland, Massachusetts, Michigan, Minnesota, Nevada, New Jersey, New Mexico, New York, Oklahoma, Rhode Island, South Carolina, South Dakota, Virginia, Washington	23
Net Metering[9]	Arizona, Arkansas, California, Colorado, Connecticut, Delaware, District of Columbia, Florida, Georgia, Hawaii, Idaho, Illinois, Indiana, Iowa, Kansas, Kentucky, Louisiana, Maine, Maryland, Massachusetts, Michigan, Minnesota, Missouri, Montana, Nebraska, Nevada, New Hampshire, New Jersey, New Mexico, New York, North Carolina, North Dakota, Ohio, Oklahoma, Oregon, Pennsylvania, Rhode Island, South Carolina, Texas, Utah, Vermont, Virginia, Washington, West Virginia, Wisconsin, Wyoming	46
Green Power Purchasing[10]	Connecticut, Illinois, Indiana, Maine, Maryland, Massachusetts, New York, Pennsylvania, Wisconsin	9
Green Pricing[11,12]	Colorado, Iowa, Montana, New Mexico, Oregon, Washington (Alabama, Alaska, Arizona, Arkansas, California, Connecticut, Delaware, District of Columbia, Florida, Georgia, Hawaii, Idaho, Illinois, Indiana, Kentucky, Louisiana, Maryland, Massachusetts, Michigan, Minnesota, Mississippi, Missouri, Nebraska, Nevada, New York, North Carolina, North Dakota, Ohio, Oklahoma, South Carolina, South Dakota, Tennessee, Texas, Utah, Vermont, West Virginia, Wisconsin, Wyoming)	6 (+ 39 states green pricing programs available)
Renewable Portfolio Standards (RPS)[13,14]	Arizona, California, Colorado, Connecticut, Delaware, District of Columbia, Hawaii, Illinois, Iowa, Kansas, Maine, Maryland, Massachusetts, Michigan, Minnesota, Missouri, Montana, Nevada, New Hampshire, New Jersey, New Mexico, New York, North Carolina, Ohio, Oregon, Pennsylvania, Rhode Island, Texas, Washington, Wisconsin (North Dakota, South Dakota, Utah, Vermont, Virginia [Florida])	30 (+ 5 state voluntary standards [and 1 state with a utility that offers an RPS])
Public Benefit Funds for Renewables and Efficiency[15,16]	California, Connecticut, Delaware, District of Columbia, Illinois, Maine, Massachusetts, Michigan, Minnesota, Montana, New Jersey, New York, Ohio, Oregon, Pennsylvania, Rhode Island, Vermont, Wisconsin	18
Public Benefits Funds for Energy Efficiency[17]	Hawaii, Nevada, New Hampshire, New Mexico, Nebraska, Texas	6
Energy Efficiency Resource Standards[18]	California, Colorado, Hawaii, Illinois, Maryland, Massachusetts, Michigan, Minnesota, Nevada, New Mexico, New York, New Jersey, North Carolina, Ohio, Pennsylvania, Texas, Vermont, Virginia, Washington	19
Appliance and Equipment Efficiency Standards[19]	Arizona, California, Connecticut, District of Columbia, Maryland, Massachusetts, New Jersey, New York, Oregon, Rhode Island, Vermont, Washington	12
Compliance with Stringent Residential Energy Building Codes[20]	Alaska, California, Florida, Georgia, Idaho, Iowa, Kentucky, Louisiana, Maryland, Massachusetts, Minnesota, Nevada, New Hampshire, New Jersey, New Mexico, North Carolina, Oregon, Pennsylvania, Rhode Island, South Carolina, Utah, Vermont, Virginia, Washington, Wisconsin	25

[1] See http://www.ncsl.org/?TabId=13240.

[2] See http://www.pewclimate.org/what_s_being_done/in_the_states/emissionstargets_map.cfm.

[3] Rubin, E.S. June 2009. "A Performance Standards Approach to Reducing CO₂ Emissions from Electric Power Plants." Coal Initiative Series. Pew Center on Global Climate Change. See http://www.pewclimate.org/white-papers/coal-initiative/performance-standards-electric.

[4] Pew Center on Climate Change. 2009. "Climate Change 101: State Action." See http://www.pewclimate.org/docUploads/Climate101-State-Jan09.pdf.

[5] See http://www.pewclimate.org/what_s_being_done/in_the_states/vehicle_ghg_standard.cfm.

[6] See http://www.pewclimate.org/docUploads/Climate101-State-Jan09.pdf.

[7] See http://www.pewclimate.org/docUploads/Climate101-State-Jan09.pdf.

[8] See http://www.pewclimate.org/what_s_being_done/in_the_states/leed_state_buildings.cfm.

[9] See http://www.dsireusa.org/summarytables/rrpre.cfm.

[10] See http://www.dsireusa.org/summarytables/rrpre.cfm.

[11] See http://apps3.eere.energy.gov/greenpower/markets/pricing.shtml?page=4.

[12] See http://www.pewclimate.org/what_s_being_done/in_the_states/west_coast_map.cfm.

[13] See http://www.dsireusa.org/summarytables/rrpre.cfm.

[14] See http://www.pewclimate.org/what_s_being_done/in_the_states/rps.cfm.

[15] See http://www.dsireusa.org/summarytables/rrpre.cfm.

[16] See http://www.pewclimate.org/what_s_being_done/in_the_states/public_benefit_funds.cfm.

[17] See http://www.pewclimate.org/what_s_being_done/in_the_states/public_benefit_funds.cfm.

[18] See http://www.pewclimate.org/what_s_being_done/in_the_states/efficiency_resource.cfm.

[19] See http://www.pewclimate.org/what_s_being_done/in_the_states/energy_eff_map.cfm.

[20] See http://www.pewclimate.org/what_s_being_done/in_the_states/res__energy_codes.cfm.

to reducing state energy bills and emissions, these efforts demonstrate the feasibility and benefits of clean energy to the larger market.

Net Metering

Most states have at least one utility that permits customers to sell electricity back to the grid, a practice known as net metering.[122] Net metering is available statewide in 18 states, and is offered by select utilities in 27 states.[123] Limits on the capacity of eligible installations range from 10 kW in Indiana to 80 MW in New Mexico.[124] While the utilization of these programs varies greatly among states and utilities, PG&E in California has reported more than 31,600 total installations with a combined capacity of 275 MW of solar and wind power as of June 2009, representing 1.3 percent of the peak load (the program is limited to 2.5 percent of the peak load).[125]

Renewable Energy Standards

A mandatory renewable energy standard (RES), sometimes referred to as a renewable portfolio standard (RPS), requires utilities to generate a certain amount of electricity or install a certain amount of capacity from renewable energy sources in a set time frame. As of November 2009, 30 states had mandatory RES programs, and an additional 5 states had voluntary RES programs.[126, 127]

In 1999, Texas passed its first RPS, which mandated that electricity providers collectively provide 2,000 MW of additional renewable energy capacity by 2009. The program exceeded expectations, and the state reached this goal in six years. During that time, wind power development in Texas more than quadrupled. In 2005, the state legislature expanded the RPS mandate to 5,880 MW by 2015, with a target of 10,000 MW for 2025. The required increase in renewable energy generation capacity is allocated between providers according to percentage market share of energy sales. Like most states with an RES in place, the Texas program includes a renewable energy credit (REC) program that allows utility companies to buy or trade RECs and use them toward compliance. In Texas, each REC represents 1 MW of installed capacity from renewable sources. If a utility earns extra RECs through its renewable energy generation, it may sell these credits to other utilities that do not have sufficient renewable energy capacity to meet the RPS requirements.[128]

Although Massachusetts has had an RPS program since the late 1990s, in July 2008 the state doubled existing requirements by mandating that its RPS requirement must grow by 1 percent per year. Under these guidelines, renewable energy will account for 15 percent of electricity generation by 2020 and 25 percent by 2030. The new legislation also divided renewable energy into two classes based on age of operating facility in order to support the continued operation of older renewable energy facilities. In Massachusetts, electricity providers may meet their obligations by producing renewable energy, purchasing a REC or paying an Alternative Compliance Payment.[129, 130]

Public Benefit Funds

Currently, 24 states have some form of public benefit funds, in which utility consumers pay a small charge to a common fund, often as part of the monthly billing cycle. The utility uses these funds to invest in energy efficiency and renewable energy projects, such as home weatherization and renewable technologies. Existing funds are anticipated to generate $7.3 billion for renewable energy by 2017.[131, 132, 133]

Tax Credits for Renewable Energy

Currently 21 states offer personal and/or corporate tax credits for investment in renewable energy.[134] Oregon's program includes both residential tax credits for small-scale renewable capacity and business tax credits up to $20 million for renewable energy equipment manufacturing, thereby cultivating both renewable energy and green jobs in the state.[135] Most other states offer some type of deduction or exemption from sales, income, corporate, or property taxes tied to renewable energy.[136]

Local Policies and Measures

Local, state, and tribal governments and territories are using funds made available under ARRA through the Energy Efficiency and Conservation Block Grant Program to develop and implement projects that improve energy efficiency and reduce energy consumption. The first 19 cities and counties received funds in July 2009 to develop energy efficiency and conservation strategies, to be supported with additional funds for implementation.[137]

Many local and tribal governments are developing their own programs to foster the development of renewable energy. The City of Madison (Wisconsin) has initiated the MadiSUN Solar Energy Program to double solar electric and solar hot water installations by 2011, which would reduce the city's CO_2 emissions by 100,000 tons. The Ogala Sioux Tribe Renewable Energy Development Authority will oversee community and commercial-scale renewable energy development on the Pine Ridge Indian Reservation in South Dakota. The Authority estimates that the reservation has the potential to develop several hundred megawatts of wind power.[138]

Local governments are also active in promoting weatherization projects in their jurisdictions. For example, the City of Portland's (Oregon) Block-by-Block Weatherization Program provides basic weatherization and education to low-income households, and hopes to weatherize 1,250 homes by 2020, reducing annual energy demand by 215 billion Btus (63,000 MWh).[139]

[122] See http://www.pewclimate.org/what_s_being_done/in_the_states/net_metering_map.cfm.

[123] See http://www.dsireusa.org/summarytables/rrpre.cfm.

[124] See http://www.dsireusa.org/summarytables/rrpre.cfm.

[125] See http://www.pge.com/mybusiness/energysavingsrebates/solar/nemtracking/index.shtml.

[126] See http://www.dsireusa.org/summarytables/rrpre.cfm.

[127] See http://www.pewclimate.org/what_s_being_done/in_the_states/rps.cfm.

[128] See http://www.seco.cpa.state.tx.us/re_rps-portfolio.htm.

[129] See http://www.pewclimate.org/what_s_being_done/in_the_states/rps.cfm.

[130] See http://www.dsireusa.org/incentives/incentive.cfm?Incentive_Code=MA05R&state=MA& CurrentPageID=1.

[131] See http://www.dsireusa.org/summarytables/rrpre.cfm.

[132] See http://www.pewclimate.org/what_s_being_done/in_the_states/public_benefit_funds.cfm.

[133] See http://www.dsireusa.org/documents/SummaryMaps/PBF_Map.ppt.

[134] See http://www.dsireusa.org/documents/SummaryMaps/TaxIncentives_Map.ppt.

[135] See http://www.dsireusa.org/incentives/index.cfm?re=1&ee=&spv=0&st=0&srp=1&state=OR.

[136] See http://www.dsireusa.org/summarytables/finre.cfm.

[137] See http://www.eecbg.energy.gov/.

[138] See http://www.indiancountrytoday.com/national/50848652.html.

[139] See http://www.smartcommunities.ncat.org/success/block.shtml.

Table 4-3 **Summary of U.S. Actions to Reduce Greenhouse Gas Emissions** (Tg CO_2 Eq.)[1]

Name of Policy or Measure	Objective and/or Activity Affected	Greenhouse Gas Affected	Type of Program	Status	Implementing Entities	Estimated Mitigation Impact for 2007	Estimated Mitigation Impact for 2010	Estimated Mitigation Impact for 2015	Estimated Mitigation Impact for 2020
Energy: Residential and Commercial[2]									
Appliances and Commercial Equipment Standards Program, Appliance Energy Efficiency Standards	Analyzes, develops, reviews and updates efficiency standards for most major household appliances and major commercial building technologies and equipment.	CO_2	Regulatory	Implemented	DOE	0.0	1.3	4.3	6.1
Building Energy Codes Program	Promotes stronger building energy codes and helps states adopt, implement, and enforce them. Recognizes that energy codes maximize energy efficiency only when they are fully embraced by users and supported through education, implementation, and enforcement.	CO_2	Regulatory	Implemented	DOE	0.0	1.3	5.8	11.3
Lighting Energy Efficiency Standards	Mandates standards that will result in phasing out the 130-year-old incandescent light bulb by the middle of the next decade and phases out less efficient fluorescent tubes. New standards will also apply to reflector lamps—the cone-shaped bulbs used in recessed and track lighting.	CO_2	Regulatory	Implemented	DOE	0.0	0.4	1.2	1.5
ENERGY STAR Labeled Products	Labels distinguish energy-efficient products in the marketplace.	CO_2	Voluntary	Implemented	EPA/DOE	64.5	82.5	113.6	141.2
ENERGY STAR for the Commercial Market	Promotes the improvement of energy performance in commercial buildings.	CO_2	Voluntary	Implemented	EPA	66.0	56.8	75.0	93.5
ENERGY STAR for the Residential Market	Promotes the improvement of energy performance in residential buildings beyond the labeling of products.	CO_2	Voluntary	Implemented	EPA	1.8	5.5	21.1	44.0
Net-Zero-Energy Commercial Building Initiative	Achieves marketable net-zero-energy commercial buildings by 2025. Encompasses all activities that support this goal, including industry partnerships, research, and tool development.	CO_2	Voluntary	Implemented	DOE	N/A	8.2	12.7	15.5
Building America	Develops cost-effective solutions that reduce the average energy use of housing by 40–100% through research partnerships with all facets of the residential building industry.	CO_2	Economic	Implemented	DOE	0.0	3.8	11.0	19.8
Energy Efficiency and Conservation Block Grants	Provides funds to units of local and state government, Native American nations, and territories to develop and implement projects to improve energy efficiency and reduce energy use and fossil fuel emissions in their communities.	CO_2	Economic	Implemented	DOE	N/A	N/A	N/A	N/A

Name of Policy or Measure	Objective and/or Activity Affected	Greenhouse Gas Affected	Type of Program	Status	Implementing Entities	Estimated Mitigation Impact for 2007	Estimated Mitigation Impact for 2010	Estimated Mitigation Impact for 2015	Estimated Mitigation Impact for 2020
Energy: Residential and Commercial[2] (Continued)									
Weatherization Assistance Program	Enables low-income families to permanently reduce their energy bills by making their homes more energy efficient.	CO_2	Economic	Implemented	DOE	N/A	6.7	7.8	8.9
Energy: Industrial									
ENERGY STAR for Industry	Enables industrial companies to evaluate and cost-effectively reduce energy use.	CO_2	Voluntary	Implemented	EPA	23.1	18.0	25.6	36.6
Save Energy Now	National initiative of the Industrial Technologies Program to drive a 25% reduction in industrial energy intensity in 10 years.	CO_2	Voluntary	Implemented	DOE	N/A	9.4	28.2	28.9
Industrial Assessment Centers	Provide in-depth assessments of a plant's site and its facilities, services, and manufacturing operations.	CO_2	Voluntary	Implemented	DOE	N/A	0.6	2.7	5.1
Industry-Specific and Cross-Cutting RDD&D	Develops and delivers advanced energy efficiency, renewable energy, and pollution prevention technologies for industrial applications through partnerships with industry, government, and nongovernment organizations.	CO_2	Information and Research	Implemented	DOE	N/A	0.0	0.8	3.4
Energy: Supply									
Solar Energy Development Program	Provides opportunities for and encourages use of federal public lands for the development of solar energy.	CO_2	Research and Information	Implemented	DOI	N/A	N/A	N/A	N/A
Wind Energy Program	Seeks to improve the cost-effectiveness of wind energy technology and lower market and regulatory barriers to the use of wind.	CO_2	Research and Information	Implemented	DOE	N/A	1.9	30.2	67.6
Geothermal Energy Development Program	Provides opportunities for and encourages use of federal public lands for the development of geothermal energy.	CO_2	Research and Information	Implemented	DOI	N/A	N/A	N/A	N/A
Energy Transmission Infrastructure	Provides for new and updated transmission for new energy development and improvement of the existing system.	CO_2	Voluntary	Initiated	DOI	N/A	N/A	N/A	N/A
Energy Smart Parks	Works to reduce carbon emissions in all aspects of national park operations and deploys renewable and efficient energy technologies throughout the national park system.	CO_2	Voluntary	Initiated	DOI/DOE	N/A	N/A	N/A	N/A

Name of Policy or Measure	Objective and/or Activity Affected	Greenhouse Gas Affected	Type of Program	Status	Implementing Entities	Estimated Mitigation Impact for 2007	Estimated Mitigation Impact for 2010	Estimated Mitigation Impact for 2015	Estimated Mitigation Impact for 2020
Energy: Supply (Continued)									
Clean Energy Initiative; Green Power Partnership; Combined Heat and Power Partnership	Remove market barriers to increased penetration of a cleaner, more efficient energy supply.	CO_2	Voluntary, Education	Implemented	EPA	17.6	19.8	44.0	73.3
Indian Education Renewable Energy Challenge	Provides opportunities for students attending tribal high schools and colleges to pursue careers in the fields of green and renewable energy.	—	Voluntary	Initiated	DOI/DOE	N/A	N/A	N/A	N/A
University-National Park Energy Partnership Program	Provides parks assistance in addressing their energy concerns.	CO_2	Voluntary	Initiated	DOI/DOE	N/A	N/A	N/A	N/A
Biorefinery Assistance	Provides loan guarantees for the development, construction, and retrofitting of commercial-scale biorefineries, and grants to help pay for the development and construction costs of demonstration-scale biorefineries.	CO_2	Economic	Implemented	USDA	N/A	N/A	N/A	N/A
Nuclear Power (AFCI, NP2010, Loan Guarantee Program, Standby Support for Certain Delays, PAA)	Provides risk insurance against construction and operational delays beyond the control of the plants' sponsors and against liability claims from nuclear incidents. Also provides loan guarantees for new plants and R&D support for advanced nuclear technologies.	CO_2	Economic, Fiscal, Regulatory, Research	Implemented	DOE	0.0	0.0	5.2	14.4
Renewable Energy Production Incentive	Provides financial incentives for electricity generated by new qualifying renewable energy generation, cost-sharing incentives for RDD&D of renewable energy technology manufacturing, and 50% matching grants for small-scale renewable projects.	CO_2	Economic, Fiscal, Research	Implemented	DOE	N/A	N/A	N/A	N/A
Rural Energy for America Program	Provides grants and loan guarantees to rural residents, agricultural producers, and rural businesses for energy efficiency and renewable energy systems, energy audits, and technical assistance; and for projects ranging from biofuels to wind, solar, geothermal, methane gas recovery, advanced hydro, and biomass.	CO_2	Economic	Implemented	USDA	2.0	0.9	0.7	1.2

Name of Policy or Measure	Objective and/or Activity Affected	Greenhouse Gas Affected	Type of Program	Status	Implementing Entities	Estimated Mitigation Impact for 2007	Estimated Mitigation Impact for 2010	Estimated Mitigation Impact for 2015	Estimated Mitigation Impact for 2020
Energy: Supply (Continued)									
Solar Energy Technologies Program	Supports R&D and deployment of cost-effective technologies toward growing the use of solar energy throughout the nation and the world. Seeks to make solar electricity cost-competitive with conventional forms of electricity by 2015.	CO_2	Research and Information	Implemented	DOE	N/A	0.2	1.5	2.5
Wind Energy Development Program	Provides opportunities for and encourages use of federal public lands for development of wind energy.	CO_2	Voluntary	Implemented	DOI	N/A	N/A	N/A	N/A
Biomass Program (BFI, Integrated Cellulosic Biorefineries and Financial Assistance, Woody Biomass Utilization, BRDI, Grants for Advanced Biofuels, Biorefinery Energy Efficiency)	Develops a portfolio of RD&D geared toward biomass feedstocks and conversion technologies. Includes development and deployment of infrastructure and opportunities for market penetration of bio-based fuels and products.	CO_2	Research and Information	Implemented	DOE	0.0	3.3	29.2	55.2
Coal Technologies (Innovations for Existing Plants, Carbon Sequestration Program, Gasification Technologies Program, CCPI, FutureGen, Demonstration)	Seeks to develop and demonstrate a portfolio of technologies that can increase operating efficiency and capture and permanently store GHGs in new commercial-scale plants or existing plants. Also includes tax credits.	CO_2	Research, Fiscal, Information	Implemented	DOE	0.0	18.6	25.3	23.1
Geothermal Technologies Program	Supports deployment, market transformation, technology diffusion, information, and technical assistance for expanding the use of geothermal resources.	CO_2	Research and Information	Implemented	DOE	N/A	0.0	1.1	1.8
Transportation									
Renewable Fuel Standard (RFS)	Implements the Energy Independence and Security Act of 2007 requirements revising the RFS. Changes include increasing the total volume of renewable fuel used in transportation to 36 billion gallons by 2022, as well as adding specific volume standards for cellulosic biofuel, biomass-based diesel, and advanced biofuel within the total volume required. Also includes new definitions and criteria for both renewable fuels and the feedstocks used to produce them, including new life-cycle (GHG) emission thresholds for renewable fuels.	All	Regulatory	New: Being Implemented	EPA	N/A	N/A	N/A	138

Name of Policy or Measure	Objective and/or Activity Affected	Greenhouse Gas Affected	Type of Program	Status	Implementing Entities	Estimated Mitigation Impact for 2007	Estimated Mitigation Impact for 2010	Estimated Mitigation Impact for 2015	Estimated Mitigation Impact for 2020
Transportation (Continued)									
Corporate Average Fuel Economy Standards	Raise the fuel economy standard for light-duty trucks for MY 2005–2010.	CO_2	Regulatory	Implemented	DOT	8.7	29.1	33.7	35.9
Corporate Average Fuel Economy Standards (Model Year 2011)	Raise the fuel economy standard for light-duty trucks and passenger cars for MY 2011 (cumulative with savings from previous rules).	CO_2	Regulatory	Implemented	EPA/DOT	N/A	N/A	37.6	43.4
National Policy to Establish Vehicle GHG Emissions and Corporate Average Fuel Economy Standards	EPA and DOT are jointly proposing GHG and fuel economy standards for passenger cars, light-duty trucks, and medium-duty passenger vehicles, covering MY 2012–2016. These standards that would achieve an average of 250 grams per mile in 2016 (with a related fuel economy standard of about 35.5 miles per gallon).	All	Regulatory	Implemented	EPA/DOT	N/A	N/A	39.0	132.0
National Clean Diesel Campaign	Works aggressively to reduce diesel emissions across the country through the implementation of proven emission control technologies and innovative strategies with the involvement of national, state, and local partners, including financial support through DERA.	CO_2 and black carbon	Voluntary, Education, Information	Implemented	EPA	N/A	N/A	N/A	N/A
SmartWay Transport Partnership	Accelerates the uptake of low-GHG technologies and strategies in the freight and consumer sectors.	CO_2	Voluntary, Information, Education	Implemented	EPA	4.2	15.8	36.8	43
Aviation Fuel Efficiency, Renewable Fuels, and Market Measures[3]	Improve aircraft/engine technology and operational procedures, enhance the airspace transportation system to reduce aviation's CO_2 emissions contribution, develop environmentally clean and affordable alternative fuels, secure energy future for sustainable aviation growth, and apply market measures, such as charges or cap and trade.	CO_2	Research	Implemented	FAA	N/A	N/A	N/A	N/A
Commercial Aviation Alternative Fuels Initiative	Develops environmentally clean and affordable alternative fuels and secures energy future for sustainable aviation growth.	CO_2	Research	Implemented	FAA	N/A	N/A	N/A	N/A

Name of Policy or Measure	Objective and/or Activity Affected	Greenhouse Gas Affected	Type of Program	Status	Implementing Entities	Estimated Mitigation Impact for 2007	Estimated Mitigation Impact for 2010	Estimated Mitigation Impact for 2015	Estimated Mitigation Impact for 2020
Transportation (Continued)									
Clean Automotive Technology Program	Searches for cost-effective advanced automotive technologies that cut GHG emissions, increase fuel efficiency, reduce health-related emissions, and are affordable for mainstream consumer and commercial vehicles, while partnering with industry to move these technologies to the road.	CO_2	Research	Implemented	EPA	N/A	N/A	N/A	N/A
Fuel Cell Technologies Program	Supports R&D of fuel cell technologies and infrastructure for electricity generation and transportation to reduce energy use, GHG emissions, and criteria pollutants	CO_2	Research	Implemented	DOE	0.0	0.0	0.0	0.0
Federal Transit Program	Provides grants and technical assistance to support public transportation systems across the country.	All	Voluntary	Implemented	DOT	N/A	N/A	N/A	N/A
Transit Investments for Greenhouse Gas and Energy Reduction	Provides funds to public transit agencies for capital investments that will assist in reducing the energy consumption or GHG emissions of public transportation systems.	All	Voluntary	Implemented	DOT	N/A	N/A	N/A	N/A
Loan Programs (Advanced Battery Loan Guarantee, Advanced Technology Vehicles Manufacturing Incentive)	Provide loan guarantees to advanced vehicle manufacturers and their component suppliers for the construction of facilities that manufacture U.S.-developed and -produced advanced vehicle batteries, lithium ion batteries, and hybrid electrical systems.	CO_2	Economic	Implemented	DOE	N/A	N/A	N/A	N/A
Congestion Mitigation and Air Quality Improvement Program	Provides states with funds to reduce congestion and to improve air quality through transportation control measures and other strategies.	CO_2	Fiscal	Implemented	DOT	N/A	N/A	N/A	N/A
Alternative Transport Systems and Use of Clean Vehicles	Replace traditional fueled vehicles with "clean" energy vehicles and reduce the size of vehicles and vehicle fleets.	CO_2	Voluntary	implemented	DOI	N/A	N/A	N/A	N/A

Name of Policy or Measure	Objective and/or Activity Affected	Greenhouse Gas Affected	Type of Program	Status	Implementing Entities	Estimated Mitigation Impact for 2007	Estimated Mitigation Impact for 2010	Estimated Mitigation Impact for 2015	Estimated Mitigation Impact for 2020
Industry (Non-CO_2)									
Coalbed Methane Outreach Program	Reduces methane emissions from U.S. coal mining operations through cost-effective means.	CH_4	Education, Information	Implemented	EPA	7.3	9.9	11	12.1
Natural Gas STAR Program	Reduces methane emissions from U.S. natural gas systems through the widespread adoption of industry best management practices.	CH_4	Voluntary	Implemented	EPA	37	27.5	35.6	46.9
Environmental Stewardship Initiative	Limits emissions of HFCs, PFCs, and SF_6 in industrial applications.	HFCs, PFCs, SF_6	Voluntary	Implemented	EPA	17.3	25.5	34.1	44.7
Voluntary Code of the Reductions of Emissions of HFC and PFC Fire Protection Agents	Minimizes non-fire emissions of HFCs and PFCs used as fire-suppression alternatives and protects people and property from the threat of fire through the use of proven, effective products and systems.	HFCs, PFCs	Voluntary	Implemented	EPA	N/A	N/A	N/A	N/A
HFC-23 Partnership	Encourages reduction of HFC-23 emissions through cost-effective practices or technologies.	HFC-23	Voluntary	Implemented	EPA	17.8	22.9	20.7	20.9
Mobile Air Conditioning Climate Protection Partnership	Identifies near-term opportunities to improve the environmental performance of mobile air conditioners and promotes cost-effective designs and improved service procedures to minimize emissions from mobile air conditioning systems.	CO_2, HFC-134a	Voluntary, Research	Implemented	EPA	0.0	0.2	13.6	24.6
GreenChill Advanced Refrigeration Partnership	Reduces ozone-depleting and GHG refrigerant emissions from supermarkets.	HFCs	Voluntary, Information, Education	Implemented	EPA	N/A	0.3	1.0	1.9
Responsible Appliance Disposal Program	Reduces emissions of refrigerant and foam-blowing agents from end-of-life appliances.	HFCs	Voluntary	Implemented	EPA	0.0	0.1	0.2	0.4
Significant New Alternatives Policy Program	Facilitates smooth transition away from ozone-depleting chemicals in industrial and consumer sectors.	HFCs, PFCs, SF_6	Regulatory, Information	Implemented	EPA	115.4	159.8	208.5	243
Voluntary Aluminum Industry Partnership	Encourages reduction of CF_4 and C_2F_6 where technically feasible and cost-effective.	PFCs	Voluntary	Implemented	EPA	10.0	8.1	8.1	8.2

Name of Policy or Measure	Objective and/or Activity Affected	Greenhouse Gas Affected	Type of Program	Status	Implementing Entities	Estimated Mitigation Impact for 2007	Estimated Mitigation Impact for 2010	Estimated Mitigation Impact for 2015	Estimated Mitigation Impact for 2020
Agriculture									
AgSTAR	Promotes practices to reduce GHG emissions from U.S. farms.	CH_4	Information, Education	Implemented	EPA/USDA	N/A	N/A	N/A	N/A
Conservation Reserve Program	Encourages farmers to convert highly erodible cropland or other environmentally sensitive acreage to native grasses, wildlife plantings, trees, filter strips, or riparian buffers to improve soil, water, wildlife, and other natural resources.	CO_2, N_2O	Voluntary	Implemented	USDA	59.6	57.1 4	53	53
Environmental Quality Incentives Program	Offers innovation grants to livestock producers and owners of working farmlands to accelerate the development, transfer, and adoption of innovative technologies and approaches, including those that deliver GHG benefits and improve the quality of nutrient management systems.	CO_2, CH_4, N_2O	Voluntary	Implemented	USDA/NRCS	3.9	6.3	10.2	14.2
Conservation Stewardship Program	This new program will provide financial and technical assistance to promote conservation on working cropland, pasture, and range land, as well as forested land that is an incidental part of an agriculture operation. Significant GHG emission reductions are possible, depending on the contracts enrolled.	CO_2, CH_4, N_2O	Voluntary	Implemented Beginning in 2009	USDA	0	0.3	0.6	0.7
Wetlands Reserve Program	Purchases easements and restores the hydrology of previously drained wetlands, frequently restoring woody wetlands and producing carbon sequestration benefits.	CO_2, CH_4, N_2O	Voluntary	Implemented	USDA	0.18	0.19	0.22	0.25
Grassland Reserve Program	Purchases easements to restore or maintain grassland to a good or excellent condition to enhance carbon sequestration.	CO_2, CH_4, N_2O	Voluntary	Implemented	USDA	0.007	0.01	0.02	0.03
Wildlife Habitat Incentives Program	Provides funds to help private landowners create habitat for specific wildlife species, frequently resulting in the establishment of woody and grass species that sequester carbon.	CO_2, CH_4, N_2O	Voluntary	Implemented	USDA	0.2	0.3	0.4	0.5

Name of Policy or Measure	Objective and/or Activity Affected	Greenhouse Gas Affected	Type of Program	Status	Implementing Entities	Estimated Mitigation Impact for 2007	Estimated Mitigation Impact for 2010	Estimated Mitigation Impact for 2015	Estimated Mitigation Impact for 2020
Forestry									
Enhancing Ecosystems Services on Forestland, Grasslands, Parks, and Wildlife Reserves	Conducts regional analyses to determine opportunities for carbon sequestration and to conserve, enhance, and restore, or assist in the adaptation of ecosystems.	CO_2	Voluntary	Initiated	DOI	N/A	N/A	N/A	N/A
Healthy Forest Initiative	Restores the health of U.S. forests, woodlands, and rangelands. Coordination among DOI, USDA, and DOE and cooperative work with states, tribes, private landowners, nongovernmental organizations, and other interested parties and potential partners is key to success. Improves air quality, particularly emissions from smoke, PM, CO_2, and NO_x.	CO_2	Voluntary	Implemented	USDA	N/A	N/A	N/A	N/A
Woody Biomass Utilization Grants Program	Focuses on creating markets for small-diameter material and low-valued trees removed from forest restoration activities, such as reducing hazardous fuels, handling insect and disease conditions, or treating forestland impacted by catastrophic weather events. Helps communities, entrepreneurs, and others turn residues into marketable forest and/or energy products.	CO_2	Economic	Implemented	USDA	N/A	N/A	N/A	N/A
Waste Management									
Stringent Landfill Rule	Reduces methane/landfill gas emissions from U.S. landfills.	CH_4	Regulatory	Implemented	EPA	9.2	9.2	9.5	9.9
Landfill Methane Outreach Program	Reduces methane emissions from U.S. landfills through cost-effective means.	CH_4	Voluntary, Information, Education	Implemented	EPA	19.1	22.7	26.4	30.8
Waste Wise	Encourages recycling, source reduction, and other progressive integrated waste management activities to reduce GHG emissions.	All	Voluntary, Information, Research	Implemented	EPA	20.1	23.4	29.9	38.1
Cross-Sectoral									
Carbon Monitoring and Sequestration	Assesses the potential for global, active, long-term containment of carbon geologic areas and the natural capacity of ecosystems to store carbon.	CO_2	Regulatory	Initiated	DOI	N/A	N/A	N/A	N/A

Name of Policy or Measure	Objective and/or Activity Affected	Greenhouse Gas Affected	Type of Program	Status	Implementing Entities	Estimated Mitigation Impact for 2007	Estimated Mitigation Impact for 2010	Estimated Mitigation Impact for 2015	Estimated Mitigation Impact for 2020
Cross-Sectoral (Continued)									
Interagency Partnership for Sustainable Communities	Encourages integrated regional planning by aligning federal policies for housing, transportation, and the environment. Aims to reduce vehicle-miles traveled, per-capita GHG emissions, and dependence on fossil fuels.	All	Voluntary, Economic, and Information	Implemented	EPA/DOT/HUD	N/A	N/A	N/A	N/A
Voluntary Reporting of Greenhouse Gases (1605(b))	Provides a means for organizations and individuals to record the results of voluntary measures to reduce, avoid, or sequester GHG emissions.	All	Voluntary	Implemented	DOE/EPA/ USDA	N/A	N/A	N/A	N/A
Climate Leaders	Assists companies with developing long-term comprehensive climate change strategies.	All	Voluntary	Implemented	EPA	N/A	N/A	N/A	N/A
Climate Showcase Communities Grant Program	Creates models of local and tribal government community actions that generate cost-effective and persistent GHG reductions and can be replicated across the country.	All	Economic, Information	Implemented	EPA	N/A	N/A	N/A	N/A
State Climate and Energy Partner Network	Motivates GHG emission reductions as one of several benefits states derive from implementing a comprehensive suite of cost-effective clean energy policies and programs.	All	Information, Education	Implemented	EPA	N/A	N/A	N/A	N/A
Climate Friendly Parks	Parks conduct emission inventories, develop strategies for reducing emissions by developing action plans, and educate park visitors about climate change and what parks are doing to address the issue.	CO_2	Voluntary	Implemented	DOI	N/A	N/A	N/A	N/A
National Action Plan for Energy Efficiency	Provides policy recommendations, reports, technical assistance, and outreach to encourage states, utilities, and stakeholders to meet electricity and natural gas demand with zero-GHG-emitting energy efficiency.	All	Information, Education	Implemented	EPA/DOE	N/A	N/A	N/A	N/A
State Energy Program	Strengthens and supports the capabilities of states to promote energy efficiency and to adopt renewable energy technologies, helping the nation achieve a stronger economy, a cleaner environment, and greater energy security.	CO_2	Economic, Information	Implemented	DOE	2.5	3.7	3.7	3.7

Name of Policy or Measure	Objective and/or Activity Affected	Greenhouse Gas Affected	Type of Program	Status	Implementing Entities	Estimated Mitigation Impact for 2007	Estimated Mitigation Impact for 2010	Estimated Mitigation Impact for 2015	Estimated Mitigation Impact for 2020
Federal Programs									
FEMP, Renewable Energy Purchases, Purchasing of Energy-Efficient Products, Fleet Conservation Requirements, Renewable Energy Goals, Efficiency Performance Standards	Promotes energy efficiency and renewable energy use in federal buildings, facilities, and operations.	All	Economic, Information, Education	Implemented	DOE	N/A	0.4	2.3	3.4

[1] Estimates of the mitigation impacts of programs are provided by the federal agency responsible for each individual program, based on the agency's experience and assumptions related to the implementation of voluntary programs. These estimates may include assumptions about the continued or increased participation of partners, development and deployment goals, and/or whether the necessary commercialization or significant market penetration is achieved.

[2] Estimates of mitigation impacts for individual policies or measures should not be aggregated to the sectoral level, due to possible synergies and interactions among policies and measures that might result in double counting.

[3] Aviation fuel efficiency is defined as the fuel burned per unit distance traveled.

[4] This value may change, depending on the outcome of an interagency process relating to the operation of the Conservation Reserve Program. The value shown is a straight-line interpolation between the values calculated for 2007 and 2012 and may not necessarily reflect the outcome of the interagency process.

AFCI = Advanced Fuel Cycle Initiative; BFI = Biofuels Initiative; BRDI = Biomass Research and Development Initiative; CCPI = Clean Coal Power Initiative; DOE = U.S. Department of Energy; DOI = U.S. Department of the Interior; DOT = U.S. Department of Transportation; C$_2$F$_6$ = hexafluoroethane; CF$_4$ = tetrafluoromethane; CH$_4$ = methane; CO$_2$ = carbon dioxide; DERA = Diesel Emissions Reduction Act; EPA = U.S. Environmental Protection Agency; EQIP = Environmental Quality Incentives Program; FAA = Federal Aviation Administration; FEMP = Federal Energy Management Program; GHG = greenhouse gas; GWP = global warming potential; HFCs = hydrofluorocarbons; HUD = U.S. Department of Housing and Urban Development; MY = model year; N$_2$O = nitrous oxide; N/A = not available; NO$_x$ = nitrogen oxides; NP = nuclear power; NRCS = Natural Resources Conservation Service; PAA = Price-Anderson Act; PFCs = perfluorocarbons; PM = particulate matter; R&D = research and development; RD&D = research, development, and demonstration; RDD&D = research, development, demonstration, and deployment; SF$_6$ = sulfur hexafluoride; USDA = U.S. Department of Agriculture.

5

Projected Greenhouse Gas Emissions

This chapter is concerned mostly with business-as-usual greenhouse gas (GHG) emission projections, but it begins with a description of the Obama administration's goals to reduce GHGs from that trajectory.

ADMINISTRATION POLICY AND GOALS

The Obama administration supports the implementation of a market-based cap-and-trade program to spur growth in the low-carbon economy and reduce GHG emissions to 83 percent below 2005 levels by 2050 (Figure 5-1). Since taking office in January 2009, the Obama administration has made investments in low-carbon and renewable technologies a priority, including through the American Recovery and Reinvestment Act of 2009 (ARRA).[1] The administration has

also worked to reduce GHGs through robust actions by the U.S. Environmental Protection Agency (EPA), the U.S. Department of Energy (DOE), and other executive agencies. While ARRA and the actions of these agencies will result in significant emission reductions, the projections in this chapter show that GHG emissions will gradually increase in the long term without additional measures, such as a cap-and-trade program.

In June 2009, the U.S. House of Representatives passed the American Clean Energy and Security Act, which includes economy-wide GHG reduction goals of 3 percent below 2005 levels in 2012, 17 percent below 2005 levels in 2020, and 83 percent below 2005 levels in 2050. Through a cap-and-trade program and other complementary measures, the bill would promote the development and deployment of new clean

[1] For the complete bill text, see http://frwebgate.access.gpo.gov/cgi-bin/getdoc.cgi?dbname=111_cong_bills&docid=f:h1enr.pdf.

energy technologies that would fundamentally change the way we produce, deliver, and use energy.

The bill would: (1) advance energy efficiency and reduce reliance on oil; (2) stimulate innovation in clean coal technology to sequester GHG emissions before they enter the atmosphere; (3) accelerate the use of renewable sources of energy, including biomass, wind, solar, and geothermal; (4) create strong market demand in the long run for these next-generation technologies, which will result in increased domestic manufacturing and enable American workers to play a central role in U.S. clean energy transformation; and (5) play a critical role in the American economic recovery and job growth—from retooling shuttered manufacturing plants to make wind turbines, to using equipment and expertise in drilling for oil to develop clean energy from underground geothermal sources, to tapping into American ingenuity to engineer coal-fired power plants that do not contribute to climate change.

In early 2010, the Senate was considering its own legislation, with similarly bold targets, to promote clean energy and reduce GHG emissions. If the Senate and House pass bills, a conference committee will be convened to resolve disagreements and negotiate a compromise bill for consideration in both legislative bodies.

In December 2009, at the Fifteenth Conference of the Parties of the United Nations Framework Convention on Climate Change (UNFCCC), as part of a Copenhagen Accord involving GHG mitigation contributions by developed and key developing countries, the Obama administration proposed a U.S. GHG emissions reduction target in the range of 17 percent below 2005 levels by 2020 and approximately 83 percent below 2005 levels by 2050, ultimately aligned with final U.S. legislation.

CHAPTER OVERVIEW

This chapter provides business-as-usual projections of U.S. GHG emissions through 2020 and beyond. These projections reflect national estimates considering population growth, long-term economic growth potential, and historical rates of technology improvement, and the projections are consistent with historic average weather. The projections are based on anticipated trends in technology deployment and adoption, demand-side efficiency gains, fuel switching, and many of the implemented policies and measures discussed in Chapter 4.[2]

Despite the recent global economic turmoil, the U.S. economy is expected to recover and emissions are expected to grow in the long term in a business-as-usual case. Even with projected growth in absolute emissions, emissions per unit of gross domestic product are expected to decline.

Figure 5-1 **Projected U.S. GHG Emissions Meeting Recently Proposed Goals Versus Business as Usual**

By 2050, the Obama administration's goal is to reduce U.S. greenhouse gas emissions approximately by 83 percent from 2005 levels, in the same range as legislation passed by the U.S. Congress.

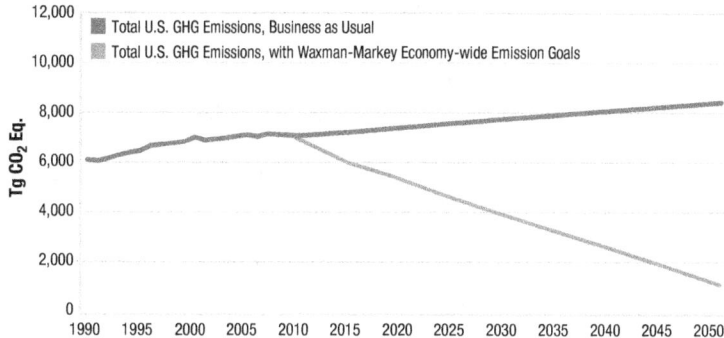

Source: U.S. EPA 2009. The historical data are derived from the *Inventory of U.S. Greenhouse Gas Emissions and Sinks: 1990–2007*; the baseline scenario is from EPA's ADAGE (Applied Dynamic Analysis of the Global Economy) model; and the decreasing emissions line includes the Waxman-Markey goals for 2012, 2020, 2030, and 2050, with intervening years interpolated.

Projections are provided by gas and by sector. In keeping with the reporting guidelines of the UNFCCC, gases included in this report are carbon dioxide (CO_2), methane (CH_4), nitrous oxide (N_2O), hydrofluorocarbons (HFCs), perfluorocarbons (PFCs), and sulfur hexafluoride (SF_6). CO_2 emissions are reported for the following sectors: electric power generation and residential, commercial, industrial, and transportation end use.

Proposed or planned policies that had not been implemented as of March 31, 2009, as well as sections of existing legislation that require implementing regulations or funds that have not been appropriated, are not included in the projections. The projections include provisions from ARRA, but do not include, for example, the vehicle fuel economy and emission standards announced by the President in May 2009 and finalized in 2010.[3]

U.S. GREENHOUSE GAS EMISSIONS: 2000–2020

Trends in Total Greenhouse Gases

DOE's Energy Information Administration (EIA) April 2009 update of the *Annual Energy Outlook 2009* (AEO 2009) provides a baseline projection of energy-related CO_2 emissions out to 2030, and reflects the provisions of ARRA, enacted in mid-February 2009 (U.S. DOE/EIA 2009h). Projected CO_2 emissions in the AEO 2009 are adjusted to match the international inventory convention.[4] EPA prepared the projections of non-energy-related CO_2 emissions and non-CO_2 emissions. Non-CO_2 emission projections are based on the report *Global Anthropogenic Non-CO_2 Greenhouse Gas Emissions: 1990–2020* (U.S. EPA/OAP 2006b). The U.S. Department of Agriculture prepared the estimates of carbon sequestration. Historical

[2] Because Chapter 5 projections and Chapter 4 mitigation effects of policies and measures are calculated using different methodologies, estimates of the total effect of policies and measures derived from each chapter are not directly comparable.

[3] For the full text of the announcement, see http://www.whitehouse.gov/the_press_office/Remarks-by-the-President-on-national-fuel-efficiency-standards/.

[4] The AEO 2009 estimate for bunker fuels was subtracted and replaced with an EPA estimate that reflects a broader definition of bunker fuels consistent with the international inventory convention. The AEO 2009 estimate of non-energy CO_2 emissions was replaced by an EPA estimate of non-energy CO_2 emissions from fuel use and all other non-energy CO_2 emissions (e.g., industrial processes). This is consistent with previous *U.S. Climate Action Reports*. An estimate of CO_2 emissions in the U.S. territories was added to the AEO 2009 number, since these emissions are not included in the AEO.

emissions data are drawn from the *Inventory of U.S. Greenhouse Gas Emissions and Sinks: 1990–2007* (U.S. EPA/OAP 2009). In general, the projections reflect long-run trends and do not attempt to consider short-run departures from those trends, with the exception that EIA explicitly models the current recession in its 2010 CO_2 emissions projection.

All GHGs in this chapter are reported in teragrams of CO_2 equivalents (Tg CO_2 Eq.), in keeping with the reporting guidelines of the UNFCCC. The conversions of non-CO_2 gases to CO_2 equivalents are based on the 100-year global warming potentials listed in the Second Assessment Report of the Intergovernmental Panel on Climate Change (IPCC 1996).

Inventory totals of GHG emissions for 2000, 2005, and 2007 and projections for 2010, 2015, and 2020 are shown in Table 5-1. From 2005 to 2020, total GHG emissions in the baseline are projected to increase by 4 percent, from 7,109 Tg CO_2 Eq. to 7,416 Tg CO_2 Eq., while the U.S. gross domestic product is projected to grow by 40 percent (US DOE/EIA 2009a).

Given the implementation of programs and measures in place as of spring 2009 and current economic projections, total gross U.S. GHG emissions are expected to drop slightly below 2005 emissions in the short term, but will rise steadily in the long term as population and total economic activity grow. Increased demand for energy services will be offset in part by shifts toward less energy-intensive industries, efficiency improvements, and increased use of renewable energy

technologies and other less carbon-intensive energy fuels. More rapid improvements in technologies that emit fewer GHGs, new GHG mitigation requirements, or more rapid adoption of voluntary GHG emission reduction programs could result in lower GHG emission levels than in the baseline projection (U.S. DOE/EIA 2009a, 2009b).

Between 2005 and 2020, CO_2 emissions in the baseline projection are estimated to increase by 1.5 percent. Over the same period, CH_4, N_2O, and PFC emissions are expected to grow by 8 percent, 5 percent, and 4 percent, respectively. A large portion of emissions growth is driven by HFCs, which are projected to more than double between 2005 and 2020, as they are more extensively used as a substitute for ozone-depleting substances. Slow growth in CO_2 emissions is driven by a combination of policies implemented to increase efficiency and the use of renewable energy, as well as increasing energy prices and economic projections that are lower than those estimated in the 2006 *U.S. Climate Action Report* (2006 CAR). Some non-CO_2 emission sources are less affected by short-term economic disruptions than CO_2 emission sources. For example, CH_4 emissions from landfills are a result of waste deposited over a long period of time, as opposed to sources based only on current economic activity.

Carbon Dioxide Emissions

Energy-related CO_2 emission estimates are taken from EIA's AEO 2009, but are adjusted to match interna-

Table 5-1 Historical and Projected U.S. Greenhouse Gas Emissions from All Sources: 2000–2020 (Tg CO_2 Eq.)

Between 2005 and 2020, total gross U.S. greenhouse gas emissions are expected to grow by 4 percent, while U.S. gross domestic product is expected to increase by 40 percent. Energy-related CO_2 emissions are expected to grow by 1.5 percent, reflecting the development and deployment of clean energy technologies, while non-CO_2 emissions growth will be driven by the shift from gases regulated under the *Montreal Protocol* to hydrofluorocarbons.

Greenhouse Gases	Historical GHG Emissions			Projected GHG Emissions		
	2000[1]	2005[1]	2007[1]	2010	2015	2020
Energy-Related Carbon Dioxide[2]	5,562	5,724	5,736	5,633	5,721	5,813
Non-energy Carbon Dioxide[3]	394	368	368	368	368	368
Methane[4]	591	562	585	590	593	605
Nitrous Oxide[4]	329	316	312	314	322	332
Hydrofluorocarbons[4]	100	116	126	147	209	279
Perfluorocarbons[4]	14	6	8	7	7	6
Sulfur Hexafluoride[4]	19	18	17	15	13	13
International Bunker Fuels (not included in totals)	100	113	110	109	108	107
U.S. Territories	36	53	51	63	74	86
Total Gross Emissions	7,008	7,109	7,150	7,074	7,233	7,416
Net Sequestration Removals[5]	–718	–1,123	–1,063	–1,238	–1,218	–1,210
Total Net Emissions	6,291	5,986	6,088	5,836	6,014	6,206

[1] Historical emissions and sinks data are from U.S. EPA/OAP 2009. Bunker fuels and biomass combustion are not included in inventory calculations.

[2] Energy-related CO_2 projections are calculated from U.S. DOE/EIA 2009h, with adjustments made to remove any non-energy CO_2.

[3] Non-energy CO_2 includes emissions from non-energy fuel use and industrial processes.

[4] Non-CO_2 emission projections are based on US EPA/OAP 2006b, adjusted to 2007 inventory emissions.

[5] Details on disaggregated historical and projected sequestration data can be found in Table 5-3.

Note: Totals may not sum due to independent rounding.

tional inventory convention, as described later in this chapter. AEO 2009 presents projections and analysis of U.S. energy supply, demand, and prices through 2030, based on results from EIA's National Energy Modeling System. The projections in AEO 2009 look beyond current economic and financial turmoil and focus on factors that drive U.S. energy markets in the longer term. Key issues highlighted in AEO 2009 include higher but uncertain world oil prices, growing concern about GHG emissions and their impacts on energy investment decisions, the increasing use of renewable fuels, the expanding production of unconventional natural gas, the shift in the transportation fleet to more efficient vehicles, and improved efficiency in end-use appliances (U.S. DOE/EIA 2009a, 2009b).

Energy-related CO_2 emissions are projected to grow by 0.1 percent per year from 2005 to 2020, as compared with 1.3 percent per year from 1990 to 2005. The growth rate between 2005 and 2020 is moderated by the impact of the current recession in the United States. In 2020, energy-related CO_2 emissions are projected to total 5,813 Tg CO_2, about 1.5 percent higher than in 2005 (U.S. DOE/EIA 2009h).

Non-energy sources of CO_2 emissions include feedstock use of energy fuels in manufacturing, natural gas production and processing, the cement industry, and waste handling and combustion. As U.S. firms voluntarily adopt recapture technologies and other mitigation practices, these emissions will slow in growth and even decline in some sectors. Because the underlying sources are so varied, no single driver of projections is available. Therefore, the estimates presented here are historical extrapolations from the U.S. GHG inventory trends (U.S. EPA/OAP 2009). Emissions from these sources are expected to remain approximately constant, as some sources have historically increased and others have decreased, at 368 Tg CO_2 Eq. in 2005 and 2020.

Non-Carbon Dioxide Emissions

Emissions other than CO_2 currently represent about 15 percent of U.S. GHG emissions. Non-CO_2 GHG emissions include CH_4 emissions from natural gas production and transmission, coal mine operations, landfills, and livestock operations; N_2O emissions from agriculture and, to a lesser degree, transportation; and HFCs, PFCs, and SF_6 gases from industrial activities including, in some cases, the life cycles of the resulting products (Table 5-2).

Non-CO_2 emission projections are based on EPA's *Global Anthropogenic Non-CO_2 Greenhouse Gas Emissions: 1990–2020* (U.S. EPA/OAP 2006b). Projections in this report are based on a combination of trends in historical emissions and source-specific modeling. These projections have been adjusted to align with the most recent U.S. GHG emissions inventory (U.S. EPA/OAP 2009). EPA is currently updating its

non-CO_2 emission projections, but the results were not available in time for this report. In some cases, non-CO_2 emissions can be less sensitive to short-term economic disruptions than energy-related CO_2 emissions. For example, CH_4 emissions from landfills result from waste deposited over a long period of time.

Methane Emissions

Between 2005 and 2020, total CH_4 emissions are estimated to increase by about 8 percent as underlying economic activity increases over the projection period, with fugitive emissions from increased natural gas production the largest influence. However, these increases are expected to be mitigated by greater control of CH_4 emissions from landfills, coal mines, and manure through increased flaring, recovery, and use. CH_4 capture-and-use projects are driven in part by the prices of electricity and natural gas, which increase over the projection period.

Nitrous Oxide Emissions

N_2O emissions are estimated to rise from 316 Tg CO_2 Eq. in 2005 to 332 Tg CO_2 Eq. in 2020—an increase of 5 percent. Emissions from agriculture, the largest

Table 5-2 **U.S. Non-CO_2 Emissions by Source and Gas** (Tg CO_2 Eq.)

Emissions other than CO_2 include methane from agriculture, landfills, and natural gas production; nitrous oxide from agriculture; and HFCs, PFCs, and SF_6 from industrial production. Of these, HFCs are expected to grow the fastest due to the ongoing CFC and HCFC phaseout and replacement with HFCs.

Gases and Sources	2005	2010	2015	2020
Methane (CH_4)	562	590	593	605
Agriculture CH_4	186	192	189	189
Landfills	128	130	128	128
Natural Gas	106	113	125	138
Coal Mines	63	61	56	56
Other	79	95	94	95
Nitrous Oxide (N_2O)	316	314	322	332
Agriculture N_2O	225	227	232	237
Mobile Combustion	37	26	26	28
Nitric and Adipic Acid Production	25	29	30	32
Other	29	32	33	34
Hydrofluorocarbons (HFCs)	116	147	209	279
ODS Substitutes (HFCs)	100	135	198	268
HCFC-22 (HFC-23)	16	12	11	11
Semiconductors	0.2	0.3	0.2	0.2
Perfluorocarbons (PFCs)	6	7	7	6
Aluminum	3	4	4	4
Semiconductors	3	4	3	3
Sulfur Hexafluoride (SF_6)	18	15	13	13
Electrical Transmission and Distribution	14	12	12	11
Magnesium	3	2	1	1
Semiconductors	1	1	1	1

CFCs = chlorofluorocarbons; ODS = ozone-depleting substance; HCFCs = hydrochlorofluorocarbons.

Note: Totals may not sum due to independent rounding.

source of N_2O emissions, are estimated to increase from 225 Tg CO_2 Eq. in 2005 to 237 Tg CO_2 Eq. in 2020. In contrast, N_2O emissions from transportation are estimated to decrease over the same time period due to improvements in vehicle emission control technologies.

HFCs, PFCs, and SF₆ Emissions

HFC emissions are estimated to increase by more than 140 percent between 2005 and 2020, from 116 Tg CO_2 Eq. to 279 Tg CO_2 Eq. Over the same period, PFC emissions are estimated to remain flat, and SF_6 emissions are estimated to decline somewhat through increased voluntary control.

HFC emissions are increasing because demand for refrigeration and air conditioning is increasing and because HFCs are predominantly used as alternatives for ozone-depleting substances, such as the hydrochlorofluorocarbons (HCFCs) that are being phased out under the *Montreal Protocol on Substances That Deplete the Ozone Layer*. Both HFCs and HCFCs are GHGs, but HCFCs are not included here consistent with UNFCCC guidelines. Growth of HFCs is anticipated to continue well beyond 2020 if left unconstrained. Other sources of HFCs, PFCs, and SF_6 in industrial production include aluminum, magnesium, and semiconductor manufacturing and, in the case of SF_6, electricity transmission and distribution. These projections assume that voluntary emission reduction goals set by these industries will be met by implementing process improvements and emission control technologies.

Bunker Fuels

Bunker fuels consist of jet fuel, residual fuel oil, and distillate fuel oil used for international aviation and marine transport. Between 1990 and 2007, CO_2 emissions from bunker fuels declined by 5 percent from 114 million metric tons of carbon dioxide (MMTCO$_2$) to 109 MMTCO$_2$. Although emissions from international flights departing the United States have increased by 14 percent, emissions from international shipping voyages have decreased by 18 percent. The projection in Table 5-1 extends the annual rate of decline (from 1990 to 2007) through 2020 (U.S. EPA/OAP 2006b).

Sequestration

Forests and agricultural soils sequester a large amount of CO_2 from the atmosphere. In 2007, American forests and soils sequestered approximately 1,063 Tg CO_2 Eq., or 15 percent of total gross U.S. GHG emissions (U.S. EPA/OAP 2009). This net removal of atmospheric CO_2 is largely the result of careful land-use decisions in the forestry and agriculture sectors, including afforestation, reforestation, forest management techniques, and increased adoption of reduced-tillage practices in agriculture. The continuation and in-

creased adoption of such practices are expected to increase net carbon sequestration within the next decade.

Net sequestration estimated in this report is –1,210 Tg CO_2 Eq. in 2020 (specific data on sequestration for each significant sink can be found in Table 5-3). Net forest land area in the United States is expected to decline by 2 percent from 2002 to 2030 (Haynes et al. 2007), largely due to conversion to urban forest and developed areas.[5] Given this low rate of projected change in U.S. forest land area, the projected increase in net forest carbon uptake is largely due to expected changes in management practices (including intensification), afforestation/reforestation, and increased adoption of sustainable forestry practices.

Intensified management of forests can lead to an increased rate of growth, which will increase the uptake of carbon. Though harvesting on forest land removes much of the above-ground carbon, there is a positive growth-to-harvest ratio in U.S. forests, meaning that more carbon is accruing than is being harvested. Because most of the timber harvested from U.S. forests is stored in wood products or disposed of in solid waste disposal facilities, significant quantities of carbon in harvested wood are stored rather than released. The reversion of cropland to forest land also increases carbon in biomass.

In the agricultural sector, changes in agricultural soil management can lead to increases in carbon sequestration. Much of the carbon accumulation that has occurred in cropped soils in the United States over the last 20 years is attributable to the Conservation Reserve Program (see Chapter 4) and land-use conversions from annual crops to perennial hay and grazing land (USDA 2008). Changes in other soil management practices have also contributed to the higher carbon stock in U.S. agricultural soils and are projected to continue through 2020 (Table 5-3). These practices include increased use of conservation tillage (particularly no-till), reduced frequency of summer fallow, and increased application of manure to cropland and pasture.

Adjustments

Adjustments, primarily to the energy-related CO_2 emissions reported in this chapter, were made to more closely adhere to UNFCCC guidelines. Emissions in U.S. territories, predominantly fuel-related, were added, and the military and civilian international use of bunker fuels was subtracted from the totals and is reported separately. Emissions from fuel use in U.S. territories are projected to increase from 53 Tg CO_2 Eq. to 86 Tg CO_2 Eq. in 2020, based on extrapolation of historical trends. Emissions from international bunker fuels will decrease from 113 Tg CO_2 Eq. in 2005 to 107 Tg CO_2 Eq. in 2020 based on extrapolation of historical trends.

[5] Updated projections for urban forest land are expected in the upcoming U.S. Forest Service 2010 Resources Planning Act (RPA) Assessment. See http://www.fs.fed.us/research/rpa/.

Sectoral Carbon Dioxide Emissions
Electric Power CO$_2$ Emissions

The largest share of U.S. energy-related CO$_2$ emissions comes from electricity generation, currently about one-third of U.S. GHG emissions. CO$_2$ emissions are projected to increase by approximately 4 percent from 2005 to 2020, or from 2,381 Tg CO$_2$ to 2,466 Tg CO$_2$ (Table 5-4). Of the CO$_2$ emissions attributable to the electricity consumed in each economic sector, most of the increase can be attributed to the residential and commercial sectors; emissions attributable to the industrial sector are expected to decrease during that time (Table 5-5).

The combination of recently enacted energy policies, such as energy efficiency measures and renewable energy incentives, and rising energy prices slows the growth in U.S. consumption of primary energy relative to historical trends. Further, when slower demand growth is combined with increased use of renewable energy technologies and fewer additions of new coal-fired conventional power plants, growth in CO$_2$ emissions due to electricity generation also is slowed relative to historical experience. The share of electricity generation that comes from fossil fuels—primarily, coal and natural gas—is expected to decline, and the share from renewables is expected to increase from 8 percent in 2005 to 15 percent in 2020.

Although a comprehensive federal policy has yet to be enacted to address climate change, growing concerns about GHG emissions appear to be affecting investment decisions in the electricity sector. In the United States, potential regulatory policies to address climate change are in various stages of development at the state, regional, and federal levels. In addition to ongoing uncertainty with respect to future demand growth

Table 5-3 **Projections of Net Carbon Sequestration Uptake** (Tg CO$_2$ Eq.)

Increases in carbon sequestration from forests will lead to an overall increase in sequestration from today's levels by 2020.

Sources of Sequestration	2000[1]	2005[1]	2007[1]	2010	2015	2020
Forests[2]	400	872	810	976	947	928
Wood Products[3]	113	104	100	104	105	109
Urban Forests[4]	82	93	98	104	115	126
Agricultural Soils[4]	111	44	45	44	42	39
Other[4,5]	11	10	10	10	9	8
Total Sequestration	718	1,123	1,063	1,238	1,218	1,210

[1] Historical values are from U.S. EPA/OAP 2009.

[2] Estimates include carbon in above-ground and below-ground biomass, dead wood, litter, and forest soils. Projections reflect adjustments to the historical value for 2005. Forest carbon stocks are calculated by the FORCARB2 model (Smith and Heath 2004), based on forest areas and volumes from the base case U.S. Forest Service 2005 *RPA Timber Assessment Update* (Haynes et al. 2007). Emissions from forest fires are implicitly included in these estimates. Historical climate is assumed in the base projection.

[3] Estimates are composed of changes in carbon held in wood products in use and in landfills, including carbon from domestically harvested wood and exported wood products (Production Accounting Approach). Projections are made using the Woodcarb II model (Skog 2008), based on base case projections of forest products production and trade from Haynes et al. 2007.

[4] Projections are estimated from historical trends.

[5] "Other" category includes landfilled yard trimmings and food scraps.

Note: Projections reflect average annual values over each period ending with the labeled year. Totals may not sum due to independent rounding.

Table 5-4 **Electricity Generation and Greenhouse Gas Emissions**

The electricity sector will experience an accelerating shift toward renewable generation through 2020.

Sources of Electricity	2005 Emissions (Tg CO$_2$ Eq.)	2005 Generation (billion kWh)	2010 Emissions (Tg CO$_2$ Eq.)	2010 Generation (billion kWh)	2015 Emissions (Tg CO$_2$ Eq.)	2015 Generation (billion kWh)	2020[5] Emissions (Tg CO$_2$ Eq.)	2020[5] Generation (billion kWh)
Fossil Fuels[1,2]	2,381	2,793	2,342	2,756	2,382	2,717	2,466	2,832
Petroleum	102	116	47	52	38	43	40	45
Natural Gas	320	684	322	700	261	561	285	620
Coal	1,958	1,992	1,962	2,003	2,071	2,113	2,129	2,166
Other[3,4]	0.4	1	12	1	12	1	12	1
Non-Fossil Fuels	0	1,102	0	1,240	0	1,423	0	1,512
Nuclear	0	782	0	809	0	831	0	876
Renewable	0	320	0	431	0	592	0	636
Non-Fossil % Share Generation	28%		31%		34%		35%	
Total Fossil and Non-Fossil Fuel Generation	3,896		3,996		4,141		4,344	

[1] Historical emissions are from U.S. EPA/OAP 2009, and historical generation data are from U.S. DOE/EIA 2009b, Table 8.2.

[2] Includes electricity-only and combined heat and power plants whose primary business is to sell electricity, or electricity and heat, to the public.

[3] Other fossil fuel emissions include emissions from geothermal power and nonbiogenic emissions from municipal waste.

[4] Due to slight differences in categories between the EPA inventory and EIA projections, the "other" category in the historical inventory is not directly comparable to the "other" category in the projections.

[5] 2020 projections are from U.S. DOE/EIA 2009h.

kWh = kilowatt-hour; Tg CO$_2$ Eq. = teragrams of carbon dioxide equivalents.

Note: Totals may not sum due to independent rounding.

Table 5-5 **U.S. Energy-Related CO_2 Emissions by Sector and Source**[1] (Tg CO_2 Eq.)

Electricity generation accounts for the largest share—about one-third—of U.S. energy-related CO_2 emissions. In 2020, energy-related CO_2 emissions are projected to total 5,813 Tg CO_2, about 2 percent higher than in 2005.

Sectors and Sources	2005[2]	2010	2015	2020
Electric Power	2,381	2,342	2,382	2,466
Petroleum	102	47	38	40
Natural Gas	320	322	261	285
Coal	1,958	1,962	2,071	2,129
Other	0.4	12	12	12
Transportation[3]	1,882	1,835	1,858	1,853
Petroleum	1,848	1,800	1,823	1,817
Natural Gas	33	35	35	36
Electricity	5	4	5	6
Industrial[3]	828	805	839	840
Petroleum	330	336	326	321
Natural Gas	382	370	389	387
Coal	116	99	124	132
Electricity	731	580	607	610
Residential[3]	358	363	346	344
Petroleum	95	93	83	78
Natural Gas	262	268	262	265
Coal	1	1	1	1
Electricity	849	879	855	882
Commercial[3]	222	225	223	224
Petroleum	50	45	42	41
Natural Gas	163	173	174	176
Coal	9	6	6	6
Electricity	797	879	916	968
U.S. Territories	53	63	74	86

[1] U.S. DOE/EIA 2009h, with adjustments for bunker fuels, non-energy CO_2 emissions, and U.S. territories.

[2] Historical emissions data are from U.S. EPA/OAP 2009.

[3] Sector total emissions do not include indirect emissions from electricity usage.

Note: Totals may not sum due to independent rounding.

and the costs of fuel, labor, and new plant construction, it appears that capacity planning decisions for new generating plants are already being affected by the potential impacts of policy changes that could limit or reduce GHG emissions.

This concern is recognized in the electric sector projections by adding a 3 percent premium on investment in carbon-intensive electric generation sources, leading to limited additions of new coal-fired capacity. Much less new coal capacity is projected in this report than in recent years (U.S. DOE/EIA 2009). Renewable generation is expected to be about half of cumulative new capacity additions through 2020, and is the largest source of new capacity additions. Key federal tax credits and a new loan guarantee program in ARRA are expected to lead to a significant expansion of renewable energy generation compared with a case without ARRA.

As a result of these factors, while electricity generation is projected to increase by 0.7 percent per year from

[6] See http://www.epa.gov/otaq/climate/regulations.htm.

2005 to 2020, CO_2 emissions from electricity generation will increase by only 0.2 percent per year.

Residential CO_2 Emissions

Energy-related CO_2 emissions from the residential sector are estimated to increase by about 1.5 percent between 2005 and 2020, including indirect emissions from electricity. U.S. energy use will grow more slowly than the U.S. population because of investments in energy efficiency, including those spurred by ARRA. Emissions from electricity should increase over the projection period, while emissions from combustion of natural gas should decrease. Since population is expanding faster in the warmer climates of the West and South, less natural gas will be used for heating and more electricity will be necessary for cooling.

Commercial CO_2 Emissions

Commercial floorspace is expected to continue to expand, following trends in economic and population growth. Improvements in efficiency can only partly offset increases in emissions from the commercial sector. Increased disposable income will continue to lead to growth in commercial floorspace in hotels, restaurants, stores, and theaters. Growth in natural gas use, predominantly for heating, will be slower than growth in electricity use because of the more rapid growth in the South and West, where heating represents a smaller share of energy use.

Industrial CO_2 Emissions

Energy-related CO_2 emissions from the industrial sector are expected to grow slowly, from 828 Tg CO_2 in 2005 to 840 Tg CO_2 in 2020. Production from energy-intensive manufacturing industries will grow more slowly than the sector average, partly accounting for the slow growth in overall emissions. Including indirect emissions from electricity, industrial emissions are expected to decline by approximately 7 percent by 2020.

Transportation CO_2 Emissions

Transportation-related CO_2 emissions are the second-largest source of U.S. emissions. As with other energy-related sources of CO_2 emissions, transportation CO_2 emission projections for this report are based on EIA's April 2009 update of AEO 2009, and adjusted to match international inventory convention (U.S. DOE/EIA 2009a). Future emissions in the transportation sector will be impacted by a number of factors.

As shown in Table 5-6, AEO 2009 projects that slower growth in income per capita and higher fuel costs will reduce the growth of personal travel, slowing the growth in demand for both highway and aviation fuels. AEO 2009 also projects that emissions from heavy-duty vehicles will grow more rapidly than light-duty vehicles as a result of increases in commercial activity (U.S. DOE/EIA 2009h). While not included in AEO 2009, future regulatory efforts, including the

Table 5-6 **Comparison of the 2002, 2006, and 2010 CAR Assumptions and Model Results for 2020**

The 2010 *U.S. Climate Action Report* reflects assumptions of lower GDP growth and higher energy prices.

Factors	Assumptions for 2020		
	2002 CAR	2006 CAR	2010 CAR
Real GDP (billion chain-weighted 2000 dollars)[1]			
Reported	18,136	17,541	15,398
Corrected	17,688	17,541	15,398
Population (millions)	325	337	343
Energy Intensity (Btu per 2000 chain-weighted dollar of GDP)[2]			
Reported	8,712	6,877	6,798
Corrected	7,398	6,877	6,798
Light-Duty Vehicle Miles Traveled (billion miles)	3,631	3,474	3,137
Refiners' Acquisition Cost of Imported Crude Oil (2000 dollars/barrel)	24.68	41.24	95.56
Wellhead Natural Gas Price (2000 dollars/thousand cubic feet)	3.26	4.49	5.67
Minemouth Coal Price (2000 dollars/short ton)	12.79	18.52	22.85
Average Electricity Price (2000 cents/kWh)	6.5	6.6	7.7
All Sector Motor Gasoline Price (2000 dollars/gallon)	1.40	1.90	3.02
Energy Consumption (quadrillion Btus)	131	121	105

[1] A table in the 2006 CAR reported that the real GDP, as reported in the 2002 CAR, was $18,136 billion in 2000 dollars (while not reported as chain-weighted these are chain-weighted dollars). This appears to be an incorrect conversion from $15,525 billion chain-weighted 1996 dollars as originally reported in the 2002 CAR. Using the deflator from the *2002 Annual Energy Outlook* (U.S. DOE/EIA 2002) for consistency, it should have been only $17,688 billion chain-weighted 2000 dollars.

[2] Energy intensity, which is total energy consumption divided by real GDP, was reported as 7,920 Btu per 1996 dollar of GDP in the original 2002 CAR. In the 2006 CAR, energy intensity for the 2002 CAR was reported as 8,712 Btu per 2000 dollar of GDP. Using a corrected real GDP, energy intensity in the 2002 CAR should be 7,398 Btu per chain-weighted 2000 dollar of GDP.

Btus = British thermal units; CAR = U.S. Climate Action Report; GDP = gross domestic product; kWh = kilowatt-hour.

Sources: U.S. DOE/EIA 2002, 2009h.

GHG tailpipe standards proposed on September 28, 2009,[6] will bring about additional reductions in future GHG emissions.

ASSUMPTIONS USED TO ESTIMATE FUTURE GHG EMISSIONS

Changes Between the 2006 CAR and the 2010 CAR, Including the Effects of New Policies and Measures

GHG emissions under the "with measures" case presented in this report are significantly lower than emission estimates in the 2006 CAR. These differences can be traced to a combination of changes in policies, energy prices, and economic growth. In the 2006 CAR, emissions increased by 19 percent from the reported 2000 levels, versus a 5.6 percent increase from 2000 levels by 2020 (4 percent from 2005 to 2020) in this 2010 CAR.

Current estimates of energy-related CO_2 emissions include the effects of a number of policies that have been implemented since the analysis was completed for the 2006 CAR. These policies include ARRA and the Energy Independence and Security Act of 2007. They also include various state vehicle technology programs and renewable portfolio standards that have been implemented since 2006 and the Regional Greenhouse Gas Initiative in the northeastern and Mid-Atlantic United States. In addition, anticipation

of future limits on GHG emissions has already shifted business practices. Analysis of the changes in energy-related CO_2 emission projections between the 2006 CAR and the 2010 CAR shows that about half of them are due to the influence of newly implemented or anticipated public policies, equivalent to about 500 Tg CO_2 in 2020.

Changes in long-term economic growth estimates and energy prices account for the remaining changes in projected emissions between the 2006 CAR and the 2010 CAR. The macroeconomic projection used for this report estimates that the U.S. economy will grow by an average of 2.3 percent per year through 2020. This is a significant departure from the 3.0 percent projection in the 2006 CAR. The anticipated price of oil in 2020 has doubled along with significant increases in the anticipated prices of natural gas, coal, and electricity (U.S. DOE/EIA 2009).

Description of NEMS and Methodology

EIA's Office of Integrated Analysis and Forecasting developed and maintains the National Energy Modeling System (NEMS). The projections in NEMS are developed using a market-based approach to energy analysis. For each fuel and consuming sector, NEMS balances energy supply and demand, accounting for economic competition among the various energy fuels and sources. The time horizon of NEMS is 2005 through 2030, approximately 25 years into the future.

NEMS is organized and implemented as a modular system. The modules represent each of the fuel supply markets, conversion sectors, and end-use consumption sectors of the energy system. NEMS also includes macroeconomic and international modules. The primary flows of information among the modules are the delivered prices of energy to end users and the quantities consumed by product, region, and sector. The delivered fuel prices encompass all the activities necessary to produce, import, and transport fuels to end users. The information flows also include other data on such areas as economic activity, domestic production, and international petroleum supply.

Each NEMS component represents the impacts and costs of existing legislation and environmental regulations that affect that sector. NEMS accounts for all combustion-related CO_2 emissions, as well as emissions of sulfur dioxide, nitrogen oxides, and mercury from the electricity generation sector. The potential impacts of pending or proposed federal and state legislation, regulations, or standards—or of sections of legislation that have been enacted but that require funds or implementing regulations that have not been provided or specified—are not reflected in NEMS.

Table 5-6 shows the underlying assumptions and results in NEMS for the year 2020 and how they have changed from the 2002 and 2006 CARs.

Key Uncertainties Affecting Projected GHG Emissions

Any projection of future emissions is subject to considerable uncertainty. In the short term (less than five years), the key factors that can increase or decrease estimated net emissions include unexpected changes in retail energy prices, shifts in the competitive relationship between natural gas and coal in electricity generation markets, changes in economic growth, abnormal winter or summer temperatures, and imperfect projection methods. Additional factors may influence emission rates over the longer term, notably technology developments, shifts in the composition of economic activity, and changes in government policies. Finally, the indirect effects and interactions among some emission sources are not fully considered in the projections. For example, the indirect effects of increased biofuel production are not considered.

Technology Development

The projections of U.S. GHG emissions take into consideration likely improvements in technology over time. For example, technology-based energy efficiency gains, which have contributed to reductions in U.S. energy intensity for more than 30 years, are expected to continue. However, while long-term trends in technology are often predictable, the specific areas in which significant technology improvements will occur and the specific new technologies that will become dominant in commercial markets are highly uncertain, especially over the long term.

Unexpected scientific and technical breakthroughs can cause changes in economic activities with dramatic effects on patterns of energy production and use. Such breakthroughs could enable the United States to considerably reduce future GHG emissions. While U.S. government and private support of research and development efforts can accelerate the rate of technology change, the effect of such support on specific technology developments remains to be seen.

Regulatory or Statutory Changes

The current projections of U.S. GHG emissions do not include the effects of any legislative or regulatory action that was not finalized before March 31, 2009. Consequently, the projections do not include any increase in the stringency of equipment efficiency standards, even though existing law requires DOE to periodically strengthen its existing standards and issue new standards for other products. Similarly, the projections do not include the Obama administration's proposed goals of reducing U.S. GHG emissions by 17 percent below 2005 levels in 2020 and 83 percent in 2050. However, the GHG projections do reflect some of the uncertainty surrounding the adoption of national climate policy and, in particular, the decreased investment in new coal power capacity that results from this uncertainty.

Energy Prices

The relationship between energy prices and emissions is complex. Lower energy prices generally reduce the incentive for energy conservation and tend to encourage increased energy use and related emissions. However, a reduction in the price of natural gas relative to other fuels could encourage fuel switching that, in turn, could reduce carbon emissions. Alternatively, coal could become more competitive vis-à-vis natural gas, which could increase emissions from the power sector.

The energy-related CO_2 projections reflect a shift in oil market assumptions, with projected oil prices substantially higher than in previous analyses (U.S. DOE/EIA 2006, 2009a, 2009b). However, energy and oil price projections are subject to significant uncertainty. Decreases in delivered energy prices could result from increased competition in the electric utility sector or improved technology. On the other hand, energy price increases could result from the faster-than-expected depletion of oil and gas resources, from political or other disruptions in oil-producing countries, or from increases in oil demand abroad.

Economic Growth

Economic growth increases the future demand for energy services, such as vehicle miles traveled, amount of lighted and ventilated space, and process heat used in industrial production. However, growth also stimulates capital investment and reduces the average age of the capital stock, increasing its average energy efficiency. The energy-service demand and energy-efficiency effects of economic growth work in opposing directions. However, the effect on service demand is the stronger of the two, so that levels of primary energy use are positively correlated with the size of the economy. The economic growth data used for this report suggest that growth will be slower (2.3 percent per year) through 2020 than projected in the 2006 CAR (3.0 percent per year), which is expected to slow emissions growth.

Weather and Natural Occurrences

Energy use for heating and cooling is directly responsive to weather variation. In the EIA projection of CO_2 emissions, normal weather is defined by the average population-weighted number of heating and cooling degree-days for the most recent 10 years of historical data. Unlike other sources of uncertainty, for which deviations between assumed and actual trends may follow a persistent course over time, the effect of weather on energy use and emissions in any particular year is largely independent from year to year. For sequestration projections, historical climate and natural disturbances are assumed. For example, emissions from forest fires are implicitly included in the estimates for forest sequestration in Table 5-3. However, the extent to which climate change could exacerbate impacts on agriculture and forestry sinks—e.g., via increased pests, different degrees of dieback, and forest fire incidence—is not included in these estimates.

LONG-TERM GREENHOUSE GAS EMISSION PROJECTIONS TO 2050

The GHG emission projections in this chapter generally extend to 2020. EPA has also developed a number of modeling tools to evaluate long-term GHG mitigation policies, often extending to 2050. One of these models is the Applied Dynamic Analysis of the Global Economy (ADAGE) model. The focus of this model is policy analysis, but it includes long-term reference

scenarios that are used as a basis to evaluate proposed policies.

Reference scenarios for long-term projections are based on the same data sources used for this report to the extent data are available. Energy-related CO_2 emission projections are based on EIA's AEO 2009 (U.S. DOE/EIA 2009h), and non-CO_2 emission projections are based on *Global Anthropogenic Non-CO_2 Greenhouse Gas Emissions: 1990–2020* (U.S. EPA/OAP 2006b). Currently, the EPA models use the projections from the main release of AEO 2009 published in March 2009, which used slightly different economic projections and did not include ARRA. Sector definitions in ADAGE differ from those used for national inventory reporting. For these reasons, ADAGE long-term projections will not match those presented in other parts of this chapter. In addition, the sources for baseline projections used in this chapter do not extend to 2050, which required taking long-term projections from other sources. AEO 2009 extends to 2030, while the *Global Anthropogenic Non-CO_2 Greenhouse Gas Emissions* report extends to 2020. Longer-term projections are drawn from emission scenarios prepared by the U.S Climate Change Science Program (U.S. CCSP 2007).

Table 5-7 presents long-term U.S. GHG emission projections used in ADAGE to evaluate policy.

Table 5-7 Long-Term U.S. GHG Emission Projections in ADAGE (Tg CO_2 Eq.)

EPA's ADAGE model projects total U.S. greenhouse gas emissions will increase gradually, by about 18 percent between 2010 and 2050.

2010	2020	2030	2040	2050
7,118	7,390	7,765	8,101	8,379

ADAGE = Applied Dynamic Analysis of the Global Economy; EIA = Energy Information Administration; EPA = Environmental Protection Agency.

Note: At the time of this report, ADAGE was calibrated to EIA's March 2009 release of its *Annual Energy Outlook* (AEO 2009) (U.S. DIE/EIA 2009h). The ADAGE projections may be different from other projections in this chapter due to differences in modeling and lag time in calibrating model baselines to new versions of AEO 2009.

6

Vulnerability Assessment,
Climate Change Impacts,
and Adaptation Measures

Recent U.S. government-led scientific assessments of climate change impacts on the United States indicate that the nation is increasingly vulnerable to current and projected changes. It is also clear that there is more emphasis than ever on adaptation measures to increase the nation's resilience and take advantage of opportunities in the face of significant change.

During the past year, the U.S. government completed a major new climate change assessment, the U.S. Global Change Research Program's (USGCRP's) *Global Climate Change Impacts in the United States* (Karl et al. 2009). This assessment received a great deal of attention from large segments of the public and is now providing the basis for significant effort to incorporate climate change into decisions made by U.S.

businesses, resource managers, and policymakers. This report is especially helpful because it assesses impacts on a regional and socioeconomic basis. Across nine regions and seven sectors it explains the current and potential U.S. impacts of climate change and illustrates adaptation measures already being adopted. As this chapter demonstrates, the motivation for adaptation is clear, and the movement to initiate and coordinate action is underway.

There has been early and significant investment by the Obama administration in developing the first overarching U.S. adaptation strategy. The Director of the Office of Science and Technology Policy highlighted the need for adaptation in recent testimony to the U.S. Senate:

[The reality of climate change] underlines the need to invest, in parallel with efforts to reduce emissions and increase the uptake of the main heat-trapping gases, in adaptation to the changes in climate that can no longer be avoided—e.g., breeding heat- and drought-resistant crop strains, bolstering defenses against tropical diseases, improving the efficiency of water use, managing ecosystems to improve their resilience, and management of coastal zones with sea-level rise in mind.[1]

Building on community, business, and resource management efforts to examine adaptation options, the President issued Executive Order 13514 on October 5, 2009.[2] In addition to reducing greenhouse gas (GHG) emissions, this order will seek to organize the national effort on climate change adaptation and ensure widespread and complementary programs across the U.S. government. The executive order states that federal agencies will participate actively in a new U.S. Interagency Climate Change Adaptation Task Force that is "developing the domestic and international dimensions of a U.S. strategy for adaptation to climate change." This strategy will contribute to the knowledge, capability, and resources to effectively respond to climate change.

Increasingly, activities to assess and address vulnerabilities are focusing on the specific information needed to make better decisions in the face of climate variability and change, and in particular to facilitate effective and appropriate adaptation initiatives. Recent scientific assessments have been more focused on defining the most relevant and useful information for decision makers and providing it in a way that is accessible for all. In other words, the process is changing to be more demand-driven, and not only identifies the expected climate impacts, but also looks at some key risks and opportunities. Some of the recent key findings follow.

GLOBAL CLIMATE CHANGE IMPACTS IN THE UNITED STATES

During the last three years (2006–2009), the U.S. government has completed a suite of focused assessments addressing high-priority climate research questions. In an open and transparent manner, this approach communicates scientific analyses to the public via a set of 21 Synthesis and Assessment Products (SAPs) developed by USGCRP (U.S. CCSP/GCRP 2006–2009). These SAPs were synthesized in a single national-scale assessment, *Global Climate Change Impacts in the United States* (GCCI), released in June 2009 (Karl et al. 2009). The report analyzed climate impacts and response options across nine U.S. regions and seven sectors, and identified the following 10 key findings:

1. Global warming is unequivocal and primarily human-induced. Global temperature has increased over the past 50 years. This observed increase is due primarily to human-induced emissions of heat-trapping gases.

2. Climate changes are underway in the United States and are projected to grow. Climate-related changes are already observed in the United States and its coastal waters. These include increases in heavy downpours, rising temperature and sea level, rapidly retreating glaciers, thawing permafrost, lengthening growing seasons, lengthening ice-free seasons in the ocean and on lakes and rivers, earlier snowmelt, and alterations in river flows. These changes are projected to grow.

3. Widespread climate-related impacts are occurring now and are expected to increase. Climate changes are already affecting water, energy, transportation, agriculture, ecosystems, and health. These impacts are different from region to region and will grow under projected climate change.

4. Climate change will stress water resources. Water is an issue in every region, but the nature of the potential impacts varies. Drought, related to reduced precipitation, increased evaporation, and increased water loss from plants, is an important issue in many U.S. regions, especially in the West. Floods, water quality problems, and impacts on aquatic ecosystems and species are likely to be amplified by climate change in most regions. Declines in mountain snowpack are important in the West and Alaska, where snowpack provides vital natural water storage.

5. Crop and livestock production will be increasingly challenged. Agriculture is considered one of the sectors most adaptable to changes in climate. However, increased heat, pests, water stress, diseases, and weather extremes will pose adaptation challenges for crop and livestock production.

6. Coastal areas are at increasing risk from sea level rise and storm surge. Sea level rise and storm surge place many U.S. coastal areas at increasing risk of erosion and flooding, especially along the Atlantic and Gulf Coasts, Pacific Islands, and parts of Alaska. Energy and transportation infrastructure and other property in coastal areas are very likely to be adversely affected.

7. Threats to human health will increase. Health impacts of climate change include heat stress, waterborne and foodborne diseases, poor air quality, extreme weather events, and diseases transmitted by insects and rodents. Robust public health infrastructure could reduce the potential for negative impacts from climate change.

8. Climate change will interact with many social and environmental stresses. Climate change will combine with air and water pollution, population growth, overuse of resources, urbanization, and other social, economic, and environmental stresses to create larger impacts than from any of these factors alone.

[1] John P. Holdren, *Climate Services: Solutions from Commerce to Communities: Hearing before the United States Senate Committee on Science, Transportation, and Commerce*, 111th Cong., 1st Sess. (2009).

[2] "Federal Leadership in Environmental, Energy, and Economic Performance." *Federal Register*, Vol. 74, No. 194. See http://www.archives.gov/federal-register/executive-orders/2009-obama.html.

9. Thresholds will be crossed, leading to large changes in climate and ecosystems. There are a variety of thresholds in the climate system and ecosystems. These thresholds determine, for example, the presence of sea ice and permafrost, and the survival of species, from fish to insect pests, with implications for society. With further climate change, the crossing of additional thresholds is expected.

10. Future climate change and its impacts depend on choices made today. The amount and rate of future climate change depend largely on current and future human-caused emissions of heat-trapping gases and airborne particles. Responses involve reducing emissions to limit future warming, and adapting to the changes that are unavoidable.

The information in all of the SAPs as well as the GCCI is intended for use by a diverse group of decision makers, stakeholders, communicators (e.g., the media), and scientists. The material addresses the nation's needs for sound scientific information that decision makers can use to develop a better understanding of climate change impacts and vulnerabilities, as well as to develop and improve the design and implementation of adaptation measures. All of the SAPs and the GCCI were extensively reviewed by scientists, federal agency officials, stakeholders, and the general public. The SAPs build on and integrate cutting-edge research and application activities, advanced over the years by the interagency research efforts in climate and global change. More information about the SAPs and GCCI may be found in Chapter 8 of this report and at www.globalchange.gov.

The information highlighted in this chapter is taken principally from the GCCI report, which synthesizes much of the analysis in the SAPs and incorporates several other assessments. It provides analyses of ongoing and potential impacts of climate variability and change, adaptability of key systems, and measures that might be taken to reduce vulnerability, including examples of adaptation measures already in evidence. This chapter highlights ongoing U.S. efforts that are generating new insights into the potential impacts of climate change on key physical and biological processes (e.g., snowpack changes, streamflow, drought, extreme events) and changing resilience and vulnerability in a range of socioeconomic sectors (e.g., energy, agriculture, water resources, coastal systems, human health, and transportation).

SOME KEY U.S. VULNERABILITIES
Water and Energy
Climate change has clearly already altered, and will continue to alter, many aspects of the water cycle in the United States, affecting where, when, and how much water is available for all uses. Changes include widespread melting of snow and ice, increasing atmo-spheric water vapor, increasing evaporation, changing precipitation patterns and intensity, changing incidence of drought, rising water temperatures, reductions in river and lake ice, and changes in soil moisture and runoff. These changes have impacts across a vast range of socioeconomic activities, such as transportation, agriculture, energy production, industrial uses, and other needs, including human consumption (Karl et al. 2009).

Some examples of regional changes already observed include drying in the Southwest, a reduction of snow-pack/snow-water equivalent in the West, an increase in the incidence of heavy precipitation events across most of the United States, increasing streamflow in the eastern United States, and reduced ice cover on the Great Lakes. All of these examples are projected to continue, along with additional emerging disruptions to the current state of the water cycle (Karl et al. 2009).

An illustration of impacts from one of these regional changes is the potential consequences of drying in the Southwest. Parts of the Southwest could see more than a 40 percent decrease in surface runoff by mid-century (even under a moderate future emissions scenario). This is occurring against a backdrop of rapid population growth in the region and in a climate that is already semi-arid. In this region (and in others), climate change places an additional stress on already overburdened water systems (Karl et al. 2009).

Many locations in the Southwest are likely to suffer from conflict over water resources by 2025, even in the absence of climate change (U.S. DOI/BOR 2005). Under drier conditions, these conflicts are likely to be more widespread or to occur sooner and with greater intensity. Water disputes already exist in several areas across the United States, including the Sacramento Bay Delta, the Rio Grande, the Klamath River in Oregon and California, the Colorado River, and the Apalachicola-Chattahoochee-Flint River system in the Southeast (Karl et al. 2009).

In addition to vulnerabilities in water supply for human consumption, water is used in the process of power production—for cooling thermal power plants, and for generating power in hydroelectric facilities. In addition, delivering and treating water require large amounts of energy; thus, the vulnerabilities of water and energy systems are tightly interconnected.

Many of the effects of climate change have clear implications for the reliable production, transmission, and use of energy itself. For example, rising temperatures are likely to increase cooling needs and reduce heating needs in different parts of the country; changing precipitation patterns may positively or negatively affect the ability to produce hydropower; and increases in hurricane intensity could impact Gulf of Mexico energy production, refining, and transportation (Karl et al. 2009). There may be changes in energy consumed for other climate-sensitive processes, such as pumping water for irrigation in agriculture (Peart et al. 1995; McCarthy et al. 2001). Depending on the magnitude of these possible energy consumption changes, it may be necessary to consider changes in energy supply or conservation practices to balance demand (Franco and Sanstad 2006; CEPA 2006).

Impacts due to climate change are more likely to be most apparent at the sub-national scale, such as the regional effects of extreme weather and reduced water availability, and increased cooling demands in areas where temperature and vulnerable populations are increasing. Overall, the national energy economy is large, and the energy industry has both the financial and the managerial resources to be adaptive (Karl et al. 2009). Of course, climate change effects on energy supply and demand will depend not only on climatic factors, but also on patterns of economic growth, land use, population growth and distribution, technological change, and social and cultural trends that shape individual and institutional actions (McCarthy et al. 2001).

Transportation

The U.S. transportation network is vital to the nation's economy, safety, and quality of life. Transportation accounts for approximately one-third of total U.S. GHG emissions. While it is widely recognized that emissions from transportation have impacts on climate change, climate will also likely have significant impacts on transportation infrastructure and operations (Karl et al. 2009; U.S. DOT 2006).

Examples of specific types of impacts include softening of asphalt roads and warping of railroad rails; damage to roads and opening of shipping routes in polar regions (McCarthy et al. 2001); flooding of roadways, rail routes, and airports from extreme events and sea level rise; and interruptions to flight plans due to severe weather (Karl et al. 2009).

Along the Gulf Coast alone, it is estimated that 3,864 kilometers (2,400 miles) of major roadways and 396 kilometers (246 miles) of freight rail lines are at risk of permanent flooding within 50–100 years as climate change and land subsidence combine to produce an anticipated relative sea level rise in the range of 1.2 meters (4 feet). In Alaska, the cost of maintaining the state's public infrastructure is projected to rise 10–20 percent by 2030 due to warming, costing the state an additional $4–$6 billion, with roads and airports accounting for about half this cost (Karl et al. 2009). In New York City, what is now a 100-year storm is projected to occur as often as every 10 years by late this century. Portions of lower Manhattan and coastal areas of Brooklyn, Queens, Staten Island, and Long Island's Nassau County would experience a marked increase in flooding frequency. Much of the critical transportation infrastructure, including tunnels, subways, and airports, lies well within the range of projected storm surge and would be flooded during such events (Karl et al. 2009).

Public Health

Climate change poses unique threats to human health, including direct threats from heat waves or storms, and indirect effects, such as heat-exacerbated air quality impacts on health, or climate-sensitive infectious diseases (Box 6-1; Karl et al. 2009). Given the complexity of the factors that influence human health, assessing health impacts related to climate change poses a significant challenge (NAS/NRC 2001). The extent and nature of climate change impacts on human health vary by region, by relative sensitivity of population groups, by the extent and duration of exposure to climate change itself, and by society's ability to adapt to or cope with the change (Rose et al. 2001).

The probability of exacerbated health risks due to climate change points to a need to maintain a strong public health infrastructure to help limit future impacts (Ebi et al. 2008). Several initiatives, especially in cities, have been implemented for reducing risk. Appropriate and focused weather and climate information from the U.S. government has been essential in these initiatives. For example, heat is already the leading cause of U.S. weather-related deaths, with more than 3,400 deaths reported between 1999 and 2003 from excessive heat. Projections for several cities indicate increasing risk of heat-related deaths with increasing temperatures, even when including the likelihood of some adaptation measures (Karl et al. 2009).

Warming will also make it more challenging to meet air quality standards that affect certain segments of the population, particularly those with existing lung conditions or those who spend more time outdoors. Under constant pollution emissions, by the middle of this century, Red Ozone Alert days (when the air is unhealthful for everyone) in the 50 largest cities in the eastern United States, are projected to increase by 68 percent due to warming alone (Karl et al. 2009).

Box 6-1 **Endangerment and Cause or Contribute Findings for Greenhouse Gases**[1]

In response to a U.S. Supreme Court decision requiring the U.S. Environmental Protection Agency (EPA) to determine whether greenhouse gases (GHGs) endanger human health or welfare, or whether the science is too uncertain to make a determination, the EPA Administrator proposed endangerment and cause or contribute findings under Section 202 (a) of the Clean Air Act in April 2009. The proposed findings then underwent a public comment period. The proposed findings stated that the total body of scientific evidence compellingly supports that GHGs threaten both public health and welfare and that emissions from U.S. vehicles cause or contribute to the problem. On December 7, 2009, EPA finalized the endangerment and the cause or contribute findings. The Administrator reached this conclusion after considering both current and projected future effects of climate change and the full range of risks and impacts to public health and welfare in the United States, as well as extensive public comments.

[1]Further information can be found at http://www.epa.gov/climatechange/endangerment.html.

Studies analyzed by the U.S. Environmental Protection Agency (EPA) show that climate change causes increases in summertime ozone concentrations over substantial regions of the country (U.S. EPA/ORD 2009). For those regions that showed climate-induced increases, the increase in the maximum daily 8-hour average ozone concentration—a key metric for regulating U.S. air quality—was in the range of 2–8 parts per billion, averaged over the summer season. The increases were substantially greater than this during the peak pollution episodes that tend to occur over a number of days each summer. Several studies suggest that climate change may increase the frequency of high-ozone events (Bell et al. 2007; Leibensperger et al. 2008). Even when considering future scenarios with large decreases in air pollution emissions, climate change partly offsets the benefit of the emission reductions (Jacob and Winner 2009). Accordingly, climate change represents a significant penalty for air quality managers working to achieve ozone air quality goals and raises concerns about adverse health outcomes.

Ecosystems

Climate is an important factor influencing the distribution, structure, function, and services of ecosystems. Ongoing climate changes are interacting with other environmental changes to affect biodiversity and the future condition of ecosystems (e.g., McCarthy et al. 2001; Parmesan and Yohe 2003; Karl et al. 2009). Many factors affect biodiversity, including climatic conditions; the influence of competitors, predators, parasites, and diseases; disturbances, such as fire; and other physical factors. Human-induced climate change, in conjunction with other stresses, is exerting major influences on natural environments and biodiversity, and these influences are generally expected to grow with increased warming (Karl et al. 2009).

Climate change is already affecting many U.S. ecosystems, including wetlands, forests, grasslands, rivers and lakes, and coastal and nearshore environments, and has led to large-scale changes in the range of species and timing of seasons and migration. Invasive weed species have also increased, as have some insect pests and pathogens. Nearshore ecosystems are under stress not only from increasing temperatures, but also from the increased acidity in the ocean. U.S. desert and dry lands are likely to become hotter and drier, feeding a self-reinforcing cycle of invasive plants, fire, and erosion. In the future, these effects are likely to increase (Karl et al. 2009).

Coasts

Approximately one-third of the U.S. population lives in counties immediately bordering the nation's ocean coasts. In addition to accommodating major cities, the coastal zone supports recreation, fishing, energy, industry, and critical transportation infrastructure. Coastal and ocean activities contribute more than one

trillion dollars to the national gross domestic product (Karl et al. 2009).

Increasing vulnerability at the coast will result from extreme events and sea level rise, and also from population changes, building practices, beach management, increasing nitrogen runoff, and many other socioeconomic factors. Sea levels have been rising by 2–3 millimeters (0.078–0.117 inches) per year along most of the U.S. coast (Zervas 2001). However, due to the feedback loops of climate change, that rise is projected to accelerate in the coming decades. Accounting for local subsidence, coastal scientists are considering the possible impacts of a 0.9-meter (3-foot) rise in sea level (or more in some locations) over the next century (Titus et al. 2009; Karl et al. 2009).

Key concerns associated with these changes include land loss, increased flooding of low-lying coastal communities, coastal erosion, barrier island migration, wetland loss, and increased salinity of aquifers and estuaries, especially during droughts. Various health impacts, including those associated with population displacement, are among the secondary effects of these changes. This increasing societal vulnerability is leading some insurance companies to raise rates or deny property coverage to communities along the Gulf and Atlantic coasts (Mills 2005).

SAMPLE U.S. RESEARCH, ASSESSMENTS, AND ACTIVITIES PERTAINING TO VULNERABILITY, IMPACTS, AND ADAPTATION

Many of the key U.S. vulnerabilities discussed above are being addressed across the government and within communities through specific programs at a variety of different geographic scales. Following is a sample cross-section of the programs being carried out by the United States at the international, federal, state, regional, and local levels to assess the impacts of and reduce vulnerability to climate change. A goal of the new U.S. Interagency Climate Change Adaptation Task Force is to create a coherent, comprehensive program of activities that allows synergies among these many and varied programs.

International Activities
NASA, USAID, and NOAA Hubs
The National Aeronautics and Space Administration (NASA), the U.S. Agency for International Development (USAID), and the National Oceanic and Atmospheric Administration (NOAA) are working to develop regional hubs around the world to apply remotely sensed information to development assistance. Based on the successful SERVIR (Regional Visualization and Monitoring System) hub in Central America, this activity will link available data streams to new applications, develop tools, and build local human and institutional capacity to use this information. These systems will sup-

port decision making in a number of areas, including climate change, land management, urban planning, food security, agriculture, and disaster mitigation.

USAID Climate Change Program[3]

USAID, often in partnership with other agencies, leads a number of activities to help build developing country capacity to understand climate change and adapt to its impacts. This includes supporting innovative applications and tools for climate and weather observation, and developing guidance on how to build the resilience of projects designed to promote economic development. USAID works to make data and guidance readily accessible and useful for development decisions at the community, national, and regional levels. USAID also provides support for cutting-edge research to develop more climate change-resilient agricultural inputs, and provides capacity building in disaster preparedness and risk reduction. USAID's approach places particular emphasis on partnerships with the private sector and on working with local and national authorities, communities, and nongovernmental organizations.

Recent climate change adaptation projects supported by USAID include:

- Community-based drought preparedness planning in Cambodia, East Timor, and Vietnam.

- An early-warning system to ensure that drought and other threats to the well-being of East African pastoralists' free-ranging livestock can be detected and addressed in a timely manner.

- A three-year initiative to help vulnerable communities in the seven Zambezi river basin countries use conservation-based farming techniques, soil conservation, water-harvesting techniques, and reforestation to adapt to climate-related threats.

- Research into the development of heat-tolerant wheat and flood-tolerant rice varieties in South Asia, where farmers are already seeing the impacts of higher temperatures and more severe flooding on crop yields.

- Community training to help farmers in Malawi diversify their livelihoods and adopt new agriculture conservation practices that reduce soil erosion, improve water quality, and sequester carbon in the soil.

- A Collaborative Research Support Program that is identifying ways to build the resilience of Andean small-holder production systems and their capacity to adapt to climate change.

- Dissemination of micro-irrigation technologies, such as foot pumps, in Mali, to address rainfall variability and increase water use efficiency.

USAID recognizes that adapting to climate change requires a hierarchy of linked efforts. USAID is working to make Earth observation information readily accessible and applicable to development decisions,

including developing innovative applications and appropriate tools, and communicating that information to stakeholders and decision makers. Through interaction with local partners and with new tools, USAID can better understand how environmental changes may affect sectors critical for development. Once those impacts are understood, stakeholders need to assess and agree on preferred adaptation options. Then, on-the-ground actions can be implemented to build the resilience of projects designed to promote economic development.

USAID is also supporting a three-year initiative implemented by the International Federation of Red Cross and Red Crescent Societies and the United Nations World Meteorological Organization, to help vulnerable communities in the seven Zambezi river basin countries use conservation-based farming techniques, soil conservation, water-harvesting techniques, and reforestation to adapt to climate-related threats.

Activities at Local to National Scales
USGCRP Assessments[4]

USGCRP, an interagency body of 13 agencies, has published a suite of 21 SAPs over the past three years (U.S. CCSP/GCRP 2006–2009). USGCRP is the focal point for the development of the detailed assessments referred to in this chapter. Each of the 21 reports focused on particular elements of climate change and U.S. vulnerabilities, and many assessment efforts of individual agencies are incorporated into these USGCRP assessments. In the spring of 2008, a technical scientific assessment was produced to provide an ongoing summary of the work (CENR 2008). In June 2009, the comprehensive national-scale GCCI assessment was released, which incorporated the results of all 21 SAPs and included a specific focus on adaptation information (Karl et al. 2009). The assessment is being discussed at dozens of local, regional, and national meetings and stakeholder forums, and is being built on to provide a foundation for more focused scientific information to support adaptation and decision making.

NASA Applied Sciences Program

This program benchmarks practical uses of NASA-sponsored observations from Earth observation systems and predictions from Earth science models. NASA implements projects that carry forth this mission through partnerships with public, private, and academic organizations working toward developing innovative approaches for using Earth system science information to provide decision support that can be adapted in applications worldwide. This program focuses on applications of national priority, such as agriculture, water resources, and air quality, and expands and accelerates the use of knowledge, science, and technologies resulting from the NASA goal of improving predictions in the areas of weather, climate, and natural hazards.[5]

---

[3] See http://www.usaid.gov/our_work/environment/climate/.

[4] See http://www.globalchange.gov/publications/reports.

[5] See http://science.hq.nasa.gov/earth-sun/applications/index.html.

EPA Global Change Research Program

EPA's Global Change Research Program (GCRP) is an assessment-oriented program that emphasizes understanding the potential consequences of climate variability and change on U.S. human health, ecosystems, and socioeconomic systems. This program has four areas of emphasis: human health, air quality, water quality, and ecosystem health. In an attempt to capitalize on expertise in the academic community, a significant portion of EPA's GCRP resources is dedicated to extramural research grants administered through the STAR (Science To Achieve Results) grants program, which supports science related to assessments of the consequences of global change and human dimensions research.[6] Another sample of research and assessment is available in the form of the Sanctuaries Condition Reports published by NOAA's Office of National Marine Sanctuaries.

NOAA Office of National Marine Sanctuaries

The Office of National Marine Sanctuaries (ONMS) manages marine areas in both nearshore and open ocean waters that range in size from less than 2.59 square kilometers (1 square mile) to almost 362,600 square kilometers (140,000 square miles). To study marine ecosystems and the human influences that affect them, in 2001 ONMS began to implement System-Wide Monitoring (SWiM). Part of SWiM includes the preparation of Condition Reports that summarize the resources in each sanctuary, pressures on those resources, the current condition and trends, and management responses to the pressures that threaten the integrity of the marine environment. Specifically, the reports include information on the status and trends of water quality, habitat, living resources, and maritime archaeological resources and the human activities that affect them. They also consider ways to observe and respond to climate-related changes in sea level, water temperature, ocean acidity, coral bleaching, invasive species, and diseases.[7]

USGS Climate Effects Science Network[8]

The U.S. Geological Survey (USGS) Climate Effects Science Network (CESN) is coordinated through all U.S. Department of the Interior (DOI) resource management bureaus. CESN integrates climate and environmental change data sets with conceptual and digital models across disciplines, including remote sensing, geography, geology, biology, and hydrology, to better understand the impacts of climate on natural resources, agriculture, and human populations on episodic to decadal and millennial time scales, local to global spatial scales, and weather to climate process scales. The goal of CESN is to develop a systems-level understanding of biogeochemical processes resulting from changes in climate, to link these changes to the sustainability of ecosystems, wildlife, subsistence cultures, and societal infrastructure, and to apply the knowl-edge gained for decision support. In 2008, USGS initiated pilot-integrated research in northern Alaska, where permafrost thaw and sea-ice melting are resulting in rapid and poorly understood changes to regional ecosystems.

U.S. National Integrated Drought Information System

More than a dozen U.S. federal agencies or offices collaborate in the U.S. National Integrated Drought Information System (NIDIS) effort to provide risk and drought management information.[6] Some of the agencies include NOAA, the U.S. Department of Agriculture (USDA), DOI, the U.S. Department of Transportation (DOT), and the U.S. Department of Energy. NIDIS was formally launched in 2006 and has significant milestones to increase capabilities in every year. It is designed to develop the leadership and networks to implement an integrated drought monitoring and forecasting system at federal, state, and local levels and to foster and support a research environment focusing on risk assessment, forecasting, and management. A key piece of the NIDIS activity will be an early-warning system for drought to provide accurate, timely, and integrated information. All of this information is incorporated into interactive systems, such as the Web portal, which provides not only timely access to the early-warning capabilities, but also a framework for public awareness and education about droughts.[9]

NSF Decision Making Under Uncertainty Centers[10]

Under the leadership of the U.S. National Science Foundation (NSF) five interdisciplinary research teams are studying important aspects of problems associated with understanding climate-related decisions under uncertainty. The increased knowledge generated by recent scientific research on the causes and consequences of climate change and variability has led to a growing need to better understand how decision makers choose among alternative courses of action. These teams are expected to produce new insights of interest to the academic community, generate significant educational benefits, and develop new tools that will benefit decision makers and a range of stakeholders. Research centers are located at Arizona State, Carnegie-Mellon, and Columbia universities. Other interdisciplinary teams are conducting research at the University of Colorado at Boulder, and Rand Corporation in Santa Monica, California.

USDA/NRCS National Water and Climate Center

The National Water and Climate Center (NWCC) leads the development and transfer of water and climate information and technology through natural resource planning support, data acquisition and management, technology innovation and transfer, partnerships, and joint ventures.[11] The NWCC develops and manages key observation and monitoring networks called SNOTEL (SNOpack TELemetry) and SCAN

[6] See http://cfpub.epa.gov/gcrp/about_ov.cfm.

[7] See http://sanctuaries.noaa.gov/science/condition/welcome.html.

[8] See http://www.usgcrp.gov/usgcrp/agencies/interior.htm.

[9] See http://www.drought.gov.

[10] See http://www.nsf.gov/news/news_summ.jsp?cntn_id=100447.

[11] See http://www.wcc.nrcs.usda.gov/.

(Soil Climate Analysis Network). These networks provide automated comprehensive snowpack, soil moisture, and related climate information designed to support natural resource assessments. They collect and disseminate continuous, standardized soil moisture and other climate data in publicly available databases and climate reports. Uses for these data include inputs to global circulation models, verifying and ground truthing satellite data, monitoring drought development, forecasting water supply, and predicting sustainability for cropping systems.

Regional Activities

NOAA Regional Integrated Sciences and Assessments Program

One of the key questions NOAA faces is how to improve the link between climate sciences and society. The Regional Integrated Sciences and Assessments (RISA) program is helping to realign the nation's climate research to better serve society. NOAA's RISA program supports research that addresses complex climate-sensitive issues of concern to decision makers and policy planners at a regional level. RISA research team members are primarily based at universities, though some are based at government research facilities, nonprofit organizations, or private-sector entities. Research areas include the fisheries, water, wildfire, and agriculture sectors, coastal restoration, and climate-sensitive public health issues.[12] The program currently supports eight regional centers. A new RISA initiative, the Southern Climate Impacts Planning Program, was initiated in 2009 in the Gulf Coast region.[13]

NOAA Regional Climate Centers

NOAA's six Regional Climate Centers (RCCs) are a federal-state cooperative effort designed to provide regional and local expertise and assistance to a wide range of customers.[14] The RCCs are engaged in the timely production and delivery of useful climate data, information, and knowledge for decision makers and other users at local, state, regional, and national levels. The RCCs support NOAA's efforts to provide operational climate services, while leveraging improvements in technology and collaborations with partners to expand quality data dissemination capabilities.

DOI Practitioner Development Program

In partnership with many stakeholders and several federal agencies (including NOAA, EPA, the Department of Defense [DOD], and USDA), the DOI Bureau of Reclamation (BOR) has implemented the Basin Study Program, which will incorporate the latest science, engineering technology, climate models, and innovative approaches to water management in the western United States.[15] The program will serve as a part of BOR's Water Conservation Initiative and as a key element in implementing the Secure Water Act. To integrate climate change into water management activities and to help improve methods of water resources planning, these partners are working together to develop and provide a Climate Change Integration Technical Training Program for western water practitioners, planners, technical specialists, and decision makers. This effort also exposes practitioners to emerging methods as they become available.

DOI Climate Change Response Centers

Under a new secretarial order, DOI has launched a coordinated strategy to address the current and future impacts of climate change on land, water, and other natural and cultural U.S. resources. A cornerstone of this strategy is the implementation of eight Climate Change Response Centers located across the country. A network of Landscape Conservation Cooperatives will also work at landscape scales to foster partnerships and assess climate impacts on such issues as wildlife migration, invasive species, and wildfire risk. This 2009 initiative will support and promote adaptation responses.

State Activities

California Climate Change Center[16]

The California Climate Change Center is investigating the range of possible changes to the state's climate and the likelihood and rate of progression of such changes.[17] Using the results of this work, the Center is assessing the potential future economic and ecological consequences of climate change for California, and examining a range of impacts and adaptation options (e.g., agriculture and water resources), as well as mitigation strategies. The center manages a robust research program with a dynamic community of California researchers from various scientific disciplines and a worldwide network of peers collaborating on climate change issues of interest to California.

SAMPLE U.S. ASSESSMENT AND ADAPTATION ACTIVITIES IN SPECIFIC SECTORS

The sample sector- and region-specific impact summaries and adaptation projects included in this section demonstrate the variety and scale of information and methods utilized within the United States. The examples are illustrative of key areas of investigation, and are not intended to be a comprehensive listing of all efforts across the nation.

Water Resources

Working Toward a Drought-Resilient U.S. Southwest

A NOAA RISA program based at the University of Arizona, titled the Climate Assessment for the Southwest (CLIMAS),[18] is developing and utilizing new information on drought to increase societal resilience to this recurrent phenomenon. The impacts of U.S. drought during the last five to seven years have included

[12] See http://www.climate.noaa.gov/cpo_pa/risa/.

[13] See http://www.southernclimate.org.

[14] See http://www.ncdc.noaa.gov/oa/climate/regionalclimatecenters.html.

[15] See http://www.usbr.gov/WaterSMART/docs/Basin%20Study%20Program.pdf.

[16] See http://www.doi.gov/archive/climatechange/SecOrder3289.pdf.

[17] See http://www.climatechange.ca.gov/research/index.html.

[18] See http://www.ispe.arizona.edu/climas/.

sustained and significant economic losses, significantly reduced reservoir levels, water emergencies, and widespread and severe wildfires. Creating a more drought-resilient society requires a fundamental shift from crisis management to risk management. Investigators studying the impacts of drought are studying the historical record, evolving demographics and population growth, water law, and ecosystem management. For example, investigators are working to develop methods to utilize seasonal climate and streamflow forecasts more effectively to mitigate the impact of drought on water supplies. It is expected that knowledge of this type will become even more valuable in the coming decades, if climate model projections of increasing aridity in continental interiors prove accurate.

Developing Strategies for Improving Water Management

If the allocation of water is already a concern in many locations around the country, the added challenges of climate change pose increased risk and vulnerability for some of these locations. An interagency report, released in 2009 by DOI, DOD, and NOAA, explored strategies to improve water management by tracking, anticipating, and responding to climate change (Brekke et al. 2009). This report describes the existing and still needed science crucial to addressing the many impacts of climate change on water resource management. It provides adaptation and planning options for water resource practitioners.

In addition, federally funded researchers are working with water and ecosystem managers as new insights and techniques become available, allowing incorporation of scientific data and information into near- and long-term planning. Interagency and partnership projects are occurring for the Colorado River, in the Columbia River Basin, and in California and many other locations.[19]

Evaluating Hydroclimatic Conditions in the Pacific Northwest

The Climate Impacts Group (CIG) at the University of Washington is using emerging knowledge to help inform decision making related to changing hydroclimatic conditions in the Pacific Northwest. CIG is utilizing its hydrologic modeling and prediction capabilities to evaluate water resource issues, including the consequences of alternative water and hydroelectric power management strategies for salmon restoration efforts and the consequences of changing water demands and changes in land cover for regional water resources.[20] CIG is one of seven similar RISA programs funded by NOAA's Climate Program Office. These programs are designed to provide the nation with experience-based knowledge about how to develop climate services.[21] They are an important element of the USGCRP's efforts to support decision making on climate-related issues.

Planning for Climate Change in New York City

New York City is an example of adaptation in the face of water resource concerns. The city's Department of Environmental Protection (DEP), which provides water for 9 million people in the New York City metropolitan area, is beginning to alter its planning to take into account the effects of climate change—sea level rise, higher temperatures, increases in extreme events, and changing precipitation patterns—on the city's water systems. Examples of measures that have emerged from the comprehensive assessment and evaluation process include relocating critical control systems to higher floors in low-lying buildings or to higher ground, building flood walls, and modifying infrastructure design criteria. The DEP is also establishing a system for reporting the impacts of extreme weather on the city's watershed and infrastructure (Karl et al. 2009).

Assessing Groundwater Availability

The depletion of groundwater at a variety of scales and the compounding effects of recent droughts emphasize the need for an updated status on the availability of the nation's groundwater resources. Assessments of the current state of the highest-stressed groundwater flow systems are necessary tools for characterizing the availability of groundwater. The USGS Groundwater Resources Program[22] is taking advantage of the quantitative work previously conducted by the Regional Aquifer-System Analysis Program and information available from USGS, DOI (e.g., BOR, the National Park Service [NPS], Bureau of Land Management [BLM]) and other federal agencies (e.g., EPA and NOAA), states, tribes, and local governments to provide an updated quantitative assessment of groundwater availability in areas of critical importance. The assessments currently underway and continuing into 2010 and beyond will (1) document the effects of human activities on water levels, groundwater storage, and discharge to streams and other surface-water bodies; (2) explore climate variability impacts on the regional water budget; and (3) evaluate the adequacy of data networks to assess impacts at a regional scale.

Developing a National-Scale View of Water Quality Impacts from Climate Change

EPA is using high-resolution simulations of future climate change over the contiguous United States from the North American Regional Climate Change Assessment Program and land-use scenarios from EPA's Integrated Climate and Land Use Scenarios project to develop hydrologic and water quality change scenarios for 20 major drainage basins across the United States. These scenarios will be based on watershed simulations conducted with the HSPF (Hydrologic Simulation Program–Fortran) and SWAT (Soil and Water Assessment Tool) watershed models, and will focus on the response of stream flow,

[19] See, for example, http://pubs. usgs.gov/circ/1331/Circ1331.pdf and http://wwa.colorado.edu/ colorado_river/.

[20] See http://www.cses. washington.edu/cig/res/hwr/hwr. shtml.

[21] See http://www.climate.noaa. gov/cpo_pa/risa/.

[22] See http://water.usgs.gov/ogw/ gwrp/.

nitrogen, phosphorus, and suspended sediment to a range of projected changes in climate and land use.

Ecosystems
Enhancing Ecosystem Management Strategies
The extent to which ecosystem conditions will be affected in the future will depend on the magnitude of climate change, the degree of sensitivity of the ecosystem to that change, the availability of adaptation options for effective ecosystem management, and the willingness to deploy those options. USGCRP addressed management strategies for facilitating ecosystem adaptation to climate variability and change in several state-of-the-art reports focused on federal lands (Baron et al. 2008; Backlund et al. 2008; Fagre et al. 2009). The goal of these adaptation strategies is to reduce the risk of adverse outcomes through activities that increase the resilience of ecological systems to climate change, and to take advantage of positive outcomes (Turner et al. 2003; Tompkins and Adger 2004; Scheffer et al. 2001; Baron et al. 2008). Because some changes in the climate system are likely to persist into the future regardless of emissions reduction, adaptation is an essential response for future protection of climate-sensitive ecosystems.

Adaptation options for enhancing ecosystem resilience include changes in processes, practices, or structures to reduce anticipated damages or enhance beneficial responses associated with climate variability and change. In some cases, opportunities for adaptation offer stakeholders multiple beneficial outcomes, such as the addition of riparian buffer strips that, for example, manage pollution loadings from agricultural land into rivers or provide a protective barrier to increases in both pollution and sediment loadings that may be associated with future climate or other environmental change (Baron et al. 2008).

Identifying Necessary Information and Tools
A range of adaptation options is possible for many ecosystems, but a lack of information or resources may impede successful implementation. In some cases, managers may not have the knowledge or information they need to address climate change impacts. In other instances, managers may understand the issues and have the relevant information but lack resources to implement adaptation options. Furthermore, even with improvement in the knowledge and communication of available and emerging adaptation strategies, the feasibility and effectiveness of adaptation will depend on the adaptive capacity of the ecological system or social entity (Baron et al. 2008).

Thus, increasing adaptive capacity will require information and tools that aid in (1) understanding the combined effects on ecosystems of climate changes and non-climate stressors, and consequent implications for achieving specific management goals;

(2) applying existing management options or developing new adaptation approaches that reduce the risk of negative outcomes; and (3) understanding the opportunities and barriers that affect successful implementation of management strategies to address climate change impacts (Baron et al. 2008).

Reducing the Frequency and Severity of Wildfires in the West
In the western United States, invasive annual grasses (e.g., cheatgrass) are increasing rapidly throughout the region. These fire-tolerant species increase fire frequency, eliminating native plants, wildlife and livestock forage, and habitat. USGS is providing science in support of decision making, including (1) mapping annual plant invasions (ground, aerial, satellite); (2) developing native plant restoration protocols; and (3) mapping historic fires to understand causes.[23] BLM, which is responsible for managing much of the federal land affected by these issues, is developing adaptation plans to restore native plant communities, ensure the necessary presence of pollinators, reduce the frequency and severity of wildfire, and "pre-adapt" these lands for climate change—planting communities in anticipation of local changes due to a changing climate. Specifically, BLM and its partners are conducting a natural habitat restoration effort for millions of acres in the Great Basin of Nevada, Oregon, Idaho, California, and Utah, and they are working with commercial seed producers to grow native seed for restoration.[24]

Public Health
Addressing Heat-Related Health Threats
Critical U.S. government information to support action to reduce the health impacts of excessive heat days would be routinely used during heat waves in many U.S. cities. This information has applicability under projected increases in the number of such events, due to climate change. For example, NOAA's National Weather Service provides temperature and heat index information for the determination of "heat warnings."

During heat waves in Philadelphia, a heat alert is issued and news organizations are provided with tips on how vulnerable people can protect themselves. The health department and thousands of block captains check on elderly residents, and public cooling places extend their hours. The city also operates a "heatline" with nurses ready to assist callers with heat-related health problems. In addition, the Cool Homes program offers assistance to elderly, low-income residents to install roof insulation and cool surfaces to save energy and lower indoor temperatures. Philadelphia's system is estimated to have saved 117 lives in its first three years of operation (Karl et al. 2009).

As another example, EPA and other federal agencies responsible for addressing excessive heat events (EHEs) developed a guidebook that provides interest-

[23] See http://www.usgs.gov/hazards/wildfires/.

[24] See http://www.blm.gov/id/st/en/prog/gbri.html.

ed public health officials with information on the risks of and impacts from EHEs, including guidance on EHE forecasting and identification (U.S. EPA/OAP 2006a). The guidebook also provides a menu of notification and response actions to consider when developing or enhancing a local EHE program based in part upon a review of various EHE response programs.

Developing Integrated Health Assessment Frameworks

EPA is also undertaking important work assessing the relationships between climate change and human health. This assessment work goes beyond basic epidemiological research to develop integrated health assessment frameworks that consider the effects of multiple stresses, their interactions, and human adaptive responses. Along with health sector assessments, conducted in conjunction with the USGCRP national assessment process, there are research and assessment activities focused on the consequences of global change on weather-related morbidity and vector- and water-borne diseases. In addition, the results from the USGCRP air quality assessments will be used to evaluate health consequences.[25]

Working Internationally to Fight Malaria

Internationally, NOAA, EPA, and NSF have collaborated to fund efforts, in partnership with developing country colleagues, to identify relationships between malaria and climate and to develop an early-warning system for malaria.[26] Clear relationships were identified and significant capacity was built, including through stakeholder input, to develop early-warning systems.

Coasts

Adapting to Rising Sea Levels Along the East Coast

In an example from the USGCRP assessment of coastal vulnerability to sea level rise (Titus et al. 2009), adaption options for coastal wetland ecosystems were outlined. Wetlands are rich ecosystems that provide protection from coastal storms as well as a number of other resources and services (Karl et al. 2009). To adapt to rising sea level without damaging vulnerable ecosystems through the implementation of "hard armoring" of the shorelines (e.g., sea walls), a system of "rolling easements" for property and ecosystems to migrate inland as the sea rises could be implemented. For example, in Massachusetts and Rhode Island, hard armoring of the shoreline is prohibited in some areas to allow the migration of ecosystems. Maryland has recently enacted a Living Shoreline Protection Act,[27] which requires its Department of the Environment to create maps delineating the areas where hard structures will be allowed or prohibited (Karl et al. 2009).

Preparing for Sea Level Rise

In recognition of significant potential impacts from climate change, the Coastal Zone Management Act of 1972 states: "Because global warming may result in a substantial sea-level rise with serious adverse effects in the coastal zone, coastal states must anticipate and plan for such an occurrence" (16 U.S.C. [U.S. Code] § 1451).[28] Property owners and federal, state, and local governments are already starting to take measures to prepare for the consequences of rising sea level. Most coastal states are working with the U.S. Army Corps of Engineers to place sand on their beaches to offset shore erosion. Property owners are elevating existing structures in many low-lying areas, to provide resilience to episodic storms as well as long-term change. Shoreline erosion along estuaries has led many property owners to defend their back yards by erecting shore protection structures, such as bulkheads, which eliminate the intertidal wetlands and beaches that would otherwise be found between the water and the dry land.

Several states have adopted policies to ensure that beaches, dunes, or wetlands are able to migrate inland as sea level rises. Some states prohibit new houses in areas likely to be eroded in the next 30–60 years (e.g., North Carolina Coastal Resources Commission). Concerned about the need to protect both natural shores and private property rights, Maine, Massachusetts, Rhode Island, South Carolina, and Texas have implemented some version of "rolling easements," in which people are allowed to build, but only on the condition that they will remove the structure if and when it is threatened by an advancing shoreline (McCarthy et al. 2001; Titus et al. 2009).

Addressing Climate's Impacts on Estuaries

EPA has created the Climate Ready Estuaries program to help address climate change in coastal areas. This effort is helping the National Estuary Programs and other coastal managers develop the technical capacity to assess climate change vulnerabilities, engage and educate local stakeholders, and develop and implement adaptation strategies.[29]

Developing Assessments and Adaptation Plans

DOI through USGS is conducting a national risk assessment due to future sea level rise for the U.S. Atlantic, Pacific, and Gulf of Mexico coasts. This includes work with NPS for coastal park units. In parallel, DOI through the U.S. Fish and Wildlife Service (FWS) is working with partners to create and implement adaptation plans for specific coastal wildlife refuges. In 2008, FWS created an adaptation plan with The Nature Conservancy and Duke Energy. Solutions articulated in the plan include (1) restore wetland hydrology: restore damage caused by artificial ditches; (2) reforest and restore shoreline: protect and restore existing natural coastal and inland habitat to facilitate species migration as sea level rises; (3) restore oyster reefs: restore oyster reefs in Pamlico Sound to protect shorelines from storms and rising seas;[30] and (4) mea-

[25] See http://cfpub.epa. gov/ncea/cfm/recordisplay. cfm?deid=203459.

[26] See http://portal.iri.columbia. edu/portal/server.pt/gateway/ PTARGS_0_0_4621_223_0_43/ http%3B/portal.iri.columbia. edu%3B9086/irips/projectview. jsp?id=31 and http://www.iwmi. cgiar.org/health/malaria/projects. htm#climatev.

[27] See http://mlis.state. md.us/2008rs/chapters_noln/ Ch_304_hb0973E.pdf.

[28] See http://coastalmanagement. noaa.gov/czm/czm_act.html.

[29] See http://www.epa.gov/cre.

[30] See http://www.dot.gov/ climate.

sure and monitor carbon sequestration: monitor the effects of management on soil carbon.

Developing Data for Informed Decision Making

Many agencies and organizations are developing data that can provide insights regarding the implications of sea level rise. Sample data include elevation models and data sets; geographic information systems; ecosystem, fish, and wetlands impact information; tidal gauge data; economic and population data; insurance information; storm surge databases; and hurricane research.

Transportation

The United States is working to provide better information to decision makers across the sector about what future climate variability and change could mean for existing and planned infrastructure and about the set of potential response strategies that might be implemented to adapt to future climate.

Incorporating Climate Change and Variability into Decision Making

DOT's Center for Climate Change and Environmental Forecasting is dedicated to fostering awareness of the potential links between transportation and global climate change, and to formulating policy options to deal with the challenges posed by climate change and variability.[18] DOT research projects are investigating the potential impacts of climate variability and change on transportation infrastructure and its operation, and provide guidance as to how transportation planners and decision makers may incorporate this information into transportation planning decisions to ensure a reliable and robust future transportation network.

"Climate Proofing" Roads

An example of adaption measures undertaken in response to climate threats to transportation includes an effort to "climate proof" a road on the island of Kosrae

in the U.S.-affiliated Federated States of Micronesia. In response to projections of increased heavy downpours and sea level rise, authorities placed the road higher and introduced improved drainage systems. The additional costs to incorporating these measures were projected to be offset by the reduced repair and maintenance costs after 15 years (Karl et al. 2009).

Energy

Prospects for adaptation to climate change effects by energy providers, energy users, and society at large are speculative, in part because of the lack of research to date, although the potentials are considerable and several examples exist of early adaptation planning. It is possible that the greatest challenges would be in connection with possible increases in the intensity of extreme weather events and possible significant changes in regional water supply regimes. But adaptation prospects depend considerably on the availability of information about possible climate change effects to inform decisions about adaptive management, along with technological change in the longer term.

One example of addressing energy vulnerabilities along the Gulf Coast in association with the oil and gas industry can be found in Port Fourchon, Louisiana. The port supports 75 percent of deepwater oil and gas production in the Gulf of Mexico, and its role is increasing. The oil imported (1 million barrels/day) and produced (300,000 barrels/day) through Port Fourchon is linked to 50 percent of the refining capacity of the nation. Only one road, Highway 1, connects Port Fourchon with the nation, and it transports machinery, supplies, and workers to the port. Responding to concerns about storm surge and flooding, related in part to climate change and rising sea levels, Louisiana is upgrading Highway 1 and raising it to about the 500-year flood level, as well as building higher bridges over Bayou LaFourche and the Boudreaux Canal (Karl et al. 2009; Savonis et al. 2008).

7

Financial Resources and Technology Transfer

Climate change is a global challenge that must be addressed in the context of a dynamic global economy. Since the United Nations Framework Convention on Climate Change (UNFCCC) was negotiated, patterns of economic growth and emissions in the developed and developing world have shifted substantially. It is now clearer than ever that no single country or group of countries can solve the threat of climate change alone, and all countries must take actions to alter their emission trajectories commensurate with the demands of science and consistent with their specific capabilities and circumstances.

In this context, the United States recognizes that poorer developing countries have urgent and growing needs for financial and technical support to adapt to the effects of climate change and to promote low-

emissions development pathways. As part of our commitment to implement the Copenhagen Accord, the United States is working with other developed countries to provide "fast-start" climate finance, approaching $30 billion during 2010–2012, to help meet the adaptation and mitigation needs of developing countries. We are also committed to a leadership role in addressing the long-term climate finance challenge.

On December 16, 2009, President Obama signed the Consolidated Appropriations Act, 2010, which more than tripled climate-related foreign assistance from the previous fiscal year (FY), to over $1 billion.

The FY 2010 enacted budget also includes a dramatic increase for adaptation assistance, including a first-ever U.S. contribution of $50 million to the United Nations

Least Developed Country Fund (LDCF) and Special Climate Change Fund (SCCF). It also includes $375 million for the World Bank-managed Climate Investment Funds, and a substantial increase in funding for U.S. Agency for International Development (USAID) climate programs. This will lead to scaled-up cooperation in many parts of the world, including Africa and Asia and among the small island states.

President Obama has also taken major steps to promote new international technology cooperation on climate and energy issues with specific regions and countries. In June 2009, the President announced a new Energy and Climate Partnership of the Americas to promote the diffusion of clean energy technologies across the Western Hemisphere. In July 2009, leaders of the 17 major economies in the Major Economies Forum on Energy and Climate (MEF) announced the establishment of a Global Partnership to speed clean energy technology deployment. They are also working to develop and implement action plans for eight key technologies. The United States has accelerated collaboration with key partners, such as China, India, the European Union, Canada, Brazil, Mexico, and Russia, to combat climate change, coordinate clean energy research and development, and support efforts to achieve a successful agreement under the UNFCCC.

Most recently, the United States announced in Copenhagen that it would ramp up U.S. climate financing to ensure a fast start to post-Copenhagen efforts, contributing its share of developed country financing approaching $30 billion by 2010, and working with other Parties to the UNFCCC to mobilize $100 billion globally by 2020 for countries in need, from various public- and private-sector sources, given the existence of robust mitigation efforts by all key Parties. As part of that larger effort, the United States pledged $1 billion through 2012 to reduce emissions from deforestation, land degradation, and other activities, as part of a multilateral donor effort of $3.5 billion. Also as part of the broader multiyear, multidonor effort, the United States pledged $85 million toward the Climate Renewables and Efficiency Deployment Initiative (Climate REDI), which will channel a total of $350 million to fund programs over five years.

This chapter describes first the general roles of U.S. government agencies in climate change financial and technical assistance, and then ongoing programs and partnerships that involve significant U.S. government resources and investments. The latter are broken out further into cross-cutting initiatives, mitigation and forests, vunerability and adaptation, trade and development assistance, and private-sector assistance. Tables 7-1 through 7-3 present U.S. financial contributions to the Global Environment Facility (GEF) and multilateral institutions in support of climate change programs and activities. Table 7-5 at the end of this chapter presents U.S. financial contributions related to U.S. implementation of the UNFCCC. The United States expects that a transition to a larger and more robust foreign assistance program will include substantial new efforts to assist countries in their efforts to address climate change in FY 2010.

U.S. FINANCIAL AND TECHNICAL ASSISTANCE BY AGENCY

U.S. government agencies provide developing and transition countries with finance and technical assistance for low-carbon and climate-resilient development. These agencies facilitate the transfer of technologies for mitigation, adaptation, capacity building, and research through official assistance, export credits, project financing, risk guarantees, and insurance to U.S. companies, as well as credit enhancements for host-country financial institutions.

U.S. Agency for International Development

As the foreign assistance arm of the U.S. government, USAID works in the areas of agriculture, the environment, economic growth, democracy and governance, conflict prevention, education, health, global partnerships, and humanitarian assistance in more than 100 countries. With headquarters in Washington, D.C., USAID provides assistance in sub-Saharan Africa, Asia and the Near East, Latin America and the Caribbean, and Europe and Eurasia. USAID's strength is its field offices in many regions of the world. In addition to partnering with other U.S. government agencies, the agency partners with private voluntary organizations, indigenous organizations, universities, American businesses, international agencies, and other governments.

Table 7-1 **U.S. Financial Contributions to the Global Environment Facility for Climate Change Activities: 2003–2010** (Millions of U.S. Dollars)

During fiscal years 2003–2010, the United States allocated $242 million for Global Environment Facility programs related to climate change.

2003	2004	2005	2006	2007	2008	2009	2010	Total
56	32	24	26	26	26	26	26	242

Note: Since 2004, funding estimates represent the budget authority available to incur obligations, and not necessarily the amount of funds outlayed or spent.
Source: Office of Management and Budget.

The types of assistance USAID provides include technical assistance and capacity building, training and scholarships, food aid and disaster relief, infrastructure construction, small-enterprise loans, budget support, enterprise funds, and credit guarantees. USAID plays a key leadership role in delivering climate change-related assistance to developing and transition countries through its Global Climate Change Program. The program is active in more than 40 developing and transition countries, integrating climate change mitigation and adaptation into a broad range of development assistance activities.

Table 7-2 **Annual U.S. Financial Contributions to Multilateral Institutions** (Millions of U.S. Dollars)

The U.S. government provides direct funding to multilateral institutions and programs in support of sustainable economic development and poverty alleviation. Although in many cases a portion of this funding supports climate change activities, it is not currently possible to identify that amount. Therefore, this table represents total U.S. government contributions to these multilateral development institutions and funds, including amounts not directly attributable to climate change activities. Table 7-3 presents U.S. funding to multilateral programs that can be directly attributed to climate change activities.

Institutions, Funds, and Programs	2005	2006	2007	2008	2009
World Bank Group	843.200	941.800	940.500	942.300	1,115.000
International Bank for Reconstruction and Development	0.000	0.000	0.000	0.000	0.000
International Development Association	843.200	940.500	940.500	942.300	1,115.000
Multilateral Investment Guarantee Agency	0.000	1.300	0.000	0.000	0.000
International Finance Corporation	0.000	0.000	0.000	0.000	0.000
Other Multilateral Institutions, Funds, and Programs					
Inter-American Investment Corporation	0.000	1.700	0.000	0.000	
Inter-American Development Bank – Multilateral Investment Fund	10.900	1.700	1.700	24.800	25.000
Asian Development Bank	0.000	0.000	0.000	0.000	0.000
Asian Development Fund	99.200	99.000	99.000	74.500	105.000
African Development Bank	4.100	3.600	3.600	2.000	0.800
African Development Fund	105.200	134.300	134.300	134.600	150.000
European Bank for Reconstruction and Development	35.100	1.000	0.000	0.000	0.000
Global Environment Facility[3]	36.000	36.000	26.000	26.000	26.000
International Fund for Agricultural Development	14.900	14.900	14.900	17.900	18.000
North American Development Bank	0.000	0.000	0.000	0.000	0.000
United Nations Development Programme[2]	108.128	110.000	108.900	97.365	100.000
United Nations Environment Programme[2]	10.912	10.262	10.159	10.415	10.500
UNFCCC Supplementary Fund (included under IPCC/UNFCCC, below)	0.000	0.000	0.000	0.000	0.000
Multilateral Scientific, Technological, and Training Programs					
OAS Development Assistance Programs[1]	4.861	4.750	4.702	5.455	5.500
United Nations World Food Programme	1,173.720	1,123.113	1,183.235	2,076.430	
UN Development Fund for Women & UNIFEM Trust Fund[2]	2.976	4.750	4.703	5.356	6.000
World Trade Organization Technical Assistance[1,2]	0.992	0.950	0.940	0.942	0.950
International Civil Aviation Organization[1,2]	0.992	0.950	0.940	0.942	0.950
Montreal Protocol Multilateral Fund[2]	33.381	32.185	34.729	30.413	32.797
International Conservation Programs[2]	0.000	0.000	0.000	0.000	0.000
Intergovernmental Panel on Climate Change/UNFCCC[2]	6.349	5.950	5.891	6.447	7.000
International Contributions for Scientific, Educational, and Cultural Activities[1]	6.789	7.000	6930	6.447	9.000
World Meteorological Organization Voluntary Co-operation Programme[1,2]	1.984	1.900	1.881	1.885	1.900
Center for Human Settlements[2]	0.149	0.150	0.149	0.992	2.000

[1] These international organizations also receive assessed contributions through the Contributions to International Organizations account.

[2] Voluntary contributions from international organizations and programs.

[3] These numbers reflect fiscal year funding—i.e. "2005" funding is FY 2005 funding. The U.S. fiscal year begins October 1st of the preceding year and ends on September 30th.

IPCC = Intergovernmental Panel on Climate Change; OAS = Organization of American States; UN = United Nations; UNFCCC = United Nations Framework Convention on Climate Change; UNIFEM = United Nations Development Fund for Women.

Sources: U.S. Department of State (http://www.state.gov/), U.S. Department of the Treasury (http://www.ustreas.gov/), UN World Food Programme (http://www.wfp.org/node/7359).

Table 7-3 **Multilateral Programs Receiving U.S. Funding for Climate Change Programs and Activities: 2005–2009** (Millions of U.S. Dollars)

During 2005–2009, the United States provided funding to three multilateral programs specifically for programs and activities related to climate change.

Multilateral Programs	2005	2006	2007	2008	2009
Montreal Protocol Multilateral Fund	21.328	21.500	21.285	18.846	21.000
Intergovernmental Panel on Climate Change/UNFCCC	6.349	5.950	5.891	6.447	7.000
International Contributions for Scientific, Educational, and Cultural Activities	5.952	6.000	5.940	5.455	8.000

UNFCCC = United Nations Framework Convention on Climate Change.
Source: U.S. Department of State (http://www.state.gov/).

U.S. Environmental Protection Agency

The U.S. Environmental Protection Agency (EPA) designs and implements innovative programs on a variety of international environmental challenges, including efforts to make transportation cleaner, reduce greenhouse gas (GHG) emissions, and improve local air quality. As a global leader in methane mitigation, EPA spearheads the Methane to Markets Partnership, an international effort that promotes methane recovery and use as a clean energy source. EPA is also working through international partnerships to retrofit diesel engines (Partnership for Clean Fuels and Vehicles); bring energy efficiency labeling, like ENERGY STAR, to other countries; improve national GHG inventories; and reduce the impacts of buildings and vehicles on the environment.

U.S. Department of Energy

In addition to providing funding support for such interagency activities as the Climate Change Technology Program, the U.S. Department of Energy (DOE) works with numerous foreign governments and institutions to promote dissemination of energy efficiency, renewable energy, and clean energy technologies and practices. Since the 2006 *U.S. Climate Action Report*, DOE has enhanced its international engagement on technology transfer through such efforts as the Carbon Sequestration Leadership Forum (CSLF) and the MEF. DOE is participating in the International Partnership for Energy Efficiency Cooperation and the International Renewable Energy Agency (IRENA), and is working through the Global Bioenergy Partnership (GBEP) to reduce the climate change impacts of biofuel development. DOE has chaired the Climate Technology Initiative (CTI) that serves the UNFCCC, and is actively involved in the Private Financing Advisory Network (PFAN), which is seeking and finding opportunities for private entities to invest in clean energy projects in developing countries.

U.S. Department of State

The U.S. Department of State (DOS) coordinates bilateral and multilateral diplomatic efforts related to climate change, including the U.S. presence at the international climate negotiations hosted by the United Nations. DOS also deploys financial resources in support of key multilateral and bilateral priorities in adaptation, clean energy, and forestry and land use.

DOS is responsible for U.S. government commitments to the LDCF and SCCF. These multilateral funds provide financing to developing countries to help them adapt to the impacts of climate change, with a specific focus on assisting the most urgent adaptation needs of least developed countries. In FY 2010, DOS will also invest funds in pilot approaches that better integrate climate change objectives into other U.S. government development activities.

DOS, in coordination with other agencies, also funds clean energy programs in support of strategic bilateral diplomatic partnerships as well as multilateral efforts. For example, working with DOE, DOS supports the MEF, which provides an avenue for supporting low-carbon technology projects and programs of interest to key emerging economies, including China, India, Brazil, South Africa, Mexico, and Indonesia. DOS funds, in conjunction with EPA technical assistance, also support the Methane to Markets Partnership.

In forestry and land use, DOS supports the World Bank Forest Carbon Partnership Facility, to help developing countries measure forest carbon stocks and design deforestation emission reduction strategies. In FY 2010, DOS will also invest funds to support capacity building for reducing emissions from deforestation and degradation (REDD+) projects in key forested developing countries.

U.S. Department of Agriculture

The U.S. Department of Agriculture (USDA) provides a broad array of technical and financial assistance to help countries carry out agriculture- and forest-sector activities that support their efforts to mitigate or adapt to the impacts of climate change. USDA activities are coordinated through the Foreign Agricultural Service and and the U.S. Forest Service's (USFS's) International Program and draw on the technical capabilities of USDA's research and conservation programs. Strong partnerships with land grant universities, environmental nongovernmental organizations (NGOs), and the private sector make this work integrated and comprehensive.

Specific activities include developing methods and protocols for measuring GHG emissions from agricultural sources, and estimating carbon fluxes from forest and agricultural systems; designing and implementing agriculture- and forest-sector components of national GHG inventories; reducing GHG emissions through improved agricultural practices; increasing carbon sequestration through improved forest management (including forest conservation, sustainable forestry, and agroforestry); and encouraging sustainable and renewable bioenergy technology and use. In addition, USDA provides technical assistance to the USAID missions and host governments to incorporate climate change strategies into country plans and carry out country-led projects to adapt to and mitigate the effects of global climate change.

The USDA Cochran and Borlaug Fellowship programs now offer training in areas related to global climate change, in response to country requests. In 2008, USFS completed a multiyear, multimillion-dollar program funded by USAID, to integrate the Incident Command System (ICS) into India's emergency response procedures, improving the country's ability to effectively coordinate and respond to large-scale climate-related disasters. A similar program was also completed in Sri Lanka. In recent years, cooperation has also occurred between USFS and partners in Mexico and China on technologies and methods related to forest carbon inventories.

National Aeronautics and Space Administration

The National Aeronautics and Space Administration (NASA) advances scientific knowledge by observing the Earth system from space; assimilating new observations into climate, weather, and other Earth system models; and developing new technologies, systems, and capabilities for its observations, including those with the potential to improve future operational systems managed by the National Oceanic and Atmospheric Administration (NOAA) and others. NASA is a major participant in the U.S. Global Change Research Program and in U.S. activities to support the Group on Earth Observations (GEO). NASA's Earth observation data are openly available to all nations, organizations, and individuals, and the agency has many active partnerships with U.S. and international agencies to facilitate the use of its data in research and operational applications.

U.S. Department of Commerce

The U.S. Department of Commerce (DOC) contributes to developing scientific data on climate change and facilitates the development and deployment of technology to mitigate and address the effects of climate change through its various agencies:

- The Patent and Trademark Office protects intellectual property rights that enable technological innovation.

- The National Institute of Standards and Technology and the National Telecommunications and Information Administration provide research into developing a smart-grid infrastructure.

- The Economic and Statistics Administration works on identifying and quantifying "green jobs."

- The International Trade Administration promotes the global deployment of U.S. climate mitigation technologies.

- The Economic Development Administration (EDA) works to help U.S. communities develop sustainable economic development plans, projects, and activities. EDA's Global Climate Change Mitigation Incentive Fund supports communities as they develop green projects, processes, and functions, helping to create new green jobs while simultaneously reducing global GHG emissions.

National Oceanic and Atmospheric Administration

NOAA, which is a DOC agency, plays a particularly critical role in international climate activities. NOAA provides weather, water, and climate services; manages and protects fisheries and sensitive marine ecosystems; conducts atmospheric, climate, and ecosystem research; promotes efficient and environmentally safe commerce and transportation; supports emergency response; and provides vital information in support of decision making. NOAA's climate mission is to: "Understand and describe climate variability and change to enhance society's ability to plan and respond." NOAA's long-term climate efforts are designed to develop and deliver a predictive understanding of variability and change in the global climate system, and to advance the application of this information for decision making in climate-sensitive sectors through a suite of research, observations and modeling, and application and assessment activities.

Millennium Challenge Corporation

Established in January 2004, the Millennium Challenge Corporation (MCC) is a U.S. government corporation that works with some of the poorest countries in the world to reduce poverty through sustainable economic growth. MCC's innovative development assistance is based on the principle that aid is most effective when it reinforces good governance, economic freedom, and investments in people. MCC recognizes that alleviating global poverty requires urgent attention to climate change and responsible environmental stewardship, and is committed to helping partner countries integrate climate change and other environmental and social considerations into their poverty reduction programs.

To date, MCC has signed grant agreements with 20 countries totaling nearly $7.2 billion in assistance. Many of these compacts include funding for projects designed to help partner countries improve natural resource management, strengthen institutional capacity, and pursue lower-carbon growth strategies. For instance, in El Salvador, MCC is funding a $67-million community development project that includes approximately $1 million for the installation of 450 solar panel systems to serve more than 2,000 poor and isolated households in the country's northern zone, and the development of watershed management plans to promote water conservation as part of a broader water supply and sanitation program. And in the Republic of Moldova, MCC is providing $2 million in assistance to promote improved watershed management as part of a larger agriculture and irrigation development program, which will contribute to climate change adaptation and increased food security.

Looking ahead, MCC anticipates increasing opportunities to help developing countries incorporate climate change and other environmental issues into their poverty reduction programs. Future compact partners—such as Jordan, Malawi, Zambia, the Philippines, and Indonesia—face cross-cutting environmental challenges, like water scarcity, deforestation, and biodiversity loss, and are particularly vulnerable to climate risks, such as droughts and extreme weather events. MCC is actively working with these countries, as with all of its partners, to help them adapt to and mitigate these risks. Moreover, MCC is committed to working closely with other donors to harmonize efforts to integrate climate change and natural resource management into development assistance.

MAJOR U.S. CROSS-CUTTING INITIATIVES

Several major U.S. initiatives that cut across agencies and sectors provide technology transfer and financial assistance in support of climate change objectives.

USAID Global Change Program

In 2009, Secretary of State Hillary Rodham Clinton noted that USAID has for years been "a leader in advancing climate, clean energy, and conservation activities in the developing world, drawing the clear and important link between solving the climate problem and promoting sustainable development globally."[1] Since 1991, USAID has spent over $3 billion on projects and programs aimed at mitigating climate change and helping vulnerable communities build their capacity and resilience, while simultaneously meeting development objectives in the energy and water sectors, urban areas, forest conservation, agriculture, and disaster assistance. USAID is committed to further bolstering its efforts in this area, in recognition of the importance of global low-carbon growth and the increasingly urgent adaptation and resilience needs of vulnerable developing countries. USAID will continue to support and augment programs to promote the transfer of clean energy technologies, provide tools to

measure and reduce GHG emissions, deploy tools for Earth observation and early-warning systems, facilitate carbon management through improved land use, and support climate vulnerability assessments and initiatives to help countries and communities increase their adaptive capacity.

World Bank Climate Investment Funds

On July 1, 2008, the World Bank Board of Executive Directors formally approved the creation of the Climate Investment Funds (CIFs). On September 26, 2008, donors gathered to pledge over U.S. $6.1 billion. The CIFs are hosted by the World Bank, but have separate, innovative governance structures that give developed and developing countries equal voice and representation, creating a consensus-based decision-making model. The World Bank implements the CIF jointly with the Regional Development Banks (the Africa Development Bank, the Asian Develpment Bank, the European Bank for Reconstruction and Development, and the Inter-American Development Bank). The CIFs were established through an inclusive and consultative process in support of the Bali Action Plan. They are designed on the basis of consultations with potential donors and recipients, the United Nations (UN) system, the GEF, civil society organizations, and the private sector. These funds will be used to pilot new approaches to governance, and address new areas of common interest, such as reducing emissions from deforestation and increasing resilience to climate change impacts.

Global Hunger and Food Security Initiative

Early in 2009, President Obama joined fellow Group of Eight (G8) leaders in committing a collective $20 billion to jump-start a new, comprehensive approach to combating global hunger. Through the Global Hunger and Food Security Initiative, the United States is focusing on agricultural-led growth to raise the incomes of the poor and increase the availability of food, while providing support to strengthen the capacity of countries to anticipate and prevent hunger-related emergencies. This commitment is particularly critical in the context of rapidly changing climate. As stated by the Intergovernmental Panel on Climate Change in its Fourth Assessment Report, the increased frequency of extreme climatic events, such as droughts and floods, will negatively affect global agricultural production, even beyond the impacts of increasing temperatures alone (IPCC 2007).

Global Earth Observation System of Systems

The United States is a founding member of the intergovernmental Group on Earth Observations (GEO), which is coordinating efforts to build the Global Earth Observation System of Systems (GEOSS) on the basis of a 10-Year Implementation Plan for 2005–2015 (GEO 2005). The purposes of GEOSS are to achieve comprehensive, coordinated, and sustained observa-

[1] January 26, 2009. Announcement of appointment of Special Envoy on Climate Change Todd Stern. See http://www.state.gov/secretary/rm/2009a/01/115409.htm.

tions of the Earth system; improve monitoring and predictions; and increase understanding of Earth processes, especially those related to weather, climate, energy, water, agriculture, disasters, health, biodiversity, and ecosystems. The emerging public infrastructure connects a diverse and growing array of instruments and systems, including the Global Climate Observing System, for monitoring and forecasting changes in the global environment. This "system of systems" supports policymakers, resource managers, science researchers, and many other experts and decision makers. For example, an umbrella framework known as "GEOSS in the Americas" has resulted in significant increases in the availability and use of environmental data and new information tools in Central and South America and the Caribbean, particularly for coastal zone management, hurricanes and flooding, air quality, water, and agricultural management.

U.S. MITIGATION PROGRAMS, EXCLUDING FOREST PROGRAMS

Major Economies Forum on Energy and Climate

The MEF brings together 17 developed and developing economies to engage in a meaningful dialogue on clean energy technology and the need to secure a broad international agreement to combat climate change. President Obama chaired a meeting of the leaders of the MEF partners in July 2009, and the group underscored its commitment to continue to work together to strengthen the world's ability to combat climate change and to facilitate agreement at the 15th Conference of the Parties to the UNFCCC (COP-15) in Copenhagen in December 2009. The group agreed to establish a Global Partnership to drive development and deployment of eight key transformational low-carbon technologies: advanced vehicles; bio-energy; carbon capture, use, and storage; energy efficiency (including buildings and industrial processes); high-efficiency and lower-emission coal technologies; smart grids; solar energy; and wind energy. Partners created action plans by late 2009 to seek to accelerate development and deployment of these technologies, and it is anticipated that MEF partners as well as other interested countries will subsequently cooperate on the implementation of the activities identified in the action plans.

Asia-Pacific Partnership on Clean Development and Climate

The Asia-Pacific Partnership on Clean Development and Climate (APP) is an effort by Australia, Canada, China, India, Japan, the Republic of Korea, and the United States to accelerate the development and commercialization of clean energy technologies and practices.[2] Partner countries work together and with their private sectors to meet energy security, national air pollution reduction, and climate change goals in ways

that promote sustainable economic growth and poverty reduction. Using a sectoral approach that breaks climate and clean development challenges into more manageable task forces (e.g., cleaner fossil energy, renewable energy and distributed generation, power generation and transmission, steel, aluminum, cement, coal mining, and buildings and appliances) helps APP Partners take advantage of readily available opportunities to increase energy efficiency and reduce GHG emissions.

In addition to such targeted, immediate actions, a sectoral focus enables APP Partners to lay the foundations for long-term market transformation. The Partnership has endorsed 175 individual cooperative activities, including 22 flagship projects that exemplify the Partnership's goals. The United States has provided funding for 40 of these projects, and has contributed approximately $75 million to further Partnership's goals since its inception.

Methane to Markets Partnership

Emerging climate science is revealing the critical importance of reducing methane emissions to mitigate climate impacts, especially in the near term. Launched in 2004, the Methane to Markets (M2M) Partnership has made great progress in accelerating the development of methane emission reduction projects around the world.[3] This international initiative also demonstrates how countries and the private sector can work together cooperatively to reduce GHG emissions, stimulate economic growth, develop new sources of clean energy, and improve local environmental quality. Building off of its domestic methane programs, EPA is working with M2M partners—30 national governments, including the European Union, and more than 900 private- and public-sector organizations (the Project Network)—to advance methane energy projects from four major sources: agricultural and food processing waste, landfills, underground coal mines, and natural gas and oil systems.

U.S. efforts under M2M are led by EPA and involve the collective efforts of six agencies and departments across the federal government. Ongoing U.S.-supported projects potentially can reduce GHG emissions by approximately 60.7 million metric tons of carbon dioxide equivalents (MMTCO$_2$ Eq.) annually, and actually reduced global warming pollution by 26.7 MMTCO$_2$ Eq. in 2008 (Figure 7-1). U.S. contributions of $38.9 million have also leveraged more than $277.9 million in investment from other partner countries, development banks, the private sector, and members of the Project Network (Figure 7-2).

In March 2010, New Delhi, India, hosted M2M's 2010 Expo. New Delhi will seek to build on the success of its first Expo, which featured more than 90 project opportunities with annual emission reduction potential of 11.5 MMTCO$_2$ Eq.

[2] See http://www.asiapacificpartnership.org/.

[3] See http://www.methanetomarkets.org/m2m2009/index.aspx.

Figure 7-1 Actual and Potential Annual Methane Emission Reductions Resulting from U.S.-Supported Projects: 2005–2008

Ongoing U.S.-supported projects are expected to reduce annual methane emissions by approximately 60.7 million metric tons of carbon dioxide equivalents ($MMTCO_2$ Eq.). In 2008, these projects reduced global warming pollution by 26.7 $MMTCO_2$ Eq.

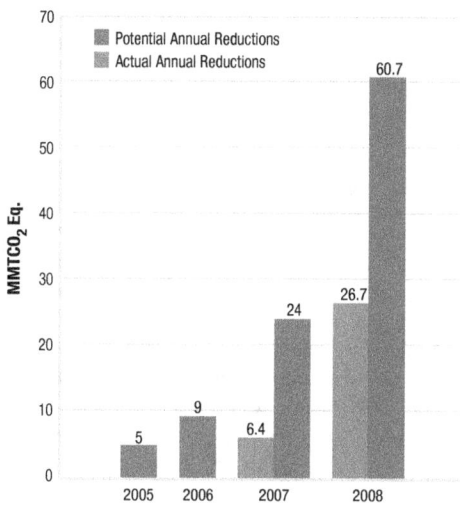

Figure 7-2 Total U.S. Government Funding and Leveraged Funding for the Methane to Markets Partnership: FY 2005–FY 2008

U.S. contributions of $38.9 million have leveraged more than $277.9 million in investment from other partner countries, development banks, the private sector, and members of the Project Network.

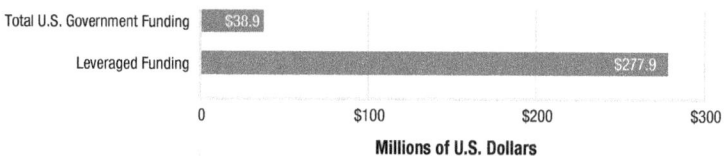

Carbon Sequestration Leadership Forum

An international climate change initiative that includes both developed and developing countries, the Carbon Sequestration Leadership Forum (CSLF) is focused on developing improved and cost-effective technologies for the separation and capture of carbon dioxide (CO_2) and its transport and long-term safe storage.[4] CSLF works to make these technologies broadly available internationally and to identify and address more comprehensive issues, such as regulation, relating to carbon capture and storage. To date, CSLF has endorsed 20 projects to evaluate and demonstrate carbon sequestration technologies.

U.S.-China Clean Energy Research Center

President Obama and President Hu Jintao formally announced the establishment of the U.S.-China Clean Energy Research Center (CERC) during the President's trip to Beijing in November 2009. Work to operationalize CERC began immediately, and further coordination of operations will continue in the months ahead. In March 2010, DOE Secretary Steven Chu announced the availability of $37.5 million in U.S. funding over the next five years to support CERC. Funding from DOE will be matched by the grantees to support $75 million in total U.S. research that will focus on advancing technologies for building energy efficiency; clean coal, including carbon capture and storage; and clean vehicles. An additional $75 million in Chinese funding will be provided. CERC will be located in existing facilities in both the United States and China.

U.S.-India Clean Energy Research Center

The U.S.-India Clean Energy Research Center aims to improve technologies to make clean energy more affordable and efficient. This center will include cooperation in wind and solar energy, second-generation biofuels, and energy efficiency, as well as unconventional sources of natural gas and clean coal technologies, including carbon capture and storage.

U.S. Climate Technology Cooperation Gateway

Sponsored by USAID and EPA, the U.S. Climate Technology Cooperation (U.S.-CTC) Gateway is an effort to increase public access to information about U.S.-sponsored international technology cooperation via an online tool.[5] The U.S.-CTC Gateway aims to facilitate climate technology cooperation for mitigation and adaptation activities in developing and transition countries consistent with U.S. responsibilities under Article 4.5 of the UNFCCC. It highlights U.S.-sponsored activities that have resulted in clear, measurable benefits, and provides information and resources that can allow users to implement climate-friendly technologies and practices throughout the world. The site is designed to be used by government officials, private-sector project developers, financiers, technology providers, NGOs, and the general public.

EPA Partnership for Clean Fuels and Vehicles— International Diesel Retrofit Projects

Through its Partnership for Clean Fuels and Vehicles, EPA has funded and provided technical assistance on several diesel retrofit projects in other countries. Projects have been completed in Mexico City, Beijing, and Bangkok. In each of these projects, EPA has worked to demonstrate state-of-the art diesel particulate filters that can remove more than 90 percent of diesel particulate matter—including black carbon, which is implicated in climate change. The projects involve substantial capacity building. In the Beijing project, Chinese entities have to date retrofitted more than 8,000 diesel trucks and buses following EPA's small demonstration project.

EPA has also funded a pilot project to retrofit two-stroke gasoline three-wheeled vehicles used as taxis ("auto rickshaws") in Pune, India. These vehicles are ubiquitous throughout Asia and contribute significantly to air quality problems in many Asian megacities.

---

[4] See http://www.cslforum.org/.

[5] See http://www.usctcgateway.gov/.

Adding electronically controlled direct-injection technology to these vehicles resulted in 30 percent fuel savings on top of substantial reductions in hydrocarbons, carbon monoxide, and particulate matter.

EPA Programs for Energy Efficiency

EPA continues to support several programs that promote energy efficiency in products and buildings.

Energy Efficiency Endorsement Labeling Programs

Drawing on experiences and resources from its successful ENERGY STAR program, EPA has worked with a number of countries, including China and India, to enhance their own voluntary energy efficiency endorsement labeling programs. For example, EPA worked with the China Standard Certification Center to harmonize product test procedures with international standards, and in India supported the technical analysis and market research for the development of labels for televisions and set-top boxes.

Vehicle Emissions and Efficiency Programs—SmartWay

EPA is working with Mexico, China, and other developing nations to export technical expertise from its successful SmartWay[SM] Transport partnership, which is touted as a model program in Australia, the European Union, Mexico, Canada, and elsewhere.[6] SmartWay promotes technologies and best practices that reduce fuel consumption throughout the freight industry and supply chain. EPA provides technical assistance, modeling tools, analysis, and "lessons learned" to enable developing nations in Asia and elsewhere to establish voluntary programs to reduce GHG and other emissions from new and legacy vehicles and to create vehicle labeling programs modeled after the SmartWay vehicle designation.

In 2008, EPA held an international workshop attended by over a dozen countries. The World Bank is currently considering a SmartWay workshop for Asia. Recently, EPA met with officials from 11 Chinese cities and the Ministry of Transportation to discuss a pilot SmartWay program in the city of Guangzhou, with the aim of expanding it to other Chinese cities. EPA also met with engineers from China's Auto Research Institute and Automotive Testing and Research Institute to share information on test methods to evaluate GHG emissions from commercial trucks.

eeBuildings

This EPA program has provided technical assistance to building owners, managers, and tenants in China and India to improve the energy efficiency of buildings.[7] In China, EPA successfully implemented an agreement with the Ministry of Housing and Urban-Rural Construction to assist with the development of a building energy benchmarking certification program, and worked with other organizations (e.g., Trade Association of Shanghai Property Managers and the China Business Council for Sustainable Development) to introduce efficiency measures. In India, in partnership with the Indian Green Buildings Council, the Bureau of Energy Efficiency, and others, EPA conducted training sessions on low- and no-cost energy efficiency measures for corporate building owners, major hotel chains, property management companies, service and equipment providers, and electric utilities.

Greenhouse Gas Inventory Improvement Project

Building on a previous successful partnership with seven nations of Central America, USAID and EPA are continuing to provide support to countries in Southeast Asia, and are exploring opportunities to collaborate with institutions and/or experts in other regions, to improve the quality and sustainability of national GHG inventories produced by developing countries. A rigorous national GHG inventory is an essential tool for informing future climate change policy decisions, designing appropriate mitigation activities, and projecting future emission trends. High-quality, transparent, consistent, and comparable inventories also provide the critical context within which mitigation actions and commitments may be monitored, reported, and verified.

The project focuses on developing long-term national inventory management systems, improving the methods and data used in the agriculture and the land use, land-use change, and forestry sectors, and training regional experts. Currently, direct assistance is being provided in collaboration with the UNFCCC Secretariat to Indonesia, Cambodia, Lao People's Democratic Republic, Malaysia, the Philippines, Thailand, and Vietnam. The project is also further refining the Agriculture and Land Use (ALU) software, which helps governments easily and accurately complete their national GHG inventories for the agriculture and the land use, land-use change, and forestry sectors.

ECO-Asia

Through the Environmental Cooperation–Asia (ECO–Asia) program, USAID develops a combination of national and regional activities in partnership with Asian governments, cities, and other organizations and agencies to promote regional dialogue in sharing and replicating innovation across Asia.[8] Key program countries include Cambodia, India, Indonesia, the Philippines, Sri Lanka, Thailand, and Vietnam.

ECO-Asia's Clean Development and Climate Program (CDCP) works to catalyze policy and finance solutions for clean energy in Asia's largest developing economies through targeted technical assistance and training, regional cooperation, and knowledge sharing. In its first two years, ECO-Asia CDCP initiated programs with partners that are expected to avoid emissions of 1.6 $MMTCO_2$ from fossil fuel consumption.

[6] See http://www.epa.gov/smartway/.

[7] See http://www.epa.gov/EEBUILDINGS/.

[8] See http://usaid.eco-asia.org/programs/cdcp/index.html.

ECO-Asia CDCP and the Asian Development Bank have also co-organized the Asia Clean Energy Forum for the past two years. This forum has become an instrumental event in Asia for policymakers, financiers, and energy experts to network, present new research findings, and share best practices. To supplement this event and other regional workshops, ECO-Asia CDCP hosts an innovative, user-generated Web site for clean energy experts to share and discuss information about clean energy solutions for Asia that help address climate change and energy security.

DOE National Laboratory Expert Technical Support

Recognizing the need for advanced technical assistance to support clean energy development across the globe, the U.S. government is establishing a network of U.S. experts to provide both targeted and cross-cutting technical assistance to developing country partners. The underlying goal of this effort is to ensure that the best U.S. technical expertise is being accessed to support global efforts to combat climate change.

USAID began working with DOE in 2009 to launch the first component of this new initiative. A team of DOE laboratory experts will be formed to provide technical assistance to USAID missions and their partner countries in the analysis, design, and implementation of GHG mitigation initiatives and clean energy projects guided by local needs and priorities. Supported activities will include sector assessments to effectively target USAID resources, country analyses for prioritizing mitigation actions, access to energy-sector and economic modelling and end-use emissions models, and assistance with analysis and design considerations for carbon markets. Ideally, these targeted efforts will aid in the development and application of analytical and technical resources to support the design and implementation of national- and sectoral-level low-carbon clean energy growth strategies in developing countries. It is anticipated that this initial effort will grow in scale and scope over the next three years, providing a framework for climate change mitigation and clean energy assistance across various U.S. government agencies and partners.

Global Bioenergy Partnership

The Global Bioenergy Partnership (GBEP) provides "a mechanism for Partners to organize, coordinate, and implement targeted international research, development, demonstration, and commercial activities related to production, delivery, conversion, and use of biomass for energy, with a focus on developing countries."[9] In response to the GBEP Steering Committee's request, the Task Force on Sustainability is working to define voluntary, science-based sustainability criteria and indicators with the intent to provide governments a toolbox to monitor trends in environmental, economic, and social sustainability. The Task Force on GHG Methodologies, chaired by the United States, has produced a Version Zero methodological framework to provide countries and institutions with a consistent and transparent method to describe measurements of the GHG footprint of biofuels. Currently in the testing phase, a Version One publication is expected in 2010.

International Partnership for Energy Efficiency Cooperation

In May 2009, the Energy Ministers of the G8, India, South Korea, China, Brazil, Mexico, and the European Commission officially launched a high-level international energy efficiency partnership that will allow participating countries to share best practices and effective policies, conduct joint activities and research, and showcase their successes in energy efficiency. The partnership will focus on developing energy efficiency tools, sharing best practices, expanding consumer awareness, and helping countries track their own progress toward national energy efficiency goals. As one of its first initiatives, the partnership will inventory the domestic energy efficiency policies of member nations.

International Renewable Energy Agency

The United States signed the IRENA Statute in June 2009.[10] The International Renewable Energy Agency (IRENA) was launched to substantially scale up global use of renewable energy through capacity building, technical assistance, sharing of best practices, and networking that does not currently exist in any other multilateral forum on energy. IRENA can help improve public policymakers' understanding of renewable energy policy and regulatory and institutional requirements for technology diffusion. Renewable energy technologies can address the multiple domestic and international objectives of energy security, climate change mitigation, economic growth and job creation, and cleaner air quality.

Climate Technology Initiative

The Climate Technology Initiative (CTI) is a multilateral cooperative activity that supports implementation of the UNFCCC by fostering international cooperation for accelerated development and diffusion of climate-friendly technologies and practices.[11] CTI was originally established at the first Conference of the Parties to the UNFCCC in 1995. Since July 2003, CTI has been operating under an implementing agreement of the International Energy Agency that includes the United States, Australia, Austria, Canada, Finland, Germany, Japan, Norway, Republic of Korea, Sweden, and the United Kingdom. Through a variety of capacity-building activities, CTI has promoted meaningful technology transfer to and among developing and transition countries. In addition to their current and future environmental benefits, these efforts are promoting near- and long-term global economic and social stability.

[9] See http://www.globalbioenergy.org/.

[10] See http://www.irena.org/.

[11] See http://www.climatetech.net.

CTI Private Financing Advisory Network

In an effort to address the growing need for clean technology financing in developing and transition countries, CTI launched PFAN in 2006 to increase access to private capital markets. PFAN is comprised of a network of practicing private finance professionals who work with project developers and other project proponents to structure projects and supporting financing proposals to meet the standard of the international financial community. These consulting services are provided at no cost to the project developer. Although there are no formal restrictions on size, CTI PFAN has thus far targeted projects with a total investment volume of $1–$50 million.

CTI PFAN is technology-neutral and will consider any clean energy, renewable energy, or energy efficiency project that can demonstrate technical and economic feasibility. Typically, PFAN projects are mitigation and distributed generation types of projects, deploying a range of such technologies as biomass, geothermal, wind, solar, biofuels, waste to energy, and small hydro. However, PFAN will also consider adaptation and technology development projects as well.

One of the techniques used by CTI PFAN to bring project developers and investors together is regionally organized financing forums that are preceded by a call for proposals. Selected projects are provided with intensive one-on-one PFAN coaching and are showcased before specially convened investor audiences. Projects identified through the calls for proposals are also considered for induction into the development pipeline. During 2009, this effort was significantly scaled up, with investor forums planned for Africa, Asia, and Latin America in early 2010.

Renewable Energy and Energy Efficiency Partnership

The United States is one of 45 countries whose governments contribute funding to the Renewable Energy and Energy Efficiency Partnership (REEEP).[12] Conceived at the World Summit on Sustainable Development in 2002, REEEP seeks to lower the policy, regulatory, and financial barriers to implementation of renewable energy and energy efficiency technologies and projects in emerging markets and developing countries; improve access to reliable clean energy services in developing and transition countries; and increase the use of local renewable resources. Since 2004, REEEP has provided support for more than 100 projects in 40 countries.

U.S. FOREST PROGRAMS AND PARTNERSHIPS

Forest Carbon Partnership Facility

The United States is a donor to the Forest Carbon Partnership Facility (FCPF), a multilateral fund at the World Bank that assists developing countries in their efforts to reduce emissions from deforestation and forest degradation (REDD). The FCPF's first objective is to build capacity in developing countries to credibly estimate forest carbon stocks and sources of forest emissions, to assist countries in building a national strategy for stemming deforestation, and to create national reference scenarios or baselines against which to measure future performance. The FCFP will then test performance-based incentive payments in a few pilot countries to set the stage for a much larger system of potential financing flows from future carbon markets. The FCFP seeks to learn lessons from first-of-a kind operations and develop a realistic and cost-effective new instrument for tackling deforestation. The experiences generated from the FCPF's pilot implementation and carbon finance experiences will provide insights and knowledge that can facilitate development of a much larger global program of incentives for REDD.

Tropical Forest Conservation Act

The Tropical Forest Conservation Act (TFCA) was enacted in 1998 to offer eligible developing countries options to relieve certain official debt owed to the United States, while at the same time to generate funds to support local tropical forest conservation activities.[13] As of August 2009, TFCA programs are being implemented in Bangladesh, Belize, Botswana, Colombia, Costa Rica, El Salvador, Guatemala, Indonesia, Jamaica, Panama (two agreements), Paraguay, Peru (two agreements), and the Philippines. The 15 agreements completed to date will directly generate more than $218 million for tropical forest conservation in these countries over the life of the agreements, and additional resources will be created through returns on investments and matching funds.

A number of other countries have expressed interest in the TFCA program. In addition to forest conservation and debt relief, TFCA is intended to strengthen civil society by creating local foundations to support small grants to NGOs and communities. The program also offers a unique opportunity for public–private partnerships. Nine of the agreements to date have included more than $14 million in cash raised by U.S.-based NGOs (and one Indonesian NGO), in addition to the approximately $135 million in appropriated debt-reduction funds contributed by the U.S. government.

Several TFCA national-level funds are exploring ways to catalyze REDD development in their respective countries. In addition, some national-level funds established through the Enterprise for the Americas Initiative, a predecessor to TFCA, are supporting the dissemination of clean technologies in several sectors.

Initiative for Conservation in the Andean Amazon

USAID's Initiative for Conservation in the Andean Amazon (ICAA) is a five-year program (2006–2011)

[12] See http://www.reeep.org/.

[13] See http://www.usaid.gov/our_work/environment/forestry/tfca.html.

that brings together the efforts of more than 20 public and private organizations currently working in the Amazon regions of Bolivia, Colombia, Ecuador, and Peru to build constituencies and agreements that promote the sustainable use and conservation of biodiversity and environmental services in the regions.[14] Beyond sharing a common language and cultural patrimony, this area includes globally unique resources and ecosystem services that are important to local, regional, and global climate, health, and economies.

ICAA activities include landscape and protected area management, improved natural resource management (timber, non-timber, and agriculture), improved environmental policy, and capacity building of local and indigenous organizations. These activities help to increase carbon sequestration and avoid GHG emissions by supporting improved resource management and minimizing conversion of natural habitats. The Woods Hole Research Center, an ICAA partner, is developing approaches to more accurately measure and monitor these GHG benefits using satellite and ground-based measurements, while also building the capacity of local NGOs to do these calculations.

Forest, Climate & Community Alliance

The Forest, Climate & Community Alliance (FCCA) is a public–private partnership between USAID and the Rainforest Alliance. A number of private-sector companies are also working with the Rainforest Alliance on this project, including Gibson Guitar Corporation, EKO Asset Management Partners, NatSource LLC, and others. Formed in 2009, the FCCA aims to increase economic opportunities for poor communities and forest management enterprises, and to combat threats to tropical forest biodiversity. The project will pilot technologies, organizational models, and local and national policies that help communities more effectively access revenue streams from payment for ecosystem services (PES) schemes, particularly carbon credits from the emerging voluntary and mandatory carbon markets. The FCCA will pilot initiatives in the countries of Ghana and Honduras and will use lessons learned from these sites to provide options for other USAID missions and cooperators.

International Small Group and Tree Planting Program

The International Small Group and Tree Planting Program (TIST) is a public–private partnership between USAID, the Institute for Environmental Innovation, and the Clean Air Action Corporation. Begun in 2005 and now in its second phase, this initiative has grown to include over 35,000 small farmers planting over 4 million trees in four countries.

The TIST program amalgamates small farmers into clusters of farmers who work together to plant and maintain trees for carbon sequestration, biodiversity enhancement, environmental benefits, and personal use. It also builds capacity in farming technologies, HIV/AIDS (human immunodeficiency virus/acquired immunodeficiency syndrome) prevention, fuel-efficient wood stoves, and other development objectives.

TIST is a PES arrangement, which pays the individual participant approximately two cents per tree for each year that it is growing on the registered plot. There is also an agreement between the Clean Air Action Corporation and the individual participant that will allow the farmers to collect 70 percent of any profits from future carbon sales from the registered trees. Carbon stocks are monitored and verified through a rigorous auditing process that uses handheld computers and global positioning system (GPS) coordinates to verify the status of each plot on an annual basis.

Center for International Forestry Project

In 2007, USAID initiated a program to build capacity within the agency on issues related to forests and climate change in general, and specifically related to the international discussion about reducing emissions from deforestation in developing countries. Implemented in collaboration with the Center for International Forestry, this program included the following three components:

1. A series of training modules on topics related to forests and climate, including forest adaptation technologies, biofuels, carbon accounting methods, forest and climate policy, and a variety of other topics.

2. Technical assistance to USAID missions in Indonesia and Liberia, to help them assess their ongoing programs and look for opportunities to make adjustments to prepare for challenges and opportunities related to forests and climate change.

3. A series of training workshops on forests and climate issues, held in Thailand and South Africa in 2009.

Forest Carbon Calculator

USAID is supporting the development of a Forest Carbon Calculator, which consists of a cutting-edge set of carbon estimation calculators for the forest sector using a Web-based tool.[15] Covering forest protection, reforestation/afforestation, agroforestry, and forest management, the tool provides estimated CO_2 emission reductions and sequestration from improved forest and land management practices. USAID is using this tool to estimate the mitigation impacts of existing programs and to explore options in the design of new programs. A public-access version of the tool will be available for the use of anyone who wishes to estimate the carbon emission reductions and sequestration of site-level activities. Future revisions of the tool are planned to allow estimation of CO_2 benefits from grasslands and agricultural management.

[14] See http://www.amazonia-andina.org/en.

[15] See http://winrock.stage.datarg.net/gcc/.

USAID/USFS Interagency Agreement

Through its long-standing agreement with USFS, USAID accesses the technical expertise of USFS to implement training, technical assistance, capacity building, and policy dialogue in the countries where USAID works. With more than 100 years of experience managing over 77 million hectares (190 million acres) of national forests and grasslands, USFS has valuable experience to contribute to the issue of climate change in the global context.

Much of the land that USFS manages is already being influenced by climate change, requiring the development of adaptation strategies. As a result, USFS is closely involved in efforts to enhance the mitigation potential of U.S. forests and is well positioned to provide technical assistance in the context of longer-term development engagement with other nations facing land management and climate change challenges. USFS is actively collaborating with numerous governments, NGOs, and the private sector to address climate change and avoided deforestation through collaborative research, policy dialogue, and technical cooperation.

In 2008, USFS experts participated in a USAID-funded assessment to identify opportunities for USAID assistance in helping Bolivia address forest and climate change issues. In 2009, USFS worked with USAID to identify and pilot models for increasing women's involvement in forest carbon management and exchanges.

Illegal Logging and Forest Trade

Illegal logging is a significant cause of forest degradation that often leads to increased carbon emissions and less resilient ecosystems. USAID is taking measures to address both the supply and the demand sides of the global forest product trade through three multiyear, multimillion-dollar public–private partnerships. For example, USAID supported the Sustainable Forest Products Global Alliance in 2008 and launched a new program in 2009 to further promote transparent supply chain management in response to new and emerging consumer country laws, such as the amended U.S. Lacey Act. The new program aims to reach out to a wider group of forest stakeholders both in the United States and abroad, to raise awareness about illegal logging, and to provide information to encourage improved supply chain management from the stump to the shelf. In the Asia region, the Responsible Asia Forest and Trade program continues to work with partners in the retail, financial, and NGO sectors to assist governments and the private sector to combat illegal logging and associated trade through the promotion of improved technologies, market strategies, management practices, and forestry policies.

Standing Forest Conservation Markets Initiative

USAID has been supporting activities in the field of PES, including the Global Assessment of Best Practices in Payment for Ecosystem Services Programs; Wildlife Conservation Society pilot projects integrating carbon emission reductions, biodiversity conservation, and community livelihoods in Madagascar and Cambodia; and Regional PES Network Meetings and site PES Incubator projects led by Forest Trends and the Katoomba Group.[16] The Standing Forest Conservation Markets Initiative, which began in 2008, builds on these recent efforts. The initial targets for this initiative are the globally important forested watersheds of the Amazon, Congo, and Mekong River Basins. Project development documentation for reduced emissions from REDD projects are currently under preparation for voluntary private-sector carbon trading, as well as for water and biodiversity markets. Pilot PES sites within these mega-diversity areas—such as the Xingu and Surui Indigenous Reserves in Brazil, and the Mbe River headlands in Gabon—are being linked to encourage south-south collaboration and learning.

The initiative also offers project development tools, such as legal and institutional assessments and a PES curriculum that can be tailored to different stakeholder groups. Training workshops to build capacity in carbon baseline measurement and market assessment were held in 2008 and 2009 in Peru, Brazil, Nepal, Cambodia, Uganda, Tanzania, Ghana, and South Africa. Finally, USAID is also providing partial support for Katoomba meetings, which aim to build regional networks and senior policymakers' interest in PES; for 2008–2010, avoided deforestation is a major theme.

U.S. VULNERABILITY AND ADAPTATION PROGRAMS

Through agencies including USAID, NASA, and NOAA, the United States continues to pursue a broad range of activities designed to help countries integrate adaptation and development concerns, understand their vulnerabilities to new climate-related risks, and build capacity at the household, community, national, and regional levels to address the urgent challenges posed by changing climatic conditions.

Adaptation Guidance Manuals

In August 2007, USAID published *Adapting to Climate Variability and Change: A Guidance Manual for Development Planning* (USAID 2007). The manual is designed to assist USAID missions and other development partners to understand, analyze, and respond to the potential impacts of climate change on development challenges, and to develop effective approaches to solving those challenges. It lays out a six-step process for assessing the potential for climatic changes to affect development efforts and engaging stakeholders in identifying alternative approaches for more climate-

---

[16] See http://www.translinks.org, http://www.oired.vt.edu/sanremcrsp/PES.php, and http://www.katoombagroup.org.

resilient development. The manual describes four pilot projects conducted by USAID to field-test the proposed methods, and also provides resources for further technical information.

In May 2009, USAID, International Resources Group, and The Coastal Resources Center at the University of Rhode Island published *Adapting to Coastal Climate Change: A Guidebook for Development Planners* (IRG and CRC-URI 2009). This coastal guide is designed to help coastal managers and other decision makers understand how climate change may affect processes in coastal areas. The guide also presents a number of well-known coastal management practices that deal with climate change, and assesses each practice or measure for its ability to address climate change or other coastal management challenges.

USAID continues to develop guidance to assist development practitioners in considering climate change impacts on development goals. However, USAID also recognizes that the methods laid out in the guidance require access to environmental, meteorological, and climatological data. To facilitate access to data, the agency has developed or leveraged a number of tools, described further in this chapter, in collaboration with other federal agencies and organizations.

Earth Observations for Decision Support
SERVIR
USAID and NASA support SERVIR, a Regional Visualization and Monitoring System that integrates satellite and other geospatial data for improved scientific knowledge and decision making by managers, researchers, students, and the general public.[17] In addition to making available previously inaccessible data, online mapping tools, and training opportunities, SERVIR provides improved monitoring of air quality, extreme weather, biodiversity, changes in land cover, and the impacts of climate change. The first SERVIR regional operational facility—for the Latin America and Caribbean region—was established in 2005 in partnership with the Water Center for the Humid Tropics of Latin America and the Caribbean (CATHALAC). In 2008, NASA and CATHALAC helped set up a second SERVIR regional operational facility for East Africa, at the Regional Center for Mapping of Resources for Development (RCMRD) in Nairobi, Kenya. The SERVIR–Africa project is building upon RCMRD's existing strengths, while augmenting its data management and training capabilities. Efforts are underway to replicate the SERVIR system model in other regions of the world, in support of climate-resilient development.

Climate Mapper
In 2008, USAID, NASA, the Institute for the Application of Geospatial Technology, the University of Colorado, and CATHALAC developed the Climate Mapper tool for SERVIR.[18] The Climate Mapper makes historical weather records and the results of climate change models accessible to a broad user community. Data are currently available for all land portions of Earth for ½-degree grid cells, which, at the equator, corresponds to roughly 50 kilometers by 50 kilometers (31 miles by 31 miles). The Climate Mapper presents historical temperature and precipitation for the base period (1961–1990), as well as modeled monthly data averaged over the decades 2031–2040 and 2051–2060. Users can assess climate change projections for the 2030s and 2050s against three-dimensional visualizations of landscape. This should enhance vulnerability assessments as development planners consider adaptation strategies and design projects to be more resilient to climate variability and change.

Climate One Stop
USAID and NASA are working with a number of other donors, NGOs, developing country institutions, and the UNFCCC to develop the Climate One Stop.[19] As its name suggests, the One Stop will be a comprehensive Web site where a variety of users—from the public to project planners and researchers—can learn about climate change, access data and tools, find climate-related news and project case studies, and link with other people and groups interested in climate change through an online community. In the beginning, the One Stop will emphasize adaptation, but it eventually will have the flexibility to grow in whatever ways the user community demands.

Regional Climate Outlook Forums
NOAA, in collaboration with USAID and the UN World Meteorological Organization (WMO), provides regional institutions in developing countries with technical support for Regional Climate Outlook Forums. These forums are a principal vehicle for providing advance information about the likely character of seasonal climate in several developing regions. They bring together climate forecasters and forecast users to develop a consensus forecast from multiple predictions and to discuss methods of dissemination and application of information. The forums also provide a unique opportunity for stakeholders to meet, share information and concerns, and forge an informal network to address common problems.

Resource-Sharing Partnerships for Sustained Ocean-Climate Observations
NOAA's Climate Program Office is working with the National Ocean Service and other NOAA offices to build sustained capacity in the Indian Ocean for ocean-climate observations and their socioeconomic application. Partnerships for New GEOSS Applications (PANGEA) builds sustainable capacity in maritime regions by providing resources and in-country, practical training by NOAA experts for local policy and budget officials, scientists, and end users. In ex-

[17] See http://www.servir.net.

[18] See http://www.servir.net/en/The_Climate_Mapper_and_SERVIR_Viz.

[19] See http://www.climate1stop.net.

change, participating national institutes provide gratis ship time and expertise to NOAA for the deployment and regular servicing of NOAA's *in situ* ocean-climate instrumentation.

PANGEA complements existing efforts to achieve a more sustainable capacity for the region by increasing both *in situ* ocean data and information as well as the more effective application of these data for more informed decision making. Capacity-building workshop topics include Planning for Climate Change in the Coastal and Marine Environment, Societal and Economic Benefits of Delivering Enhanced Ocean Observing System Data, Tools for Assessing Community Resilience, and How to Select and Evaluate Adaptation Measures for Managing Impacts From Climate Change.

NOAA's current partners include Indonesia's Ministry of Marine Affairs and Fisheries and Agency for the Assessment and Application of Technology, India's Ministry of Earth Sciences, and the Agulhas and Somali Current Large Marine Ecosystem for nine East African and Western Indian Ocean nations. The PANGEA model is equally applicable elsewhere and is being explored for other ocean basins.

International Research Institute for Climate and Society

NOAA provides approximately $9 million in annual funding to the International Research Institute for Climate and Society (IRI).[20] IRI's mission is to enhance society's capability to understand, anticipate, and manage the impacts of seasonal climate fluctuations, in order to improve human welfare and the environment, especially in developing countries. IRI's international efforts involve research in climate prediction, monitoring, and analysis targeted to address problems of climate risk in agriculture and food security, water resources, public health, disasters, and such cross-cutting issues as drought management. A combination of scientific rigor, problem-centered analysis, and partner teamwork is beginning to yield successful approaches to climate risk management. IRI has several ongoing projects in Africa, Asia, and Latin America and has launched an effort to raise institutional and societal awareness of climate vulnerability and risk as an arena for action.

Coral Triangle Initiative on Coral Reefs, Fisheries, and Food Security

The Coral Triangle Initiative (CTI) on Coral Reefs, Fisheries, and Food Security is a new multilateral partnership to safeguard the region's extraordinary marine and coastal biological resources.[21] Recognized as the global center of marine biological diversity, the Coral Triangle region embraces six countries—Indonesia, Malaysia, Papua New Guinea, the Philippines, Timor Leste, and the Solomon Islands. Home to some 363 million people, one-third of whom are directly dependent on coastal and marine resources for their livelihoods, the region faces immediate risk from a range of factors, including overfishing, unsustainable fishing methods, land-based sources of pollution, and climate change.

The CTI was officially launched in December 2007 during UNFCCC COP-13 in Bali, and was endorsed by the six heads of state at the CTI Summit in May 2009. Over the first five years of the initiative, USAID and DOS plan to invest at least $42 million in the regional initiative, in addition to providing bilateral support to Indonesia and the Philippines. Investments will strengthen the capacity of community groups, governments, and public and private organizations to conserve the Coral Triangle's globally important biodiversity, improve fisheries management, and enhance resiliency and adaptation to climate change.

Famine Early Warning System Network

The USAID-funded Famine Early Warning Systems Network (FEWS NET) is a 25-year-old program that analyzes remotely sensed data and ground-based meteorological, crop, markets and trade, and livelihoods information to provide early-warning and vulnerability information on emerging or evolving food security conditions.[22] The network is comprised of a multidisciplinary team that includes a management unit at USAID, contractual relationships with NASA, NOAA, USDA, the U.S. Geological Survey (USGS), a private-sector contractor operating a FEWS NET headquarters team, and USAID's 23 offices in food-insecure countries of the developing world.

FEWS NET includes a unique initiative to identify climate change impacts on specific food-insecure countries. This initiative has focused largely on Africa to date, and has produced a large volume of research that identifies how climate change is already affecting large parts of eastern Africa, including significant impacts on food security conditions. The broad conclusions of this research include the following: (1) climate change is already producing drought and eroding livelihoods in eastern Africa, and these drought impacts are likely to precede impacts from projected temperature rises; (2) climate change models fail to model precipitation accurately, are commonly wrong, and are inadequate tools for evaluating changes in precipitation, especially over land surfaces; (3) conversely, data sets based on precipitation observations, when properly constructed and analyzed, can provide one of the best tools for tracking decadal climatic variability; and (4) climate change impacts on food security must be assessed in an interdisciplinary investigation that can identify and trace the multiple interactions of food security with climate.

FEWS NET is also committed to the capacity and institution building of national and regional early-warning and response networks. Anticipating an increase in the threat of food insecurity in previously relatively un-

[20] See http://portal.iri.columbia.edu/portal/server.pt.

[21] See http://www.cti-secretariat.net/.

[22] See http://www.fews.net.

challenged countries, FEWS NET will expand to cover up to 30 new countries over the next five years. New climate change research will also be conducted in additional African regions and Central Asia.

RANET Program

USAID and NOAA are working with a range of humanitarian and meteorological organizations to build the capacity of regional, national, and sub-national institutions to provide useful weather and climate information to rural communities and populations in remote regions of Africa, Asia, the Pacific, and Central America. RANET (Radio and Internet for the Communication of Hydro-Meteorological and Climate-Related Information for Development) utilizes and develops applications on a variety of communication platforms, such as satellite data broadcasts, satellite telephony, FM (frequency modulation) community radio, HD (high-definition) digital e-mail networks, mobile phones, and Web-based systems.[23] RANET works with national partners to use these tools to deliver forecasts, bulletins, imagery, seasonal assessments, educational and training information, and data to rural and remote areas. The program stresses training, system development, and ongoing operation of communication networks to support country programs, and emphasizes local ownership of all systems.

In 2010 RANET will continue current activities, which include operation of satellite broadcasts, the RANET Alert Watcher Short-Message Service notification system, development of the Community Reporter Program, and creation of various technical training modules. Significant new activities will be development of GEONETCast capacity for current RANET country programs, continuation of the RANET Chatty Beetle pilot, and launch of a unified content management system designed to allow countries to publish information across multiple communications platforms.

Disaster Preparedness Activities

USAID is supporting a number of projects around the world to help vulnerable communities anticipate weather hazards and build resilience to climate change. In 2008, USDA, USAID, NOAA, and other partners began work to develop a Global Flash Flood Guidance and Early Warning System, to improve lead time for early warnings of flash floods and allow for rapid mobilization of response activities. USAID has also supported technical assessments and forums for decision makers and technical personnel from the Hindu Kush–Himalayan region to strengthen capacities on flash flood management, promote collaboration, and develop regional approaches to flood management and flood early-warning systems. USAID, NOAA, and USGS also support the Asia Flood Network, which aims to lessen flood hazard vulnerability in Asia by building regional and national capacities in climate, weather, and hydrological forecasting and warning, and by improving the dissemination of forecasts and warnings to at-risk populations.

Other programs receiving USAID support include community-based drought preparedness planning in Cambodia, East Timor, and Vietnam; training for Mekong Delta officials and affected populations on the use of flood mapping; the introduction of water management schemes that will enhance the sustainability of water supply activities in southern Swaziland; and a five-year project to support a community-based approach to flood monitoring and forecasting in Bangladesh. Finally, in partnership with the United Nations World Food Programme, USAID is supporting the development of a drought insurance program to protect against livelihood loss in Ethiopia.

International Technical Assistance Program

The Department of the Interior (DOI) International Technical Assistance Program (ITAP) leverages funds from other federal agencies, international organizations, and foreign governments to provide capacity building in other countries using the diverse expertise of technical experts from DOI bureaus.[24] Currently, DOS and USAID provide the largest share of DOI-ITAP funds, which supports training and technical assistance efforts in Bahrain, Costa Rica, the Dominican Republic, El Salvador, Guatemala, Jordan, Morocco, Nicaragua, Oman, the Philippines, the Republic of Georgia, and Tanzania.

DOI-ITAP's capacity-building efforts span the range of DOI expertise, including protected area management, environmental education, natural and cultural resource law enforcement, mining policy and regulations, and endangered species management. Recent proposals have centered on DOI's unique positioning as a leader in climate change adaptation, mitigation, and education and outreach. DOI-ITAP is seeking to apply the combined expertise of its bureaus to establish programs addressing climate change in at least 10 countries worldwide.

U.S. TRADE AND DEVELOPMENT FINANCING

The U.S. government facilitates the transfer of technologies by providing official assistance, export credits, project financing, risk and loan guarantees, and investment insurance to U.S. companies, as well as credit enhancements for host-country financial institutions. These activities help leverage direct investment by decreasing the risks associated with long-term, capital-intensive projects or projects in nontraditional sectors. Supported projects include a number of clean technology projects that further climate change objectives.

Overseas Private Investment Corporation

The Overseas Private Investment Corporation's (OPIC's) core mission is to support economic devel-

[23] See http://www.ranetproject.net/.

[24] See http://www.doi.gov/intl/itap/overview.html.

opment by facilitating U.S. private investment in developing countries and transition economies. OPIC provides project financing and political risk insurance for U.S. company projects covering a range of investments, including many independent power projects in developing countries. OPIC also supports a variety of private investment funds that make direct equity and equity-related investments in new, expanding, and privatizing companies in emerging market economies. OPIC evaluates all project applications on the basis of their contribution to economic development to ensure the successful implementation of the organization's core developmental mission, and prioritizes the allocation of scarce resources to projects on the basis of their developmental benefits. With its expanded commitment to the clean technology and renewable energy sector (i.e., via its existing Finance, Political Risk Insurance, and Investment Funds programs), OPIC anticipates significant increased future support of clean technology and renewable energy projects in OPIC-eligible countries.

OPIC Finance and Political Risk Insurance Programs
Support for renewable energy and clean technology projects is a priority area for OPIC. OPIC's Finance and Political Risk Insurance programs have considered over 100 proposals each year for 2008 and 2009 to determine projects suitable for OPIC support. OPIC is currently tracking renewable energy and clean technology projects that total nearly $2 billion for potential support, in the areas of energy-efficient buildings, solar power, hydroelectric power, wind power, geothermal power, waste-to-energy, and wastewater treatment. OPIC has a substantial portfolio of existing renewable energy and clean technology projects. Table 7-4 presents illustrative examples from OPIC's annual reports.

Investment Funds
In FY 2010 OPIC expects to commit $523 million in finance, investment funds, and insurance for renewable energy and clean technology projects, nearly a threefold increase over FY 2009. Target geographies include all OPIC-eligible emerging market countries throughout Asia, Latin America, Central and Eastern Europe, and the Middle East. Target sectors include those with proven renewable energy technologies (e.g., solar, wind, small hydro); energy efficiency systems and equipment; and water, waste, and emission control treatment.

OPIC financing will be provided to invest in existing companies and/or projects in the aforementioned renewable energy and environmental sectors in OPIC-eligible countries. Such investments will help to catalyze the sustainable development of emerging markets in which OPIC operates. By the end of 2010, OPIC anticipates signed formal commitments with the funds. Furthermore, OPIC expects the majority of such funds to have already begun investing in existing companies and/or projects within the aforementioned renewable energy and environmental sectors in OPIC-eligible countries.

Export-Import Bank
The Export-Import Bank (Ex-Im), the export credit agency of the United States, provides financial support to exporters of U.S. equipment and services through its insurance, working capital, and loan guarantee programs. Ex-Im also features an Environmental Exports Program (EEP) that provides enhanced financial support for renewable energy and other environmentally beneficial exports. Under the EEP, Ex-Im provides special support for exports of air, water, and soil pollution cleanup; ecological and forestry management;

Table 7-4 **Sample Renewable and Clean Energy Projects Supported by OPIC**

The Overseas Private Investment Corporation (OPIC) supports renewable energy and clean technology projects in developing countries and economies in transition. Current OPIC finance and insurance exposure in these projects totals more than $1 billion. This table presents some examples of this support.

Country	U.S. Sponsor/ Insured Investor	Project Name	Project Description	Committed Amount/ Maximum Coverage Amount Issued	Project Category
India	Southern Energy Partners LLC	SEP Energy India Pvt. Ltd.	Wind Energy—Tamil Nadu	$1,150,000	Insurance
Dominica	Dominica Private Power Ltd.	Dominica Electricity Services Ltd. ("DOMLEC")	Hydro and geothermal electric services	$3,033,000	Insurance
Colombia	Citibank, N.A.	ISAGEN S.A.	Hydropower generation	$310,000,000	Finance
Algeria	GE Ionics, Inc.	Fixed Rate Funding & Liquidity Ltd. (HWD SPA)	Reverse osmosis water desalination plant	$200,000,000	Finance
Serbia	DV Technologies	DV Technologies d.o.o. Belgrade	Hydroelectric power plant—SHPP Vrutci	$153,000	Insurance

Source: Overseas Private Investment Corporation annual reports (http://www.opic.gov/).

renewable and alternative energy projects, including photovoltaic, wind, biomass, fuel cells, waste-to-energy, hydroelectric, clean coal, and geothermal projects; products to measure or monitor air or water quality; equipment to reduce emissions or effluents; environmental impact assessments and ecological studies; environmental training services; and products designed to improve energy efficiency.

Ex-Im also offers foreign buyers extended repayment terms of up to 18 years to cover the purchase of U.S. goods and services for renewable energy projects. This special support is available for exports to wind energy, geothermal energy, hydropower, tidal, wave power, solar photovoltaic, solar thermal, ocean thermal, sustainable biomass, and certain bioenergy projects. In FY 2010, Ex-Im expects to commit nearly $183 million in loan guarantees, insurance, and working capital guarantees to support U.S. renewable energy exports to various foreign countries, which greatly exceeded amounts authorized for these exports in previous years ($23.1 million in FY 2009, $30.4 million in FY 2008, $2.7 million in FY 2007, $9.8 million in FY 2006, and $16.8 million in FY 2005).

USAID Development Credit Authority

The Development Credit Authority (DCA) is a broad financing authority that allows USAID to use credit guarantees to pursue any of the development purposes specified under the Foreign Assistance Act of 1961, as amended. DCA seeks to provide USAID the flexibility to make more rational choices about appropriate financing tools used in development, including individual or combined loans, guarantees, and grants. DCA activities are designed and managed by USAID's overseas missions.

Credit guarantees offer several distinct and attractive advantages over other forms of assistance, and leverage and maximize USAID resources by providing access to local private capital, sharing risk to encourage lending, mobilizing local private capital, and enhancing "the demonstration effect" so that other financial institutions will be induced to enter the market. USAID/El Salvador used a DCA guarantee to enable Citibank de El Salvador to lend to businesses that wanted to invest in energy-efficient upgrades. One guaranteed loan for $300,000 allowed a family-owned milk processing company to invest in a new filtration system and wastewater treatment facility. With these upgrades, the company now meets the environmental compliance standards under the Central America–Dominican Republic–United States Free Trade Agreement, and is certified by the U.S. Food and Drug Administration.

USAID Global Development Alliance

USAID's Global Development Alliance (GDA) business model links U.S. foreign assistance with the resources, expertise, and creativity of governments, business, and civil society. Through public–private partnerships, USAID and its partners combine their assets to address pressing development problems, achieving a solution that would not be possible for any individual partner alone.

Through 2009, USAID has cultivated more than 900 alliances with 1,700 partners to extend development assistance programming. For example, USAID and Firestone are planning to launch a GDA that will reinvigorate Liberia's rubber industry and support the development of a clean energy power plant in Monrovia through a rubber tree cultivation and recycling program. The partnership is expected to increase export-driven trade and growth, while boosting Liberia's clean energy initiatives.

In 2005, USAID created a new obligating instrument—the collaboration agreement—to provide funding directly to a nontraditional partner. For example, PFAN, an alliance of private-sector companies and investors described earlier in this chapter, receives funding from USAID to help developers of climate-friendly technologies in developing countries find financing.

U.S. Trade and Development Agency

The U.S. Trade and Development Agency (USTDA) is a foreign assistance agency that delivers its program commitments through overseas grants and contracts with U.S. firms that are designed to support activities that will have a strong and measurable development impact on emerging markets and offer opportunities for commercial participation by U.S. firms. USTDA's approach to foreign assistance generates mutually beneficial results through the formation of long-term business relationships that foster sustainable development, facilitate local private-sector growth, improve trade and investment climates, and advance U.S. commercial innovations.

USTDA's efforts are helping to realize the Obama administration's goal of reversing the effects of climate change, creating green jobs, and helping move emerging economies to a low-carbon energy base. For example, USTDA programs are providing countries with the resources necessary to discern what technologies are needed to implement a low-carbon development strategy as well as manage the technologies if and when they are implemented. In this way, USTDA is a bridge and a catalyst in paving the way for the acceptance and deployment of more environmentally sound energy production in emerging economies. USTDA has been a catalyst for new geothermal power generation in Turkey and Ethiopia; deployment of methane-capturing coalbed methane (CBM) power generation in China; and the construction of new CBM power development in Botswana. Going forward, USTDA is investing in solar, wind, energy efficiency, and new advanced technologies to help emerging economies expand their use of clean renewable energy and optimize current energy resources.

U.S. PRIVATE-SECTOR ASSISTANCE

The private sector has a crucial and complementary role to play in the successful transfer of technical know-how and climate-friendly technologies to developing and transition countries. The private sector is well placed to provide much of the human and financial capital for effective deployment of these technologies. While public-sector funding for climate change activities continues to grow, U.S. foreign direct investment by the private sector still comprises the vast majority of funding going to climate change and related activities in developing and transition countries.

The Obama administration also recognizes the importance of creating incentives for the development of low-carbon energy sources. Yet, most information relating to financing and implementing private-sector projects is proprietary. Furthermore, the classification of domestic exports in available databases uses broad categories that do not allow for the specific and comprehensive identification of clean or lower-carbon technologies and products. As a result, only public-sector funding is included in the tables in this chapter, with the intention of providing increased transparency.

Table 7-5 2008 U.S. Direct Financial Contributions Related to Implementation of the UNFCCC (Millions of U.S. Dollars)

RECIPIENT COUNTRY/ REGION	MITIGATION						ADAPTATION			TOTALS
	Energy	Transport	Forestry	Agriculture	Waste Management	Industry	Capacity Building	Coastal Zone Management	Other Vulnerability Studies	
WORLD										228.58
Africa										44.68
Africa Regional	2.40	0.00	2.50	0.00	0.00	0.00	0.00	0.00	0.00	4.90
Angola	0.00	0.00	0.00	0.00	0.00	0.00	0.00	0.00	0.00	0.00
Benin	0.00	0.00	0.00	0.00	0.00	0.00	0.00	0.00	0.00	0.00
Botswana	0.00	0.00	0.00	0.00	0.00	0.00	0.00	0.00	0.00	0.00
Burkina Faso	0.00	0.00	0.00	0.00	0.00	0.00	0.00	0.00	0.00	0.00
Burundi	0.00	0.00	0.00	0.00	0.00	0.00	0.00	0.00	0.00	0.00
Cameroon	0.00	0.00	0.00	0.00	0.00	0.00	0.00	0.00	0.00	0.00
Cape Verde	0.00	0.00	0.00	0.00	0.00	0.00	0.00	0.00	0.00	0.00
Central Africa Regional	0.00	0.00	15.25	0.00	0.00	0.00	0.00	0.00	0.00	15.25
Central African Republic	0.00	0.00	0.00	0.00	0.00	0.00	0.00	0.00	0.00	0.00
Chad	0.00	0.00	0.00	0.00	0.00	0.00	0.00	0.00	0.00	0.00
Comoros	0.00	0.00	0.00	0.00	0.00	0.00	0.00	0.00	0.00	0.00
Congo (DROC)	0.25	0.00	0.75	0.00	0.00	0.00	0.00	0.00	0.00	1.00
Congo (ROC)	0.00	0.00	0.00	0.00	0.00	0.00	0.00	0.00	0.00	0.00
Cote d'Ivoire	0.00	0.00	0.00	0.00	0.00	0.00	0.00	0.00	0.00	0.00
Djibouti	0.00	0.00	0.00	0.00	0.00	0.00	0.00	0.00	0.00	0.00
East Africa Regional	0.00	0.00	0.00	0.00	0.00	0.00	0.00	0.00	0.00	0.00
Equatorial Guinea	0.00	0.00	0.00	0.00	0.00	0.00	0.00	0.00	0.00	0.00
Eritrea	0.00	0.00	0.00	0.00	0.00	0.00	0.00	0.00	0.00	0.00
Ethiopia	0.00	0.00	0.00	0.00	0.00	0.00	0.00	0.00	0.00	0.00
Gabon	0.00	0.00	0.00	0.00	0.00	0.00	0.00	0.00	0.00	0.00
Gambia	0.00	0.00	0.00	0.00	0.00	0.00	0.00	0.00	0.00	0.00
Ghana	0.00	0.00	0.00	0.00	0.00	0.00	0.00	0.00	0.20	0.20
Guinea	0.00	0.00	0.00	0.00	0.00	0.00	0.00	0.00	0.00	0.00
Guinea-Bissau	0.00	0.00	0.00	0.00	0.00	0.00	0.00	0.00	0.00	0.00
Kenya	0.27	0.00	0.38	0.00	0.00	0.00	0.16	0.63	0.00	1.44
Lesotho	0.00	0.00	0.00	0.00	0.00	0.00	0.00	0.00	0.00	0.00
Liberia	4.95	0.00	2.00	0.00	0.00	0.00	0.00	0.00	0.00	6.95
Madagascar	0.00	0.00	8.02	0.00	0.00	0.00	0.04	0.16	0.00	8.22
Malawi	0.00	0.00	0.95	0.00	0.00	0.00	0.00	0.00	0.00	0.95
Mali	0.00	0.00	0.08	0.00	0.00	0.00	0.00	0.00	0.00	0.08
Mauritania	0.00	0.00	0.00	0.00	0.00	0.00	0.00	0.00	0.00	0.00
Mauritius	0.00	0.00	0.00	0.00	0.00	0.00	0.00	0.00	0.00	0.00
Mozambique	0.00	0.00	0.00	0.00	0.00	0.00	0.07	0.28	0.00	0.35
Namibia	0.00	0.00	0.00	0.00	0.00	0.00	0.00	0.00	0.00	0.00
Niger	0.00	0.00	0.00	0.00	0.00	0.00	0.00	0.00	0.00	0.00
Nigeria	0.00	0.00	0.00	0.00	0.00	0.00	0.00	0.00	0.00	0.00
Rwanda	0.00	0.00	0.00	0.00	0.00	0.00	0.00	0.00	0.00	0.00
Sahel	0.00	0.00	0.00	0.00	0.00	0.00	0.00	0.00	0.00	0.00
Senegal	0.00	0.00	0.75	0.00	0.00	0.00	0.04	0.16	0.00	0.95
Sierra Leone	0.00	0.00	0.00	0.00	0.00	0.00	0.00	0.00	0.00	0.00
Somalia	0.00	0.00	0.00	0.00	0.00	0.00	0.00	0.00	0.00	0.00
South Africa	0.00	0.00	0.00	0.00	0.00	0.00	0.00	0.00	0.00	0.00
Southern AfricaRegional	0.30	0.00	0.00	0.00	0.00	0.00	0.00	0.00	0.00	0.30
Sudan	0.00	0.00	0.00	0.00	0.00	0.00	0.00	0.00	0.00	0.00
Swaziland	0.00	0.00	0.00	0.00	0.00	0.00	0.00	0.00	0.00	0.00
Tanzania	0.00	0.00	0.00	0.00	0.00	0.00	0.00	0.00	0.00	0.00
Uganda	0.00	0.00	2.00	0.00	0.00	0.00	0.00	0.00	0.00	2.00
West Africa Regional	0.00	0.00	1.27	0.00	0.00	0.00	0.17	0.66	0.00	2.10
Zimbabwe	0.00	0.00	0.00	0.00	0.00	0.00	0.00	0.00	0.00	0.00

RECIPIENT COUNTRY/ REGION	MITIGATION						ADAPTATION			TOTALS
	Energy	Transport	Forestry	Agriculture	Waste Management	Industry	Capacity Building	Coastal Zone Management	Other Vulnerability Studies	
Asia/Near East										73.85
Afghanistan	17.54	0.00	0.00	0.00	0.00	0.00	0.00	0.00	0.00	17.54
Algeria	0.00	0.00	0.00	0.00	0.00	0.00	0.00	0.00	0.00	0.00
Asia/Near East Regional	2.31	0.00	0.89	0.00	0.00	0.00	0.00	0.00	0.00	3.20
Bahrain	0.00	0.00	0.00	0.00	0.00	0.00	0.00	0.00	0.00	0.00
Bangladesh	2.00	0.00	5.00	0.00	0.00	0.00	0.00	0.00	0.00	7.00
Bhutan	0.00	0.00	0.00	0.00	0.00	0.00	0.00	0.00	0.00	0.00
Brunei	0.00	0.00	0.00	0.00	0.00	0.00	0.00	0.00	0.00	0.00
Burma (Myanmar)	0.00	0.00	0.00	0.00	0.00	0.00	0.00	0.00	0.00	0.00
Cambodia	0.00	0.00	0.94	0.00	0.00	0.00	0.00	0.00	0.00	0.94
China	5.69	0.00	0.60	0.00	0.00	0.00	0.00	0.00	0.00	6.29
East Timor	0.00	0.00	0.00	0.00	0.00	0.00	0.00	0.00	0.00	0.00
Egypt	0.00	0.00	0.00	0.00	0.00	0.00	0.00	0.00	0.00	0.00
Federated States of Micronesia	0.00	0.00	0.00	0.00	0.00	0.00	0.00	0.00	0.00	0.00
Fiji	0.00	0.00	0.00	0.00	0.00	0.00	0.00	0.00	0.00	0.00
French Polynesia	0.00	0.00	0.00	0.00	0.00	0.00	0.00	0.00	0.00	0.00
Hong Kong	0.00	0.00	0.00	0.00	0.00	0.00	0.00	0.00	0.00	0.00
India	7.68	0.00	0.00	0.00	0.00	0.00	0.00	0.00	0.00	7.68
Indonesia	0.00	0.00	5.60	0.00	0.00	0.00	0.30	1.20	0.00	7.10
Iraq	0.00	0.00	0.00	0.00	0.00	0.00	0.00	0.00	0.00	0.00
Jordan	0.10	0.00	0.00	0.00	0.00	0.00	0.00	0.00	0.00	0.10
Kiribati	0.00	0.00	0.00	0.00	0.00	0.00	0.00	0.00	0.00	0.00
Korea (ROK)	0.00	0.00	0.00	0.00	0.00	0.00	0.00	0.00	0.00	0.00
Laos	0.00	0.00	0.00	0.00	0.00	0.00	0.00	0.00	0.00	0.00
Lebanon	0.00	0.00	0.00	0.00	0.00	0.00	0.00	0.00	0.00	0.00
Macao	0.00	0.00	0.00	0.00	0.00	0.00	0.00	0.00	0.00	0.00
Malaysia	0.00	0.00	0.00	0.00	0.00	0.00	0.00	0.00	0.00	0.00
Maldive Islands	0.00	0.00	0.00	0.00	0.00	0.00	0.00	0.00	0.00	0.00
Marshall Islands	0.00	0.00	0.00	0.00	0.00	0.00	0.00	0.00	0.00	0.00
Mongolia	0.00	0.00	0.05	0.00	0.00	0.00	0.00	0.00	0.00	0.05
Morocco	0.00	0.00	0.07	0.00	0.00	0.00	0.00	0.00	0.00	0.07
Nauru	0.00	0.00	0.00	0.00	0.00	0.00	0.00	0.00	0.00	0.00
Nepal	0.00	0.00	2.00	0.00	0.00	0.00	0.00	0.00	0.00	2.00
New Caledonia	0.00	0.00	0.00	0.00	0.00	0.00	0.00	0.00	0.00	0.00
North Korea	0.00	0.00	0.00	0.00	0.00	0.00	0.00	0.00	0.00	0.00
Oman	0.00	0.00	0.00	0.00	0.00	0.00	0.00	0.00	0.00	0.00
Pakistan	0.00	0.00	0.00	0.00	0.00	0.00	0.00	0.00	0.00	0.00
Palau	0.00	0.00	0.00	0.00	0.00	0.00	0.00	0.00	0.00	0.00
Papua New Guinea	0.00	0.00	0.00	0.00	0.00	0.00	0.05	0.20	0.00	0.25
Philippines	3.73	0.00	7.80	0.00	0.00	0.00	0.98	3.90	0.00	16.41
Qatar	0.00	0.00	0.00	0.00	0.00	0.00	0.00	0.00	0.00	0.00
Reunion	0.00	0.00	0.00	0.00	0.00	0.00	0.00	0.00	0.00	0.00
Samoa	0.00	0.00	0.00	0.00	0.00	0.00	0.00	0.00	0.00	0.00
Saudi Arabia	0.00	0.00	0.00	0.00	0.00	0.00	0.00	0.00	0.00	0.00
Seychelles	0.00	0.00	0.00	0.00	0.00	0.00	0.00	0.00	0.00	0.00
Singapore	0.00	0.00	0.00	0.00	0.00	0.00	0.00	0.00	0.00	0.00
South Asia Regional	2.70	0.00	0.00	0.00	0.00	0.00	0.00	0.00	0.00	2.70
Southeast Asia Regional	0.00	0.00	0.00	0.00	0.00	0.00	0.00	0.00	0.00	0.00
Sri Lanka	0.52	0.00	0.00	0.00	0.00	0.00	0.00	0.00	0.00	0.52
Syria	0.00	0.00	0.00	0.00	0.00	0.00	0.00	0.00	0.00	0.00
Taiwan	0.00	0.00	0.00	0.00	0.00	0.00	0.00	0.00	0.00	0.00
Thailand	0.00	0.00	0.00	0.00	0.00	0.00	0.00	0.00	0.00	0.00

RECIPIENT COUNTRY/ REGION	MITIGATION						ADAPTATION			TOTALS
	Energy	Transport	Forestry	Agriculture	Waste Management	Industry	Capacity Building	Coastal Zone Management	Other Vulnerability Studies	
Tokelau Islands	0.00	0.00	0.00	0.00	0.00	0.00	0.00	0.00	0.00	0.00
Tunisia	0.00	0.00	0.00	0.00	0.00	0.00	0.00	0.00	0.00	0.00
Turkey	0.00	0.00	0.00	0.00	0.00	0.00	0.00	0.00	0.00	0.00
Turks & Caicos Islands	0.00	0.00	0.00	0.00	0.00	0.00	0.00	0.00	0.00	0.00
Vanuatu	0.00	0.00	0.00	0.00	0.00	0.00	0.00	0.00	0.00	0.00
Vietnam	0.00	0.00	0.00	0.00	0.00	0.00	0.40	1.60	0.00	2.00
West Bank/Gaza	0.00	0.00	0.00	0.00	0.00	0.00	0.00	0.00	0.00	0.00
Yemen	0.00	0.00	0.00	0.00	0.00	0.00	0.00	0.00	0.00	0.00
Europe/Eurasia										29.23
Albania	0.46	0.00	0.00	0.00	0.00	0.00	0.00	0.00	0.00	0.46
Armenia	8.55	0.00	0.00	0.00	0.00	0.00	0.00	0.00	0.00	8.55
Azerbaijan	0.00	0.00	0.00	0.00	0.00	0.00	0.00	0.00	0.00	0.00
Belarus	0.00	0.00	0.00	0.00	0.00	0.00	0.00	0.00	0.00	0.00
Bosnia & Herzegovina	0.00	0.00	0.00	0.00	0.00	0.00	0.00	0.00	0.00	0.00
Bulgaria	0.00	0.00	0.00	0.00	0.00	0.00	0.00	0.00	0.00	0.00
Central Asia Regional	0.95	0.00	0.00	0.00	0.00	0.00	0.00	0.00	0.00	0.95
Croatia	0.00	0.00	0.00	0.00	0.00	0.00	0.00	0.00	0.00	0.00
Cyprus	0.00	0.00	0.00	0.00	0.00	0.00	0.00	0.00	0.00	0.00
Czech Republic	0.00	0.00	0.00	0.00	0.00	0.00	0.00	0.00	0.00	0.00
Estonia	0.00	0.00	0.00	0.00	0.00	0.00	0.00	0.00	0.00	0.00
Europe & Eurasia Regional	1.12	0.00	0.00	0.00	0.00	0.00	0.00	0.00	0.00	1.12
Georgia	3.17	0.00	0.25	0.00	0.00	0.00	0.00	0.00	0.00	3.41
Hungary	0.00	0.00	0.00	0.00	0.00	0.00	0.00	0.00	0.00	0.00
Kazakhstan	0.36	0.00	0.00	0.00	0.00	0.00	0.00	0.00	0.00	0.36
Kosovo	6.62	0.00	0.00	0.00	0.00	0.00	0.00	0.00	0.00	6.62
Kyrgyzstan	0.98	0.00	0.00	0.00	0.00	0.00	0.00	0.00	0.00	0.98
Latvia	0.00	0.00	0.00	0.00	0.00	0.00	0.00	0.00	0.00	0.00
Lithuania	0.00	0.00	0.00	0.00	0.00	0.00	0.00	0.00	0.00	0.00
Macedonia	0.00	0.00	0.00	0.00	0.00	0.00	0.00	0.00	0.00	0.00
Moldova	0.40	0.00	0.00	0.00	0.00	0.00	0.00	0.00	0.00	0.40
Montenegro	0.00	0.00	0.00	0.00	0.00	0.00	0.00	0.00	0.00	0.00
Poland	0.00	0.00	0.00	0.00	0.00	0.00	0.00	0.00	0.00	0.00
Romania	0.00	0.00	0.00	0.00	0.00	0.00	0.00	0.00	0.00	0.00
Russia	0.27	0.00	0.43	0.00	0.00	0.00	0.00	0.00	0.00	0.70
Serbia	0.00	0.00	0.00	0.00	0.00	0.00	0.00	0.00	0.00	0.00
Slovak Republic	0.00	0.00	0.00	0.00	0.00	0.00	0.00	0.00	0.00	0.00
Slovenia	0.00	0.00	0.00	0.00	0.00	0.00	0.00	0.00	0.00	0.00
Tajikistan	0.77	0.00	0.00	0.00	0.00	0.00	0.00	0.00	0.00	0.77
Turkmenistan	0.22	0.00	0.00	0.00	0.00	0.00	0.00	0.00	0.00	0.22
Ukraine	4.70	0.00	0.00	0.00	0.00	0.00	0.00	0.00	0.00	4.70
Uzbekistan	0.00	0.00	0.00	0.00	0.00	0.00	0.00	0.00	0.00	0.00
Yugoslavia	0.00	0.00	0.00	0.00	0.00	0.00	0.00	0.00	0.00	0.00
Latin America/ Caribbean										56.42
Anguilla	0.00	0.00	0.00	0.00	0.00	0.00	0.00	0.00	0.00	0.00
Antigua & Barbuda	0.00	0.00	0.00	0.00	0.00	0.00	0.00	0.00	0.00	0.00
Argentina	0.00	0.00	0.00	0.00	0.00	0.00	0.00	0.00	0.00	0.00
Aruba	0.00	0.00	0.00	0.00	0.00	0.00	0.00	0.00	0.00	0.00
Bahamas	0.00	0.00	0.00	0.00	0.00	0.00	0.00	0.00	0.00	0.00
Barbados	0.00	0.00	0.00	0.00	0.00	0.00	0.00	0.00	0.00	0.00
Belize	0.00	0.00	0.00	0.00	0.00	0.00	0.02	0.08	0.00	0.10
Bermuda	0.00	0.00	0.00	0.00	0.00	0.00	0.00	0.00	0.00	0.00

RECIPIENT COUNTRY/ REGION	MITIGATION						ADAPTATION			TOTALS
	Energy	Transport	Forestry	Agriculture	Waste Management	Industry	Capacity Building	Coastal Zone Management	Other Vulnerability Studies	
Bolivia	0.00	0.00	3.30	0.00	0.00	0.00	0.00	0.00	0.00	3.30
Brazil	0.00	0.00	7.32	0.00	0.00	0.00	0.00	0.00	0.00	7.32
British Virgin Islands	0.00	0.00	0.00	0.00	0.00	0.00	0.00	0.00	0.00	0.00
Caribbean Regional	0.00	0.00	0.00	0.00	0.00	0.00	0.40	1.60	0.00	2.00
Cayman Islands	0.00	0.00	0.00	0.00	0.00	0.00	0.00	0.00	0.00	0.00
Central America Regional	0.00	0.00	0.00	0.00	0.00	0.00	0.60	2.40	0.05	3.05
Chile	0.00	0.00	0.00	0.00	0.00	0.00	0.00	0.00	0.00	0.00
Colombia	0.00	0.00	0.00	0.00	0.00	0.00	0.00	0.00	0.00	0.00
Costa Rica	0.00	0.00	0.00	0.00	0.00	0.00	0.00	0.00	0.00	0.00
Dominica Islands	0.00	0.00	0.00	0.00	0.00	0.00	0.00	0.00	0.00	0.00
Dominican Republic	0.00	0.00	0.00	0.00	0.00	0.00	0.10	0.40	0.00	0.50
Ecuador	0.00	0.00	1.08	0.00	0.00	0.00	0.44	1.76	0.00	3.28
El Salvador	0.00	0.00	0.00	0.00	0.00	0.00	0.06	0.24	0.00	0.30
French Guiana	0.00	0.00	0.00	0.00	0.00	0.00	0.00	0.00	0.00	0.00
Grenada Islands	0.00	0.00	0.00	0.00	0.00	0.00	0.00	0.00	0.00	0.00
Guadeloupe	0.00	0.00	0.00	0.00	0.00	0.00	0.00	0.00	0.00	0.00
Guatemala	0.00	0.00	1.80	0.00	0.00	0.00	0.01	0.02	0.00	1.83
Guyana	0.00	0.00	0.00	0.00	0.00	0.00	0.00	0.00	0.00	0.00
Haiti	0.00	0.00	0.00	0.00	0.00	0.00	4.01	16.05	0.00	20.06
Honduras	0.00	0.00	0.50	0.00	0.00	0.00	0.52	2.00	0.00	3.02
Jamaica	0.00	0.00	0.00	0.00	0.00	0.00	0.10	0.46	0.00	0.56
Latin America Regional	0.00	0.00	6.50	0.00	0.00	0.00	0.00	0.00	0.00	6.50
Martinique	0.00	0.00	0.00	0.00	0.00	0.00	0.00	0.00	0.00	0.00
Mexico	0.35	0.00	0.00	0.00	0.00	0.00	0.05	0.20	0.00	0.60
Montserrat Islands	0.00	0.00	0.00	0.00	0.00	0.00	0.00	0.00	0.00	0.00
Nicaragua	0.00	0.00	1.00	0.00	0.00	0.00	0.00	0.00	0.00	1.00
Panama	0.00	0.00	0.00	0.00	0.00	0.00	0.40	1.60	0.00	2.00
Paraguay	0.00	0.00	0.00	0.00	0.00	0.00	0.00	0.00	0.00	0.00
Peru	0.00	0.00	1.00	0.00	0.00	0.00	0.00	0.00	0.00	1.00
Saint Kitts-Nevis	0.00	0.00	0.00	0.00	0.00	0.00	0.00	0.00	0.00	0.00
Saint Lucia Islands	0.00	0.00	0.00	0.00	0.00	0.00	0.00	0.00	0.00	0.00
Saint Pierre & Miquelon	0.00	0.00	0.00	0.00	0.00	0.00	0.00	0.00	0.00	0.00
Saint Vincent & Grenadines	0.00	0.00	0.00	0.00	0.00	0.00	0.00	0.00	0.00	0.00
Suriname	0.00	0.00	0.00	0.00	0.00	0.00	0.00	0.00	0.00	0.00
Trinidad & Tobago	0.00	0.00	0.00	0.00	0.00	0.00	0.00	0.00	0.00	0.00
Uruguay	0.00	0.00	0.00	0.00	0.00	0.00	0.00	0.00	0.00	0.00
Venezuela	0.00	0.00	0.00	0.00	0.00	0.00	0.00	0.00	0.00	0.00
Other Global Programs	3.28	0.00	0.00	0.00	0.00	0.00	0.00	0.00	21.13	24.41

8

Research and Systematic Observation

nvestments in climate science over the past several decades have greatly increased understanding of global climate change, including its attribution mainly to human influences. The air and oceans are warming, mountain glaciers are disappearing, sea ice is shrinking, permafrost is thawing, the great land ice sheets on Greenland and Antarctica are showing signs of instability, and sea level is rising. The consequences for human well-being are already being felt: more heat waves, floods, droughts, and wildfires; tropical diseases reaching into the temperate zones; vast areas of forest destroyed by pest outbreaks linked to warming; alterations in patterns of rainfall on which agriculture depends; and coastal property increasingly at risk from the surging seas. All of these impacts are being experienced in the United States and globally, as extensively documented in the recently released U.S. Global Change Research Program report *Global Climate Change Impacts in the United States* (Karl et al. 2009).

RESEARCH ON GLOBAL CHANGE

The latest and best scientific information forms the bedrock on which effective policy to combat and cope with climate change must be built. To assist the government and society as a whole with understanding, mitigating, and adapting to climate change, U.S. government agencies deploy a wide range of powerful science and technology resources. Each agency has different sets of key specialists and capabilities, different networks and relationships with the external research community, and separate program and budget authorities.

U.S. Global Change Research Program

The U.S. Global Change Research Program (USGCRP) brings together into a single interagency program the essential capacities for research and observations that are widely distributed across U.S. government agencies.[1] An essential component of success in delivering the information necessary for decision making is coordination of the programmatic and budgetary decisions of the 13 agencies that make up the USGCRP.

Growing out of interagency activities and planning beginning in about 1988, creation of the USGCRP energized cooperative interagency activities, with each agency bringing its strength to the collaborative effort. In 1990, USGCRP received congressional support under the Global Change Research Act (GCRA). The act called for the development of a research program "...to understand, assess, predict, and respond to human-induced and natural processes of global change," and guided federally supported global change research for the next decade.[2] In 2001, President Bush established the Climate Change Research Initiative (CCRI) to investigate uncertainties and set research priorities in climate change science, aiming to fill gaps in understanding within a few years.[3] In the following year, it was announced that the USGCRP and CCRI together would become the Climate Change Science Program (CCSP). The USGCRP label remained attached to many of the program's activities; now, consistent with the statutory language of the GCRA, the whole effort is going forward in the Obama administration as the USGCRP.

The USGCRP is managed by a director from one of the participating agencies (currently from the National Aeronautics and Space Administration [NASA]), with the help of the USGCRP Integration and Coordination Office (ICO) and interagency working groups that plan future research and cross-cutting activities, such as communications, decision support, and information and data concerns. The Office of Science and Technology Policy (OSTP) and Office of Management and Budget work closely with the ICO and the working groups to establish research priorities and funding plans to ensure the program is aligned with the administration's priorities and reflects agency planning.

The fiscal year 2011 budget provides $2.1 billion for USGCRP/CCSP programs—an increase of about 3 percent, over the 2009 level (excluding American Recovery and Reinvestment Act of 2009 [ARRA] funds).[4] USGCRP programs also received $604 million in ARRA funding based on preliminary agency allocations, including $237 million for NASA climate activities. ARRA funding also includes $170 million for the National Oceanic and Atmospheric Administration's (NOAA's) climate modeling activities. The 2011 budget supports research activities, including the goals set forth in the program's strategic plan. These activities can be grouped under the following areas: increase society's knowledge of Earth's past and present climate variability and change; improve understanding of natural and human forces of climate change; accelerate the capability to model and predict future conditions and impacts; assess the nation's vulnerability to current and anticipated impacts of climate change; and improve the nation's ability to respond to climate change by providing climate information and decision-support tools that are useful to policymakers and the general public.

Although the USGCRP continues to support a variety of research activities to gain more detailed predictive understanding of climate change, increased emphasis is being placed on bridging the significant gaps between estimating how much climate may change and the effects these changes may have on ecosystem services, water resources, natural resource utilization, human health, and societal well-being. The USGCRP is making a strong commitment to provide information that will reduce vulnerabilities and improve resilience to variability and change.

A recent National Academy of Sciences National Research Council report recommended restructuring the USGCRP around "...the end-to-end climate change problem, from understanding causes and processes to supporting actions needed to cope with the impending societal problems of climate change" (NAS/NRC 2009). The USGCRP is committed to supporting a balanced portfolio of fundamental and application-oriented research activities, from expanded modeling efforts to studies of coupled human-natural systems and institutional resilience. Plans are being developed to boost adaptation research; to bolster the capacity to monitor change and its impacts (including not only enhancing monitoring networks on land and at sea but also strengthening our system of Earth observation satellites); to produce the integrated assessments of the pace, patterns, and regional impacts of climate change that will be needed by decision makers as input into their deliberations on the metrics and goals to be embraced for both mitigation and adaptation; and to make climate data and information accessible to those who need it.

Adaptation Research

Knowledge is currently limited about the ability of communities, regions, and sectors to adapt to a changing climate. To address this shortfall, research on climate change impacts and adaptation must address complex human dimensions, such as economics, management, governance, behavior, and equity. Interdisciplinary research on adaptation that takes into account the interconnectedness of the Earth system and the complex nature of the social, political, and economic environment in which adaptation decisions must be

---

[1] See http://www.usgcrp.gov/.

[2] See http://www.gcrio.org/gcact1990.html.

[3] See http://www.climatescience.gov/about/ccri.htm.

[4] See http://www.gpoaccess.gov/usbudget/.

made is central to this effort. Given the relationships between climate change and extreme events, the communities of researchers, engineers, and other experts who work on reducing risks from natural and human-caused disasters have an important role to play in framing climate change adaptation strategies and in providing information to support decision making during implementation. For example, assessments of emergency preparedness and response systems, insurance systems, and disaster-relief capabilities are an important component of a society's adaptive capacity.

Integrated Assessment

Preparing for and adapting and responding to the impacts of climate change at the national level must begin with an integrated assessment that cuts across regional and sectoral lines. Any national assessment activity must engage localities and sectors to aggregate information into a national picture of climate impacts, and should also use this engagement to gather information on the "demand side" of adaptation, where people live and work, to re-orient research and observation investments. While national policy may be warranted for addressing certain climate change issues, individuals, public- and private-sector organizations, local communities, states, and regions will need to respond to many challenges as well. Future USGCRP activities will serve these different scales and stakeholders, providing the information and capabilities they need to prepare for and adapt and respond to future conditions.

Climate Services

Coordinated climate information and services are needed to assist decision making across public and private sectors. Local planners need information on likely changes in precipitation amount and flooding rains; farmers and farm cooperatives need information on changes in season length and temperature, not just for their own farms, but for those of their local and distant competitors; coastal zone managers need information on likely changes in sea level, storms, and estuarine temperatures; water resource managers need information on likely changes in snowpack and run-off, and the chance of floods and drought; community health planners need information on changes in the location of freezing conditions and the frequency of extreme heat waves; industry needs information on changes in extremes that might affect their businesses and shipping; and economic analysts need information across the region.

Just as the nation's climate research efforts require and benefit from interagency and academic partnerships, so too will the development and communication of climate change information benefit citizens and researchers. Providing this information requires sustained federal agency partnerships and collaboration with climate service providers and end users.

While much work has been done to evaluate the need for climate services and a National Climate Service, the Obama administration believes that additional assessment and analysis of existing climate-service capabilities and user needs for climate services are necessary. A National Climate Service and, more broadly, our nation's approach to delivering climate services will require that such analysis and assessment are ongoing, science-based, user-responsive, and relevant to all levels of interest—e.g., local, regional, national, and international. Such a framework must also be able to adapt to new developments in the scientific understanding of climate change and resultant impacts to serve the needs of decision makers and the public.

To address the nation's need for reliable and accurate climate information, OSTP is working to convene a task force with representation from a diverse group of key agencies whose charge will be to examine national assets, existing data and information gaps, and costs related to the development of a cohesive framework for delivering accurate climate-related information to the public. This process is intended to result in a more detailed functional and organizational approach for delivering climate services to the nation, in concert with a broad authorizing framework.

SYSTEMATIC OBSERVATIONS

Long-term, high-quality observations of the global environmental system are essential for defining the current state of the Earth's system, discovering past trends, and measuring its variability. This task requires both space- and surface-based observation systems. The term *climate observations* can encompass a broad range of environmental observations, including (1) routine weather observations, which, when collected consistently over a long period of time, can be used to help describe a region's climatology; (2) observations collected as part of research investigations to elucidate chemical, dynamic, biological, or radiative processes that contribute to maintaining climate patterns or to their variability; (3) highly precise, continuous observations of climate system variables collected for the express purpose of documenting long-term (decadal-to-centennial) change; and (4) observations of climate proxies, collected to extend the instrumental climate record to remote regions and back in time to provide information on climate change for millennial and longer time scales. A full documentation of all U.S. systematic climate observational activities can be found in *The United States National Report on Systematic Observations for Climate for 2008: National Activities with Respect to the Global Climate Observing System (GCOS) Implementation Plan* (U.S. CCSP 2008). The following input will summarize, and where appropriate, update that information.

Past reports have categorized observing systems as being either satellite-based or *in situ*, to indicate a ground-based system. However, the term *in situ* is somewhat misleading, as it appears to include all non-satellite measurements (i.e., remote sensing from the ground, balloons, or aircraft), which it does not. For example, atmospheric observations necessary for climate studies include both real *in situ* measurements (which directly sample the air mass surrounding the instrument) as well as non-satellite remote-sensing measurements. Therefore, this report distinguishes observing systems as being either non-satellite or satellite in nature.

Satellite observations provide a unique perspective of the global integrated Earth system and are necessary for good global climate coverage. Non-satellite observations are required for the measurement of parameters that cannot be estimated from space platforms (e.g., biodiversity, groundwater, carbon sequestration at the root zone, and subsurface ocean parameters). Non-satellite observations also provide long time series of observations required for the detection and diagnosis of global change, such as surface temperature, precipitation and water resources, weather and other natural hazards, the emission or discharge of pollutants, and the impacts of multiple stresses on the environment due to human and natural causes.

NOAA has provided leadership in the *in situ* long-term measurement of atmospheric greenhouse gases (GHGs) for 40 years. Flask, tower, and aircraft measurements are routinely made by NOAA's Earth System Resources Laboratory, and ocean carbon inventory and exchange observations are routinely made by NOAA's Pacific Marine Environmental Laboratory. NOAA maintains the global observational networks and field programs on which society will increasingly depend for reliable information.

One critical challenge to the Earth observation field is to maintain existing observation capabilities in a variety of areas. For example, maintaining the observational record of stratospheric ozone is essential in discerning the effects of climate change on the nature and timing of ozone recovery. Other key areas include radiative energy fluxes of the Sun and Earth, atmospheric carbon dioxide (CO_2), global surface temperature, and global sea level. Efforts to create a long-term record of global land cover, started by Landsat in the 1970s, are currently being prepared for the transition to a Landsat Data Continuity Mission (LDCM) being planned by NASA and the U.S. Geological Survey (USGS).[5] The LDCM is currently planned for launch in December 2012 and is expected to have a 5-year mission life with 10-year expendable provisions.

The USGS Landsat 35-year record of the Earth's surface is available to users at no charge. Under a transition toward the National Land Imaging Program, USGS is pursuing an aggressive schedule to provide users with electronic access to any Landsat scene held in the USGS-managed national archive of global scenes dating back to Landsat 1, launched in 1972. As of February 2009, any Landsat archive scene selected by a user will be automatically processed, at no charge. In addition, newly acquired scenes meeting a cloud cover threshold of 20 percent or below will be processed and placed online for at least three months, after which time they will remain available for selection from the archive.

The National Polar-orbiting Operational Environmental Satellite System (NPOESS)[6] program has undergone a major restructuring in order to put this critical program on a more sustainable pathway toward success. The satellite system is a national priority that is essential to meeting both civil and military weather-forecasting, storm-tracking, and climate-monitoring requirements. The restructured program, now known as the Joint Polar Satellite System (JPSS), will consist of platforms based on the NPOESS Preparatory Project (NPP) satellite scheduled for launch in September 2011. The JPSS is designed to monitor global environmental conditions, and collect and disseminate data related to weather, atmosphere, oceans, land, and near-space environment with a planned launch in 2015. The first JPSS satellite will host some of the sensors that were planned for the first NPOESS satellite, including the Visible Infrared Imaging Radiometer Suite, the Cross-track Infrared Sounder, the Advanced Technology Microwave Sounder, the Ozone Mapping and Profiling Suite, and the Cloud and Earth Radiant Energy System instrument.

Remotely sensed observations continue to be a cornerstone of the overall U.S. climate-observing program as coordinated by the USGCRP. The Cloud-Aerosol Lidar and Infrared Pathfinder Satellite Observation (CALIPSO) satellite mission and CloudSat radar instruments are providing an unprecedented examination of the vertical structure of aerosols and clouds over the entire Earth. These data—when combined with data from the Aqua, Aura, and Parasol satellites orbiting in formation (the "A-Train")—will enable systematic pursuit of key issues, including the effects of aerosols on clouds and precipitation, the strength of cloud feedbacks, and the characteristics of difficult-to-observe polar clouds. The increasing volume of data from remote-sensing and non-satellite observing systems presents a continuing challenge for USGCRP agencies to ensure that data management systems are able to handle the expected increases.

To meet the long-term needs for the documentation of global changes, the United States integrates observations from both research and operational systems. The United States supports the need to improve global observing systems for climate, and to exchange information on national plans and programs that contribute to the global capacity in this area.

Providing for wide access to information from the Global Earth Observation System of Systems

---

[5] See http://landsat.gsfc.nasa.gov/ and http://landsat.usgs.gov.

[6] See http://www.ipo.noaa.gov/.

[7] See http://www.earthobservations.org/.

(GEOSS)[7] for applications that benefit society has been a focus of efforts coordinated by the intergovernmental Group on Earth Observations (GEO)[8] and the U.S. Group on Earth Observations (USGEO).[9] The United States continues to be a very active participant in the development of GEOSS. The purpose of GEOSS is to achieve *comprehensive, coordinated, and sustained* observations of the Earth system in order to improve monitoring of the changing state of the planet, to increase understanding of complex Earth processes, and to enhance the prediction of the impacts of environmental change, including climate change. Finally, GEOSS provides the overall conceptual and organizational framework to build toward integrated global Earth observations to meet user needs and to support decision making in an increasingly complex and environmentally stressed world. It is a "system of systems" consisting of existing and future Earth observation systems, supplementing—not supplanting—the mandates and governance arrangements of those systems, such as GCOS.[10] The established Earth observation systems, through which many countries cooperate as members of the United Nations (UN) Specialized Agencies and as contributors to international scientific programs, provide essential building blocks for GEOSS.[11]

Potential benefits of Earth observations were detailed in the U.S. Integrated Earth Observation System 10-year strategic plan that covered climate and eight other related areas—agriculture, disasters, ecology, energy, health, integration, ocean resources, water resources, and weather (IWGEO and NSTC/CENR 2005). The importance of this global collaborative effort was evident at the November 2009 GEO-VI Plenary Meeting, and a regional implementation of GEOSS, known as GEOSS in the Americas, continues to be an effort strongly supported by the United States.[12] The first significant GEOSS in the Americas project involved the shifting of the Geostationary Operational Environmental Satellites GOES-10 satellite in 2006 to a new orbit, to greatly improve environmental satellite coverage of the Western Hemisphere, especially over South America.[13] By significantly enhancing satellite detection of such natural hazards as severe storms, floods, drought, landslides, and wildfires, the shift helps protect lives and property in both South America and the United States, and allows for improved prediction, response, and follow-up and expanded understanding of Earth system processes. Planning is underway to continue this South America coverage after GOES-10's service is completed, by using GOES-12. Since 2006, other collaborative activities in the region have been taken on, including the North American Drought Monitoring Program[14] and a coastal zone management effort discussed at a special forum in concert with the 2009 GEO-VI Plenary Meeting.

Documentation of U.S. Climate Observations

The United States supports a large number of remote-sensing satellite platforms, as well as a broad network of Earth-based global atmospheric, ocean, and terrestrial observation systems that are essential to climate monitoring. The United States contributes to the development and operation of several global observing systems, both research and operational, that collectively provide a comprehensive measure of climate system variability and climate change processes. These systems are a baseline Earth-observing system and include NASA, NOAA, and USGS Earth-observing satellites and extensive non-satellite observational capabilities. The USGCRP also supports several ground-based measurement activities that provide the data used in studies of the various climate processes necessary for better understanding of climate change. U.S. observational and monitoring activities contribute significantly to several international observing systems, including the GCOS, principally sponsored by the World Meteorological Organization (WMO); the Global Ocean Observing System (GOOS),[15] sponsored by the UN Educational, Scientific, and Cultural Organization's Intergovernmental Oceanographic Commission;[16] and the Global Terrestrial Observing System (GTOS), sponsored by the UN Food and Agriculture Organization.[17] The latter two have climate-related elements being developed jointly with GCOS. For example, NOAA has provided leadership in the *in situ* long-term measurement of atmospheric GHGs for 40 years. Flask, tower, and aircraft measurements are routinely made by NOAA's Earth System Resources Laboratory, and ocean carbon inventory and exchange observations are routinely made by NOAA's Pacific Marine Environmental Laboratory.

U.S. priorities for advancement of the atmospheric, oceanic, and terrestrial observing components of GCOS include (1) reducing the uncertainty in the global carbon inventory (in the atmospheric, oceanic, and terrestrial domains), sea-level change, and sea surface temperature; (2) continuing support for existing non-satellite atmospheric networks in developing nations; and (3) planning for surface and upper-air GCOS reference observations consistent with the USGCRP Synthesis and Analysis Report 1.1 (Karl et al. 2006). As such, the GOOS will make incremental advances, building out to 60 percent completion in 2010, in which 50 surface drifters will be equipped with salinity sensors for satellite validation and salinity budget calculations, particularly in the polar regions; a new reference array will be added across the Atlantic basin, to measure changes in the ocean's overturning circulation—an indicator of possible abrupt climate change; a pilot U.S. coastal carbon-observing network will enter sustained service, to help quantify North American carbon sources and sinks and to measure ocean acidification caused by CO_2 sequestration in

[8] See http://www.earthobservations.org/.

[9] See http://www.usgeo.gov/.[10] See http://www.gosic.org/ios/GCOS-main-page.htm.

[11] See http://earthobservations.org.

[12] See http://www.ssd.noaa.gov/PS/SATS/GOES/TEN/.

[13] See http://www.strategies.org/EOPA.html.

[14] See http://www.ncdc.noaa.gov/oa/climate/monitoring/drought/nadm/.

[15] See http://www.ioc-goos.org/.

[16] See http://ioc-unesco.org/.

[17] See http://www.fao.org/gtos/.

the ocean; and dedicated ships will target deployments of Argo and surface drifters in undersampled regions of the world oceans. The North American Carbon Program (NACP), under the auspices of the USGCRP, aims heavily at terrestrial uncertainties that are supported by a growing set of atmospheric measurements to place constraints on flux estimates.[18] These observations are ongoing and involve participation from numerous agencies and institutions.

Non-Satellite Atmospheric Observations

The United States supports 75 stations in the GCOS Surface Network, 21 stations in the GCOS Upper Air Network,[19] and 4 stations in the Global Atmospheric Watch (GAW).[20] These stations are distributed geographically, as prescribed in the GCOS and GAW network designs. The data (metadata and observations) from these stations are shared according to GCOS and GAW protocols.

Since publishing its last report to the UN Framework Convention on Climate Change (UNFCCC), the United States has continued to field and commission a system known as the U.S. Climate Reference Network (USCRN).[21] The USCRN is designed to answer the question: How has the U.S. climate changed over the past 50 years at national, regional, and local levels? Since beginning in 2002, and as of September 2008, 114 USCRN stations have been commissioned in the continental United States. In 2008, work was begun on expanding that network to include 29 additional commissioned USCRN stations in Alaska; as of September 2009, 2 new USCRN stations had been installed in Alaska. The USCRN concept is also being applied toward expanding reference surface observing on an international basis as resources allow. After a few years of planning, an effort is underway to install a USCRN station at the Russian Arctic observing station in Tiksi as part of a U.S.–Russia bilateral effort.

In addition, the Cooperative Observer Program (COOP) is the nation's largest and oldest weather network, with nearly 10,250 observations taken daily, mostly by volunteers, over the course of the past 121 years. The COOP is the primary source for monitoring U.S. climate variability, including measuring weekly to interannual time frames on national, regional, and local scales. These data are also the primary basis for assessments of century-scale climate change. The network is in stable locations of urban, suburban, and rural settings in flat, mountainous, and beach areas where people live, work and play. The COOP is designed to provide weather, water, and climate information. Due to the density of this observation network, the information collected by the COOP can clarify how the U.S. climate has changed in the past 120 years or more, on a national, regional, and local level. The USCRN installed the final station in 2008, and uses historic data from the COOP network to develop

pseudo-normals. Each year these data help to inform decisions related to Federal Disaster Declarations based on weather, insurance industry claims, water resource management, drought declarations, transportation issues, legal issues, computing model guidance to daily weather forecasts, normals and extremes, and energy consumption.

The Atmospheric Radiation Measurement (ARM) Climate Research Facility (ACRF) is a scientific user facility for obtaining continuous, long-term measurements of radiative fluxes, cloud and aerosol properties, and related atmospheric characteristics in diverse climate regimes.[22] The ACRF paradigm of long-term continuous measurements is essential to the evaluation and enhancement of climate models that must simulate the evolution of atmospheric properties for long continuous periods, from decades to centuries. The ACRF expands its geographic coverage through deployments of a mobile facility and includes aerial measurements that complement the ground measurements. In 2008, the mobile facility was deployed to China to examine aerosol indirect effects. In 2009, the mobile facility began a two-year deployment to the Azores to study processes controlling the radiative properties and microphysics of marine boundary layer clouds, a high-priority science question. An International Polar Year (IPY)[23] experiment was conducted using combinations of ground and aerial measurements. Data from the IPY experiment will be used as a case study by the Global Energy and Water Cycle Experiment Cloud Systems Study.[24]

The U.S. GCOS program's primary mission is support of non-satellite reference observational efforts, including developing the GCOS Reference Upper-Air Network (GRUAN). The GRUAN is intended to aid in enhancing the quality of upper-tropospheric and lower-stratospheric water vapor measurements at a subset of 30-40 global stations. The GRUAN began operation on January 1, 2009, and is led by the GRUAN Lead Center in Lindenberg, Germany; seven U.S. stations (including five ACRF sites, one NOAA/National Center for Atmospheric Research site, and one site from Howard University) have been invited to be part of the initial configuration of stations. The GRUAN is a key contributing network to GCOS; contributes to the GEOSS goal of "understanding, assessing, predicting, mitigating, and adapting to climate variability and change"; and is a key element supporting the Global Space-Based Inter-Calibration System effort. Long-term surface-based reference climate sites are essential for creating a continuous and homogeneous climate data record, such as those used by the Intergovernmental Panel on Climate Change and the UNFCCC, in global climate assessments. A reference climate data record is also essential for use by least-developed nations for local and regional planning related to protecting and monitoring water resources

[18] See http://www.nacarbon.org/.

[19] See http://www.ncdc.noaa.gov/oa/hofn/guan/guan-intro.html.

[20] See http://www.wmo.ch/web/arep/gaw/gaw_home.html.

[21] See http://www.ncdc.noaa.gov/crn/.

[22] See http://www.arm.gov/acrf/.

[23] See http://www.ipy.org/.

[24] See http://www.gewex.org/gewex_overview.html.

(e.g., drought forecasting), for understanding the effects of climate change on human health, and for understanding, assessing, predicting, mitigating and adapting to climate variability and change. Additionally, this kind of data record is a key element in reducing uncertainties in global temperature and precipitation variances, providing reference ground-truth data to aid in the evaluation of climate model simulations and in the provision of quality data for the calibration and validation of satellite data.

While it is difficult to list all observing campaigns and systems, several others should be noted for their global climate significance. The Southern Hemisphere ADditional OZonesondes (SHADOZ)[25] provides a consistent data set from balloon-borne ozonesondes for ground verification of satellite tropospheric ozone measurements at 12 sites across the tropical and subtropical regions of the Southern Hemisphere. Another key system along these lines is the AErosol RObotic NETwork (AERONET),[26] which is a federation of ground-based remote-sensing aerosol networks established in part by NASA and France's Centre National de la Recherche Scientifique.

AERONET provides a long-term, continuous, and readily accessible public domain database of aerosol optical properties for research and characterization of aerosols, validates satellite retrievals, and provides synergy with other databases. AERONET collaboration provides a series of globally distributed observations of spectral aerosol optical depth, inversion products, and precipitable water in diverse aerosol regimes. The collaborative effort between NASA's Advanced Global Atmospheric Gases Experiment (AGAGE)[27] and NOAA's Flask Monitoring Network has been instrumental in measuring the composition of the global atmosphere continuously since 1978. The AGAGE is distinguished by its capability to measure globally and at high frequency most of the important gases in the *Montreal Protocol on Substances That Deplete the Ozone Layer* and almost all of the significant non-CO_2 gases in the *Kyoto Protocol* to mitigate climate change. Also, both NASA and NOAA demonstrate great collaborative research efforts in this key climate monitoring activity.[28]

AERONET retrievals of atmospheric particulate absorption will continue to be utilized in climate-forcing studies and in the validation of current and future satellite missions, such as the Glory satellite (late 2010 launch), which will measure aerosol light absorption from space. Network expansion will continue, with focus on regions that are not adequately sampled and that are important for understanding of global climate change, such as Asia. An experimental effort is underway to investigate the possibility of measuring sunlight reflected off the moon to make aerosol measurements at night. In addition, an experimental algorithm is under development for measurements of atmospheric CO_2. In the future, light detection and ranging (LIDAR) data will be used in studies of the influence of polar stratospheric clouds on ozone formation over the South Pole, Arctic haze impacts on polar climate, and generation of climatological aerosol and cloud properties at several Micro Pulse Lidar (MPL)[29] Network (MPLNET)[30] sites. To enhance data value, MPL instrument designs and hardware will be continually improved. In addition, several new MPLNET data products will be made available to the research community.

Non-Satellite Ocean Observations

The climate requirements of GOOS are the same as those for GCOS. Also like GCOS, GOOS is based on a number of non-satellite and space-based observing components. The United States supports the Integrated Ocean Observing System's (IOOS's) surface and marine observations through a variety of components, including fixed and surface-drifting buoys, subsurface floats, and volunteer observing ships.[31] It also supports the Global Sea Level Observing System (GLOSS) through a network of sea level tidal gauges.[32]

The United States currently provides satellite coverage of the global oceans for sea-surface temperatures, surface elevation, ocean-surface vector winds, sea ice, ocean color, and other climate variables. The first element of the climate portion of GOOS, completed in September 2005, is the global drifting buoy array, which is a network of 1,250 drifting buoys measuring sea-surface temperature and other variables as they flow in the ocean currents. At present, the United States is the world leader in implementing the non-satellite elements of GOOS for climate, and sponsors the majority of the IOOS Global Component, which is the U.S. contribution to the international GOOS program and the ocean baseline of the GEOSS. The United States sponsors nearly half of the platforms presently deployed in the global ocean (3,860 of 7,723), with 72 other countries providing the remainder. The United States has historically contributed about half of the international system, and has been a leader in fostering an international systems approach to the implementation of GOOS.

Expanding in coverage, currently 60 percent of the initial GOOS design is complete. The demand for ocean data and the products and forecasts derived from these data require international cooperation with other nations to complete deployment as soon as possible.

In 1998, an international consortium presented plans for Argo, a global array of 3,000 autonomous instruments that would revolutionize the collection of critical, climate-relevant information from the upper 2 kilometers (km) (1.2 miles [mi]) of the world's oceans.[33] These instruments drift at depth, periodically

[25] See http://croc.gsfc.nasa.gov/shadoz/.

[26] See http://aeronet.gsfc.nasa.gov/data_frame.html.

[27] See http://agage.eas.gatech.edu/.

[28] See http://www.strategies.org/EOPA.html.

[29] See http://www.arm.gov/instruments/instrument.php?id=mpl.

[30] See http://www.mplnet.com/.

[31] See http://www.ocean.us/what_is_ioos.

[32] See http://www.gloss-sealevel.org/.

[33] See http://www.argo.ucsd.edu/.

rising to the sea surface, collecting data along the way, and report their observations in real time via satellite communications. The initial deployment objective of 3,000 instruments distributed homogenously throughout the world's oceans has been attained, and Argo now provides over 100,000 high-quality temperature and salinity profiles annually, along with global-scale velocity data, all without a seasonal bias. The Argo array has been deployed through the collaboration of more than 40 countries plus the European Union.

A guiding principle of Argo is that the program should benefit everyone. Thus, the data are openly and immediately available to anyone wishing to use them. Argo data coupled with global-scale satellite measurements from radar altimeters have made possible huge advances in the representation of the oceans in coupled ocean-atmosphere models for climate forecasts and the routine analysis and forecasting of the state of the subsurface ocean. Argo data are being used in an ever-widening range of research applications that have led to new insights into how the ocean and atmosphere interact in extreme as well as normal conditions. Two examples are the processes in polar winters when the deep waters that fill most of the ocean basins are formed, and the transfer of heat and water to the atmosphere beneath tropical cyclones. Both conditions are crucial to global weather and climate, and could not be observed by ships.

The present generation of instruments has a design life of four years when profiling to 2-km (1.2-mi) depth every 10 days. Maintaining the array will require annual deployments of around 800 floats. Having deployed the array and built the data delivery system, the challenge is to maintain the full array for a decade in a pre-operational "sustained maintenance" phase, including ensuring the availability of the platform resources to maintain the array. This will allow Argo's design to be optimized and its value fully demonstrated. The United States has committed to maintaining half of the array, and other contributing nations are striving to continue the array's strong international nature.

Continued upgrading of the GLOSS tidal gauge network from 43 to 170 stations is planned for the period 2006–2010. Ocean carbon inventory surveys in a 10-year repeat survey cycle help determine the anthropogenic intake of carbon into the oceans. Plans for advancement of the global Tropical-Atmosphere-Ocean (TAO) network of ocean buoys include an expansion of the network into the Indian Ocean (the Pacific Ocean has a current array of 70 TAO buoys).[34] From 2005 to 2007, 8 new TAO buoys were installed in the Indian Ocean in collaboration with partners from India, Indonesia, and France. Plans call for a total of 38 TAO buoys in the Indian Ocean by 2013, which will form the Research Moored Array for African-

Asian-Australian Monsoon Analysis and Prediction (RAMA) network. RAMA is a multinationally supported element of the Indian Ocean Observing System, a combination of complementary satellite and non-satellite measurement platforms for climate research and forecasting purposes. NASA is currently investing in the development of new prototype geodetic instruments for deployment later this decade to support the creation of a next-generation geodetic network for the improvement of the terrestrial reference frame.

IOOS is the U.S. coastal observing component of the GOOS. It is envisioned as a coordinated national and international network of observations, data management, and analyses that systematically acquires and disseminates data and information on past, present, and future states of the oceans. A coordinated IOOS effort is being established by NOAA via a national IOOS Program Office. The IOOS observing subsystem employs both remote and non-satellite sensing. Remote sensing includes satellite-, aircraft- and land-based sensors, power sources, and transmitters. Non-satellite sensing includes platforms (ships, buoys, gliders, etc.), non-satellite sensors, power sources, sampling devices, laboratory-based measurements, and transmitters.

Non-Satellite Terrestrial Observations

For terrestrial observations, GCOS and GTOS have identified permafrost thermal state and permafrost active layer as key variables for monitoring the state of the cryosphere. The United States operates a long-term "benchmark" glacier program to intensively monitor climate, glacier motion, glacier mass balance, glacier geometry, and stream runoff at a few select sites. The data collected are used to understand glacier-related hydrologic processes and improve the quantitative prediction of water resources and glacier-related hazards, and the consequences of climate change. Long-term, mass balance monitoring programs have been established at three widely spaced U.S. glacier basins that clearly sample different climate-glacier-runoff regimes.

SNOTEL and SCAN Networks

The SNOTEL (SNOpack TELemetry) and SCAN (Soil Climate Analysis Network) monitoring networks provide automated comprehensive snowpack, soil moisture, and related climate information designed to support natural resource assessments. SNOTEL operates more than 660 remote sites in mountain snowpack zones of the western United States.[35] SCAN, which began as a pilot program, now consists of more than 120 sites.[36] These networks collect and disseminate continuous, standardized soil moisture and other climate data in publicly available databases and climate reports. Uses for these data include inputs to global circulation models, verifying

---

[34] See http://www.pmel.noaa.gov/tao/proj_over/pubs/overview.html.

[35] See http://www.wcc.nrcs.usda.gov/snow/.

[36] See http://www.wcc.nrcs.usda.gov/scan/.

and ground truthing satellite data, monitoring drought development, forecasting water supply, and predicting sustainability for cropping systems.

AmeriFLUX Network

The AmeriFLUX network endeavors to establish an infrastructure for guiding, collecting, synthesizing, and disseminating long-term measurements of CO_2, water, and energy exchange from a variety of ecosystems.[37] Its objectives are to collect critical new information to help define the current global CO_2 budget, to enable improved projections of future concentrations of atmospheric CO_2, and to enhance the understanding of carbon fluxes, net ecosystem production, and carbon sequestration in the terrestrial biosphere.

North American Carbon Program

NACP, a major focus of the USGCRP, is a multidisciplinary research program established to obtain the scientific understanding of North America's carbon sources, sinks, and changes in carbon stocks needed to meet societal concerns and provide tools for decision makers. NACP is supported by a number of federal agencies through a variety of intramural and extramural funding mechanisms and award instruments. NACP relies upon a rich and diverse array of existing observational networks, monitoring sites, and experimental field studies in North America and its adjacent oceans. The program's goals are to develop quantitative scientific knowledge, robust observations, and models to determine: the emissions and uptake of CO_2, methane (CH_4), and carbon monoxide (CO); the changes in carbon stocks; and the factors regulating these processes for North America and adjacent ocean basins. NACP also aims to develop the scientific basis to implement full carbon accounting on regional and continental scales. This is the knowledge base needed to design monitoring programs for natural and managed CO_2 sinks and emissions of CH_4; to support long-term quantitative measurements of fluxes, sources, and sinks of atmospheric CO_2 and CH_4; and to develop forecasts for future trends.

Glacier Monitoring

The United States operates a long-term "benchmark" glacier program to intensively monitor climate, glacier motion, glacier mass balance, glacier geometry, and stream runoff at a few select sites. The data collected are used to understand glacier-related hydrologic processes and improve the quantitative prediction of water resources, glacier-related hazards, and the consequences of climate change. The approach has been to establish long-term mass balance monitoring programs at three widely spaced U.S. glacier basins to clearly sample different climate-glacier-runoff regimes—South Cascade Glacier in Washington State, and Gulkana and Wolverine Glaciers in Alaska. Mass balance data are available beginning in 1959 for the South Cascade Glacier, and beginning in 1966 for the Gulkana and Wolverine Glaciers.[38]

Land Cover Characterization Program

This program was begun in 1995 to develop land cover and other land characterization databases to address national and international requirements that were becoming increasingly sophisticated and diverse. To meet these requirements, USGS develops multi-scale land cover characteristics databases used by scientists, resource managers, planners, and educators (Global and National Land Cover Characterization), and contributes to the understanding of the patterns, characteristics, and dynamics of land cover across the United States and the globe (Urban Dynamics and Land Cover Trends). The program also conducts research to improve the utility and efficiency of large-area land cover characterization and land cover characteristics databases.

The initial goal to develop a global 1-km (0.6-mi) land cover characteristics database was achieved in 1997. Current efforts focus on revising the database utilizing input from users around the world. USGS, the University of Nebraska-Lincoln, and the European Commission's Joint Research Centre generated the initial 1-km (0.6-mi) resolution Global Land Cover Characteristics database using NOAA Advanced Very-High-Resolution Radiometer data from April 1992 through March 1993. The database was built on a continent-by-continent basis using standard map projections and 1-km (0.6-mi) nominal spatial resolution; each continental database contains unique elements based on the geographic aspects of the specific continent. The continental databases are combined to make seven global data sets, each representing a different landscape based on a particular classification legend.[39]

National Ecological Observatory Network

NEON is a planned continental-scale research platform for discovering and understanding the impacts of climate change, land-use change, and invasive species on ecology.[40] NEON would be a national observatory, not a collection of regional observatories. It is currently in the planning and development stages.

NEON would consist of distributed sensor networks and experiments, linked by advanced cyber infrastructure to record and archive ecological data for at least 30 years. Using standardized protocols and an open data policy, NEON would gather long-term data on ecological responses of the biosphere to changes in land use and climate, and on feedbacks from the geosphere, hydrosphere, and atmosphere. NEON would be designed to serve as a U.S. terrestrial contribution to GEOSS.

Space-Based Observations

Space-based, remote-sensing observations of the atmosphere–ocean–land system have evolved substantially since the early 1970s, when the first operational weather satellite systems were launched. Over the last decade, satellites have proven their observational capa-

---

[37] See http://public.ornl.gov/ameriflux/.

[38] See http://ak.water.usgs.gov/glaciology/.

[39] See http://edcdaac.usgs.gov/glcc/glcc.html.

[40] See http://www.neoninc.org/.

bility to accurately monitor nearly all aspects of the total Earth system on a global basis. Currently, satellite systems monitor the evolution and impacts of El Niño and La Niña, weather phenomena, natural hazards, and vegetation cycles; the ozone hole; solar fluctuations; changes in snow cover, sea ice and ice sheets, ocean surface temperatures, and biological activity; coastal zones and algal blooms; deforestation and forest fires; urban development; volcanic activity; tectonic plate motions; aerosol and three-dimensional cloud distributions; water distribution; and other climate-related information.

A number of U.S. satellite operational and research missions form the basis of a robust national remote-sensing program that fully supports the requirements of GCOS (U.S. DOC/NOAA 2001). These include instruments on the GOES and Polar-orbiting Operational Environmental Satellites (POES),[41] the series of Earth Observing System (EOS) satellites,[42] the Landsats 5 and 7,[43] and the Jason satellite[44] measuring sea-surface height, wind speed, and waves. Additional satellite missions in support of GCOS include (1) the Active Cavity Radiometer Irradiance Monitor for measuring solar irradiance;[45] (2) the EOS Terra, Aqua, and Aura series; (3) the Sea-viewing Wide Field-of-view Sensor (SeaWiFS) for studying ocean and productivity, as well as aerosols;[46] (4) the Shuttle Radar Topography Mission;[47] and (5) the Tropical Rainfall Measuring Mission for measuring rainfall, clouds, sea-surface temperature, radiation, and lightning.[48] A major upgrade to the GOES system, known as GOES-R, is under development, with a first launch scheduled for 2015.[49]

POES

Since 1979, the NOAA POES system has provided the nation with the longest time series of essential climate variables, including atmospheric temperature, water vapor, clouds, ozone, vegetation, and sea and land surface temperature.

GOES

Since the 1980s GOES has provided essential information on the diurnal cycle of clouds, and has been used as a key data set for the International Satellite Cloud Climatology Project. GOES has also been used to study the diurnal cycle of sea surface temperature.

Jason Altimeter Series

Global sea level rise is the most obvious manifestation of climate change in the ocean. It directly threatens coastal infrastructure through increased erosion, more frequent storm-surge flooding, and loss of habitat through drowned wetlands. The only feasible way to accurately determine global sea level rise is through satellite altimetry, the systematic collection of sea level observations, gathered today by the ongoing Jason series of satellite missions. These observations suggest that sea level rise is accelerating; in particular, the val-

ue of approximately 3.1 millimeters (0.12 inches) per year from altimeters over the past 15 years is almost twice the estimate of approximately 1.7 mm (0.07 in) per year from tide gauges over the past century.

The Jason series is being transitioned as a research endeavor from NASA and the Centre National d'Etudes Spatiales (the French Space Agency) to NOAA and the European Organisation for the Exploitation of Meteorological Satellites (EUMETSAT, NOAA's operational satellite counterpart in Europe) for joint implementation as a sustained operational capability. NOAA and EUMETSAT have already assumed responsibility for the ground system and operation of the Jason-2 satellite launched in June 2008. Additionally, with funding requested in the President's fiscal year (FY) 2010 budget (OMB 2009), NOAA plans to begin developing Jason-3, a joint mission with EUMETSAT. Assuming resources are made available on schedule, a joint Jason-3 could be launched in late 2013, in time to overlap at least the last 6 months of the design life of Jason-2, thus helping ensure the continuity of the climate record of global sea level.

Ocean Surface Vector Wind Measurements

NASA's QuikSCAT (or Quick Scatterometer) scans the ocean surface to measure wind speed and direction, providing observations over the open ocean where other tools, such as buoys, ships, and reconnaissance flights, are sparse or unavailable. Over time, these data have proved useful to NOAA for marine and tropical cyclone forecasts. In November 2009, NASA announced performance degradation of the QuikSCAT satellite.

NOAA and NASA continue to pursue short-term and long-term strategies to replace these space-based scatterometry measurements. NOAA is using data from the Advanced Scatterometer (ASCAT) onboard the Meteorological Operational (MetOp) satellite operated by EUMETSAT. NOAA and NASA have signed a letter of intent to collaborate on science and obtain Oceansat-2 data from the Indian Space Research Organisation. NOAA and NASA have also had discussions with the Japanese Aerospace Exploration Agency (JAXA) about flying a U.S.-developed scatterometer instrument on a future JAXA satellite.[50]

The Afternoon Train ("A-Train")

A collaboration between NASA and the space agencies of Canada and France, the A-Train (Afternoon Train) is a key Sun-synchronous, Earth-orbiting satellite formation that studies the atmosphere.[51] The A-Train constellation consists of five satellites flying in close proximity to each other. A sixth satellite, the Orbiting Carbon Observatory (OCO), was to have been added to the A-Train constellation in early 2009, but experienced a launch failure in February 2009.

[41] See http://www.oso.noaa.gov/poes/.

[42] See http://eospso.gsfc.nasa.gov/.

[43] See http://landsat.usgs.gov/.

[44] See http://topex-www.jpl.nasa.gov/mission/jason-1.html.

[45] See http://acrim.jpl.nasa.gov/.

[46] See http://oceancolor.gsfc.nasa.gov/SeaWiFS/.

[47] See http://www2.jpl.nasa.gov/srtm/.

[48] See http://trmm.gsfc.nasa.gov/.

[49] See http://www.goes-r.gov/.

[50] See http://www.jaxa.jp/projects/sat/gcom/index_e.html.

[51] See http://nasascience.nasa.gov/earth-science/a-train-satellite-constellation.

The first satellite in the A-Train constellation, Aqua, was launched in 2002; the second satellite, Aura, was launched in July 2004; and the CloudSat, CALIPSO, and PARASOL satellites were launched in April 2006. The A-Train satellites cross the equator within a few minutes of one another at around 1:30 p.m. local solar time. By combining the different sets of observations from the A-Train, a better understanding of atmospheric composition, clouds, and aerosols has led and is leading to major advances in atmospheric knowledge. More details on the five A-Train components follow.

NASA Aura—The NASA Aura satellite was launched with four instruments to extensively monitor the composition of the atmosphere.[52] Two of these instruments, the Microwave Limb Sounder and High-Resolution Dynamics Limb Sounder, obtain highly resolved altitude profiles of the stratosphere and upper troposphere for understanding photochemical and dynamical processes in these altitude ranges. The Tropospheric Emission Spectrometer obtains column and partial altitude profiles for ozone and tropospheric trace gases, while the Ozone Monitoring Instrument obtains nearly daily global ozone column maps, as well as columns for other important air quality parameters. Aura observes the atmosphere to answer the following three high-priority environmental questions: (1) Is the Earth's ozone layer recovering? (2) Is air quality getting worse? and (3) How is the Earth's climate changing?

PARASOL—A French Centre National d'Etudes Spatiales microsatellite project,[53] PARASOL has improved the characterization of cloud and aerosol microphysical and radiative properties. This advance has substantially increased our understanding of the radiative impact of clouds and aerosols, which in turn has led to improving numerical modeling of these processes in general circulation models.

CALIPSO and CloudSat—NASA's highly complementary CALIPSO[54] and CloudSat[55] satellites provide new, three-dimensional perspectives of how clouds and aerosols form, evolve, and affect weather and climate. Both satellites fly in formation as part of the NASA A-Train constellation, providing the benefits of near simultaneity and thus the opportunity for synergistic measurements made with complementary techniques.

NASA Aqua—The NASA Aqua satellite is designed to acquire precise atmospheric and oceanic measurements that provide a greater understanding of these components in the Earth's climate.[56] Other instruments on Aqua, such as the Moderate Resolution Imaging Spectroradiometer (MODIS) instrument, provide regional-to-global land cover, sea surface temperature, and ocean color. Data from the A-Train instruments will help answer these important questions: (1) What are the aerosol types, and how do satellite observations match global emission and transport models? (2) How do aerosols contribute to the Earth radiation budget, and to what extent are they a climate forcing? (3) How does cloud layering affect the Earth radiation budge? (4) What is the vertical distribution of cloud water/ice in cloud systems? and (5) What is the role of polar stratospheric clouds in ozone loss and denitrification of the Arctic vortex?

AIRS—Additional advances have been achieved by exploiting new thermal sounder measurements from the Atmospheric InfraRed Sounder (AIRS). Global retrievals from AIRS have proven valuable to understand the distribution as well as the transport mechanisms of CO, CH_4, and CO_2 in the middle troposphere with respective precisions of 10, 1.5, and 0.5 percent. NOAA has incorporated the lessons learned from AIRS into operational carbon products from the EUMETSAT Infrared Atmospheric Sounding Interferometer (IASI), which launched aboard the MetOp-A satellite in 2006. NOAA is planning to continue these products with the NPOESS Cross-track Infrared Sounder (CrIS). The currently scheduled IASI and CrIS missions will allow the creation of a 20-year record of satellite thermal sounder-derived carbon trace gases, along with self-consistent ozone, temperature, moisture, and cloud information.

Other Recent NASA Missions

Other recent NASA missions include the following:

ICESat—The Ice, Cloud, and Land Elevation Satellite, launched in 2003, has been measuring surface elevations of ice and land, vertical distributions of clouds and aerosols, vegetation-canopy heights, and other features with unprecedented accuracy and sensitivity.[57] The primary purpose of ICESat has been to acquire time series of ice-sheet elevation changes for determining the present-day mass balance of the ice sheets, to study associations between observed ice changes and polar climate, and to improve estimates of the present and future contributions of ice melt to global sea level rise.

SORCE—The Solar Radiation and Climate Experiment satellite, also launched in 2003, is equipped with four instruments that measure variations in solar radiation much more accurately than previous measurements and observe some of the spectral properties of solar radiation for the first time.[58]

GRACE—The Gravity Recovery and Climate Experiment (GRACE) twin satellites celebrated their seventh anniversary in orbit in March 2009, completing a successful primary mission that has provided improved ongoing estimates of the Earth's gravity field.[59] In conjunction with other data and models, GRACE has provided observations of terrestrial water storage changes, ice-mass variations, ocean-bottom pressure changes, and sea level variations.

[52] See http://aura.gsfc.nasa.gov/.

[53] See http://smsc.cnes.fr/PARASOL/.

[54] See http://www.nasa.gov/mission_pages/calipso/main/.

[55] See http://cloudsat.atmos.colostate.edu/.

[56] See http://aqua.nasa.gov/.

[57] See http://icesat.gsfc.nasa.gov/.

[58] See http://lasp.colorado.edu/sorce/index.htm.

[59] See http://www.csr.utexas.edu/grace/.

Planned Missions

NASA plans to launch the following four missions over the next several years:

Glory—2010 launch: Glory will measure black carbon soot and other aerosols, as well as total solar irradiance.

Aquarius—2010 launch: Aquarius will measure global sea surface salinity.

NPOESS Preparatory Project—2011 launch: This project will demonstrate advanced technology for atmospheric sounding, providing continuity of weather and climate observations following EOS-PM (Terra) and EOS-AM (Aqua). It will supply data on atmospheric and sea surface temperatures, humidity soundings, land and ocean biological productivity, and cloud and aerosol properties.

Global Precipitation Measurement—2013 launch: The Global Precipitation Measurement mission will study global precipitation (rain, snow, and ice).

Data Management

Data management is an important aspect of any systematic observing effort. U.S. agencies have unique mandates for climate-focused and -related systematic observations, and for the attendant data processing, archiving, and use of the important information from these observing systems.

Integrated Earth Observations

Cooperative efforts by USGCRP and USGEO agencies are moving toward providing integrated and more easily accessible Earth observations. Currently operating USGCRP systems for data management and distribution highlighted in the 2007 *Our Changing Planet* report include NASA's Global Change Master Directory and Earth Observing System Data and Information System, and the U.S. Department of Energy's (DOE's) Carbon Dioxide Information Analysis Center (USGCRP and SGCR 2006). DOE's Carbon Dioxide Information Analysis Center (CDIAC) provides comprehensive, long-term data management support, analysis, and information services to the global climate research community and the general public. The CDIAC data collection is designed to answer questions pertinent to both the present-day carbon budget and temporal changes in carbon sources and sinks. The data sets provide quantitative estimates of anthropogenic CO_2 emission rates, atmospheric concentration levels, land-atmosphere fluxes, ocean-atmosphere fluxes, and oceanic concentrations and inventories. The data holdings also support the NACP. NOAA's Climate Services Portal aggregates NOAA climate data, products, and services and serves as a touchstone for inquiries, interactive public dialogue, education, climate assessment information, and user requests.

NOAA's National Climatic Data Center's (NCDC's) Climate Data Online site provides climate data from multiple stations around the world. Plans for 2007 and 2008 included the IPY participation through a focus on polar climate observations via NCDC's World Data Center for Meteorology.[60] Finally, efforts are being explored to improve climate data integration in the Pacific Islands region and produce more useful, end-user-driven climate products.

PRICIP AND PaCIS

The Pacific Region Integrated Climatology Information Products (PRICIP) project[61] is an example of a region-wide collaborative activity under the auspices of the Pacific Climate Information System (PaCIS).[62] PRICIP is a regional path-finding activity with the goal of developing a national comprehensive coastal climatology program. It aims to improve our understanding of patterns and trends of storm frequency and intensity—"storminess"—within the Pacific region and develop a suite of integrated information products that can be used by emergency managers, mitigation planners, government agencies, and decision makers in key sectors, including water and natural resource management, agriculture and fisheries, transportation and communication, and recreation and tourism.

PRICIP is exploring how the climate-related processes that govern extreme storm events are expressed within and between three thematic areas: heavy rains, strong winds, and high seas. It involves analyses of historical records collected throughout the Pacific region, and the integration of these climatological analyses with near-real-time observations to put the current weather into a longer-term perspective. PaCIS provides a programmatic framework to integrate ongoing and future climate observations, operational forecasting services, and climate projections, research, assessment, data management, outreach, and education to address the needs of American Flag and U.S.-Affiliated Pacific Islands.

Integrated Data and Environmental Applications Center

NOAA's Integrated Data and Environmental Applications (IDEA) Center helps meet critical regional needs for ocean, climate, and ecosystem information to protect lives and property, support economic development, and enhance the resilience of Pacific Island communities in the face of changing environmental conditions.[63] This region-wide data integration activity (1) integrates regional observations, research, assessment, and services, and provides a prototype for a next-generation NOAA data center; (2) strengthens the delivery of ocean, climate, and ecosystem data products and information services to the diverse Pacific Island user community; (3) supports NOAA research and service programs in the Pacific; (4) supports the emergence of regional ocean- and climate-observing systems and information services that are responsive to the needs of Pacific Island com-

---

[60] See http://www.ncdc.noaa.gov/oa/wdc/.

[61] See http://www.pricip.org/.

[62] See http://www.ideademo.org/pacis.

[63] See http://www.ideademo.org/.

munities, governments, and businesses via the evolving PaCIS program; (5) supports integrated ecosystem science and services needed for Pacific Island ocean and coastal resource management programs; and (6) continues NOAA/U.S. leadership in the emergence of a global environmental observing system (e.g., GCOS, GOOS, IOOS, and GEOSS).

National Integrated Drought Information System

Droughts have far-reaching impacts on many aspects of our daily lives, from water management to health to energy consumption and conservation. To mitigate these impacts, NOAA, other federal and state agencies, partners, and countries developed the plan for the National Integrated Drought Information System (NIDIS). NIDIS is a dynamic and accessible drought-risk information system that was created in response to extended drought conditions, especially in the western United States, over the past decade.

In 2007, the United States unveiled a new, interactive Web site called the U.S. Drought Portal (USDP) that allows the public and civic managers to monitor U.S. drought conditions, get forecasts, assess the impacts of drought on their communities, and learn about possible mitigation measures.[64] This Web site is useful internationally as nations work to coordinate drought preparedness, response, mitigation, and recovery activities, and it fits in well with drought-related bilateral activities the United States is engaged in with partners in Canada and Mexico.

In 2008, NOAA, along with its partners, including the U.S. Department of Agriculture (USDA), began to institute geographic information system (GIS) mapping capabilities into the USDP. In 2009, the program will work to integrate enhanced GIS capabilities into the USDP. Additionally, communities will be unveiled in the portal, serving as a location for subject matter experts to share improvements in drought monitoring, forecasting, and mitigation. These communities will also serve as a coordinating and communications mechanism for NIDIS regional pilot projects. NIDIS was featured at the 2007 GEO-IV Plenary Session as a major contribution to GEOSS.

State of the Climate Report

Produced in partnership with WMO and numerous national and international partners, the annual *State of the Climate Report–Using Earth Observations to Monitor the Global Climate*,[65] consists of operational monitoring, analysis, and reporting on atmosphere, ocean, and land surface conditions from the global to local scale. By combining historical data with current observations, this report places today's climate in historical context and provides perspectives on the extent to which the climate continues to vary and change, as well as the effect that climate is having on societies and the environment.

More than 150 scientists from over 30 countries are now part of an annual process of turning raw observations collected from the global array of observing systems into information that enhances the ability of decision makers to understand the state of the Earth's climate and its variation and change during the past year, with context provided by decades to centuries of climate information. Many observational and analytical systems are unique to countries or regions of the world. Nevertheless, through this effort, the information from each system is openly shared, which is essential to transitioning data to operational use and filling critical gaps in current knowledge about the state of the global climate system. A State of the Climate Report is distributed through publication in the *Bulletin of the American Meteorological Society* each year,[66] and is translated into other languages and distributed to all 187 WMO member nations. The report seeks to provide details on as many of the essential climate variables (ECVs) as possible, as identified in the GCOS Second Adequacy Report.[67] Since this report began monitoring ECVs in 2001, and in line with the recently published 2008 edition, the number of reported ECVs has more than doubled to nearly 25.

Earth Observing System Data and Information System

NASA's Earth Observing System Data and Information System (EOSDIS) provides convenient mechanisms for locating and accessing products of interest either electronically or via orders for data on media. EOSDIS facilitates collaborative science by providing sets of tools and capabilities, such that investigators may provide access to special products (or research products) from their own computing facilities. EOSDIS has an operational EOS Data Gateway (EDG) that provides access to the data holdings at all the Distributed Active Archive Centers (DAACs) and participating data centers from other U.S. and international agencies. Currently, there are 14 EDGs around the world that permit users to access Earth science data archives, browse data holdings, select data products, and place data orders.

Eight NASA DAACs, representing a wide range of Earth science disciplines, comprise the data archival and distribution functions of EOSDIS. The DAACs carry out the responsibilities for processing certain data products from instrument data, archiving and distributing NASA's Earth science data, and providing a full range of user support. More than 2,100 distinct data products are archived at and distributed from the DAACs. These institutions are custodians of Earth science mission data until the data are moved to long-term archives. They ensure that data will be easily accessible to users.

NASA and NOAA have initiated a pilot project to develop a prototype system for testing candidate ap-

[64] See http://drought.gov.

[65] See http://www.noaa.gov/features/climate/climatemonitoring2.html.

[66] See http://www.ametsoc.org/PUBS/bams/.

[67] An archive of these reports from 2000 to 2007 can be found at http://www.ncdc.noaa.gov/oa/climate/research/state-of-climate.

proaches for moving MODIS data into long-term NOAA archives. This pilot project is part of the evolution of NOAA's Comprehensive Large Array-data Stewardship System (CLASS). Acting in concert with their users, DAACs provide reliable, robust services to those whose needs may cross traditional discipline boundaries, while continuing to support the particular needs of their respective discipline communities. The DAACs are currently serving a broad and growing user community at an increasing rate. CLASS is NOAA's online facility for the distribution of NOAA and U.S. Department of Defense (DOD) POES data, NASA mission data, NOAA GOES data, and derived data. CLASS is an electronic library of NOAA environmental data that provides capabilities for finding and obtaining such satellite data.[68]

Global Observing System Information Center

The transition of the Global Observing System Information Center (GOSIC) from a developmental activity at the University of Delaware to an operational global data facility at NOAA's NCDC was completed on behalf of and with the concurrence of the global observing community in October 2006. GOSIC provides information; facilitates easier access to data and information produced by GCOS, GOOS, and GTOS and their partner programs; provides explanations of the various global data systems, as well as an integrated overview of the myriad global observing programs, which includes online access to their data, information, and services; and offers a search capability across international data centers, to enhance access to a worldwide set of observations and derived products.[69]

TECHNOLOGY FOR GLOBAL CHANGE

The United States has committed not only to improving the science to better understand global climate change, but also to promoting the development and deployment of technologies to reduce GHG emissions. These efforts are targeted at increasing energy end-use efficiency and supplying energy with greatly reduced GHG emissions to meet the nation's goals of reducing GHG emissions and stabilizing GHG atmospheric concentrations at a level that avoids dangerous human interference with the climate system. To address these challenges, the Obama administration and Congress are working together to spur a revolution in clean energy technologies.

U.S. Climate Change Technology Program

The U.S. Climate Change Technology Program (CCTP) was established administratively in 2002, and authorized by the Energy Policy Act of 2005.[70] CCTP developed its August 2005 *Vision and Framework for Strategy and Planning* (CCTP 2005) and September 2006 *Strategic Plan* (CCTP 2006) to guide and prioritize the federal government's climate technology efforts.

CCTP's strategic vision has six complementary goals: (1) reducing emissions from energy end use and infrastructure, (2) reducing emissions from energy supply, (3) capturing and sequestering CO_2, (4) reducing emissions of other GHGs, (5) measuring and monitoring emissions, and (6) bolstering the contributions of basic science. DOE serves as the lead agency for the CCTP effort. Twelve agencies participate in the interagency coordination efforts of CCTP. Eight of these fund activities are included in the CCTP portfolio.

In FY 2009, approximately $5.2 billion was appropriated for CCTP activities. ARRA provided over $25 billion in additional funding for CCTP research and development (R&D) activities across a broad portfolio of GHG mitigation options, including high-performance buildings; efficient manufacturing; advanced vehicles; clean biofuels; wind, solar, geothermal, and nuclear power; carbon capture and sequestration; advanced energy storage; a more intelligent electric grid; and techniques for reducing emissions and/or increasing uptake of CO_2 in agriculture and forestry. ARRA also provided $400 million for establishing the Advanced Research and Projects Agency–Energy (ARPA-E) within DOE to overcome the long-term and high-risk technological barriers to the development of clean energy technologies.[71]

Energy End Use and Infrastructure

Major sources of GHGs are closely tied to the use of energy in transportation, residential and commercial buildings, and industrial processes. Improving energy efficiency and reducing the intensity of GHG emissions in these sectors can significantly reduce overall GHG emissions. In addition, improving the infrastructure of the electricity transmission and distribution grid can reduce GHG emissions by making power generation more efficient and by providing expanded use and grid access of low-emission electricity from renewable energy technologies, including wind, solar, and geothermal power.

Key research activities include DOE's nationwide plan to modernize the electric grid, enhance the security of the U.S. energy infrastructure, and ensure reliable electricity delivery to meet growing demand. The R&D program is focused on technologies that reduce GHG emissions and contribute to energy independence and economic growth by improving the reliability, efficiency, flexibility, functionality, and security of the nation's electricity delivery system. The emphasis is on development of advanced transmission technologies, including more efficient cables and conductors to reduce energy loss; strengthening the reliability of the electric grid by enhancing real-time visualization tools; and developing a "smart grid" system with enhanced intelligence and connectivity.

Several U.S. agencies (DOE, Department of Transportation [DOT], DOD, Environmental Protection Agen-

[68] See http://www.nsof.class.noaa.gov/saa/products/welcome.

[69] See http://gosic.org.

[70] See http://www.climatetechnology.gov/.

[71] See http://arpa-e.energy.gov/.

cy [EPA], and NASA) are working on cost-effective automotive technologies that increase fuel efficiency and produce ultra-low pollution and GHG emissions. Under the Clean Automotive Technology Program, EPA facilitates collaboration with the automotive industry through innovative research to achieve ultra-low-pollution emissions, increase fuel efficiency, and reduce GHGs. By developing cost-effective technologies, the program encourages manufacturers to produce cleaner and more fuel-efficient vehicles. The DOT Federal Transit Administration's National Fuel Cell Bus Program develops and demonstrates fuel cell transit bus technology.[72] In addition, DOE's Vehicle Technologies Program supports R&D to make vehicles more efficient and capable of operating on non-petroleum fuels.[73]

Other DOT programs include efforts to improve travel activity, reduce vehicle miles traveled, and enhance vehicle and system operations. Aviation yields GHG emissions that have the potential to influence global climate. To identify opportunities for GHG emission reductions in the aviation sector, DOT's Federal Aviation Administration (FAA) recently launched the Aviation Climate Change Research Initiative. Currently, measuring and tracking fuel efficiency from aircraft operations provide the data for assessing the improvements in aircraft and engine technology, operational procedures, and the airspace transportation system that reduce aviation's contribution to CO_2 emissions. The FAA's Commercial Aviation Alternative Fuels Initiative is a government–private-sector coalition that focuses the efforts of commercial aviation to engage the emerging alternative fuels industry.[74] With support from NASA, the FAA recently launched the Continuous Lower Energy Emissions and Noise Program to advance maturing engine and aircraft technologies for quick fusion into the fleet in order to achieve increases in fuel efficiency (which is directly related to CO_2 emissions) and reduction in nitrogen oxide emissions (which affects distributions of ozone and methane—both of which are GHGs). These strategies to improve the transportation system can reduce GHG emissions, lead to environmental benefits, reduce oil use, improve America's energy security, and benefit the economy.

Reducing energy consumption and transforming the carbon footprint of the built environment through the development of technologies that will enable cost-competitive, zero-energy buildings, and supporting the advancement of clean and efficient industrial technologies and processes are other areas of research that could yield significant emission reductions both domestically and globally. At the July 9, 2009, Group of Eight (G8) meetings in L'Aquila, Italy, the Major Economies Forum countries (G8 + China, India, South Africa, Brazil, Mexico, and Indonesia) announced a Global Partnership to drive transforma-

tional low-carbon, climate-friendly technologies. A commitment was made to dramatically increase and coordinate public-sector investments in research, development, and demonstration (RD&D) of these technologies, with a view to doubling such investments by 2015, while recognizing the importance of private investment, public-private partnerships and international cooperation, including regional innovation centers. The United States will lead on "efficiency," which includes both buildings and industrial sector efficiency. Technology Action Plans and roadmaps will be developed along with recommendations for further progress. Drawing on global best practice policies, the Global Partnership will undertake to remove barriers, establish incentives, enhance capacity building, and implement appropriate measures to aggressively accelerate deployment and transfer of key existing and new low-carbon technologies, in accordance with national circumstances.

Energy Supply

Global and domestic energy supplies are dominated by fossil fuels that emit CO_2 when burned. The transition to a low-carbon energy future will require the availability of cost-competitive low- or zero-carbon energy supply technologies.

Renewable energy includes a range of different technologies that can play an important role in reducing GHG emissions. The United States currently invests considerable resources in wind, wave, tidal, hydropower, solar photovoltaics, and biomass technologies. In FY 2009, CCTP-related investments in renewable energy technologies included a combined $800 million. For example, DOE is helping meet America's increasing energy needs by working with wind industry partners to develop clean, domestic, innovative wind energy technologies that can compete with conventional fuel sources. DOE's Wind and Hydropower Technologies Program efforts have culminated in some of industry's leading products today and have contributed to record-breaking industry growth.[75] DOE's Biomass Program also conducts R&D in four key areas of technology required to produce biomass feedstocks and convert them to useful biofuels and value-added products: feedstocks, processing and conversion, integrated biorefineries, and infrastructure.[76]

USDA's Biomass Research and Development Initiative addresses feedstock development, biofuels and bio-based product development, and biofuel development analysis.[77] All projects are implemented in accordance with a life-cycle perspective that considers both direct and indirect environmental and economic impacts. USDA's Rural Development program provides (1) loan guarantees for the development, construction, and retrofitting of commercial-scale biorefineries; (2) grants to help pay for the development and construction costs of demonstration-scale biore-

[72] See http://www.nrel.gov/hydrogen/proj_fc_bus_eval.html.

[73] See http://www1.eere.energy.gov/vehiclesandfuels/.

[74] See http://www.caafi.org/.

[75] See http://www1.eere.energy.gov/windandhydro/.

[76] See http://www1.eere.energy.gov/biomass/.

[77] See http://www.brdisolutions.com/default.aspx.

fineries; (3) payments to biorefineries to replace fossil fuels used to produce heat or power with renewable biomass; and (4) loan guarantees to rural residents, agricultural producers, and rural businesses for energy efficiency and renewable energy systems, energy audits, and technical assistance for projects ranging from biofuels to wind, solar, geothermal, methane gas recovery, advanced hydro, and biomass.[78]

Advanced fossil -based power and fuels are areas of particular interest for the United States. With coal likely to remain one of the nation's most widely used energy resources for the foreseeable future, the United States is actively funding applied R&D of advanced coal technologies that improve efficiency and reduce the intensity of CO_2 emissions. These activities are conducted through such programs as the Clean Coal Power Initiative, a cost-shared partnership between the government and industry to develop and demonstrate advanced coal-based power generation technologies.[79]

Concerns about resource availability, energy security, air quality, and climate change suggest a larger role for nuclear power as an energy supply choice. A key mission of DOE's nuclear energy R&D program is to plan and conduct applied research in advanced reactor and fuel and waste management technologies. The aim of these efforts is to enable nuclear energy to be used as a safe, advanced, cost-effective source of reliable energy that will help address climate change by reducing GHG emissions. The Generation IV Nuclear Energy Systems program is investigating the next-generation reactor and fuel-cycle systems, which represent a significant leap in economic performance, safety, and proliferation resistance.[80] Fusion energy is a potential major new source of energy that, if successfully developed, could be used to produce electricity and possibly hydrogen. Fusion has features that make it an attractive option from both environmental and safety perspectives. However, the technical hurdles of fusion energy are very high, and with a commercialization objective of 2050, its impact will not be felt until the second half of the century.

Carbon Capture and Sequestration

Carbon capture and sequestration (CCS) is a central element of CCTP's strategy, because for the foreseeable future, fossil fuels will continue to be the world's most widely used forms of energy. Global energy models suggest that with current global coal use patterns, it will not be possible to stabilize atmospheric GHG concentrations at acceptable levels. Thus, a realistic approach is to find ways to "sequester" the CO_2 produced when fossil fuels—especially coal—are used. The term *carbon sequestration* describes a number of technologies and methods to capture, transport, and store CO_2 or remove it from the atmosphere. These include capturing carbon (or CO_2), geologic storage of CO_2, and terrestrial sequestration in natural environs.

Advanced techniques to capture gaseous CO_2 from energy and industrial facilities and store it permanently in geologic formations are under development. In 2008, the G8 nations called for advancing CCS internationally, resulting in 20 major demonstrations by 2020; the United States agreed to sponsor at least 10 of these. Central to these U.S. demonstrations is DOE's core Carbon Sequestration Program, which emphasizes technologies that capture CO_2 from large point sources and store the emissions in geologic formations capable of holding vast amounts of CO_2. The Carbon Sequestration Program is complemented by other DOE programs that seek to significantly reduce the overall cost of integrated plants that will produce electricity and other co-products while capturing and sequestering CO_2.[81] The focus is on CO_2 capture from both new and existing coal plants.

In 2003, DOE launched a nationwide network of seven Regional Carbon Sequestration Partnerships that include 43 U.S. states, four Canadian provinces, three Native American nations, and over 350 organizations.[82] The partnerships' main focus is on determining the best approaches for sequestration in their regions and taking the initial steps to develop the infrastructure that will be needed for eventual large-scale deployment. This includes examination of regulatory needs. Small-scale validation testing of 35 sites involving terrestrial and geologic sequestration technologies began in 2005. During 2009–2012, CO_2 injection will begin for nine large-scale geologic storage tests that will be carried out by the seven regional partnerships. These tests will be of sufficient scale to allow the partnerships to address the kinds of challenges that will be encountered for commercial projects.

Terrestrial sequestration—removing CO_2 from the atmosphere and sequestering it in trees, soils, or other organic materials—has proven to be a low-cost means for long-term carbon storage. The DOE-supported Carbon Sequestration in Terrestrial Ecosystems consortium provides research on mechanisms that can enhance terrestrial sequestration. In addition, USDA's Agricultural Research Service operates the Greenhouse Gas Reduction through Agricultural Carbon Enhancement Network (GRACEnet) at 31 locations around the country to measure and predict carbon sequestration and GHG emissions across a range of agricultural systems, land and animal management practices, soils, and climate zones.[83] Elements of GRACEnet include the development and use of standardized measurement methods, process model development, data base development, and the development of guidelines for producers.

Other Greenhouse Gases

A main component of the U.S. strategy is to reduce other GHGs, such as CH_4, nitrous oxide (N_2O), sulfur hexafluoride (SF_6), and fluorocarbons.

[78] See http://www.rurdev.usda.gov/.

[79] See http://www.netl.doec.gov

[80] See http://www.ne.doe.gov/GenIV/neGenIV1.html.

[81] See http://www.netl.doe.gov/technologies/carbon_seq/.

[82] See http://fossil.energy.gov/sequestration/partnerships/index.html.

[83] See http://www.ars.usda.gov/research/programs/programs.htm?np_code=204&docid=17271.

Improvements in methods and technologies to detect and either collect or prevent CH_4 emissions from various sources—such as landfills, coal mines, natural gas pipelines, and oil and gas exploration operations—can prevent this GHG from escaping to the atmosphere. Reducing CH_4 emissions may also have a positive benefit in reducing local ozone problems, as CH_4 is a long-lived ozone precursor. In agriculture, improved management practices for fertilizer applications and livestock waste can reduce CH_4 and N_2O emissions appreciably.

Hydrofluorocarbons (HFCs), perfluorocarbons (PFCs), and SF_6 are all high global warming potential (GWP) gases. HFCs and PFCs are used as substitutes for ozone-depleting chlorofluorocarbons and are used in or emitted during complex manufacturing processes. Advanced methods to reuse, recycle, and reduce the leakage of these chemicals and to use lower GWP alternatives are being explored.

Programs aimed at reducing particulate matter have led to significant advances in fuel combustion and emission control technologies to reduce U.S. black carbon aerosol emissions. Reducing emissions of black carbon, soot, and other chemical aerosols can have multiple benefits, including better air quality and public health and, in some cases, can reduce radiative forcing.

Measuring and Monitoring

To meet future GHG emission measurement requirements, a wide array of sensors, measuring platforms, monitoring and inventorying systems, and inference methods are being developed. Many of the baseline measurement, observation, and sensing systems used to advance climate change science are being developed as part of CCSP. CCTP's efforts focus primarily on validating the performance of various climate change technologies, such as in terrestrial and geologic sequestration.

The U.S. Department of Commerce's National Institute of Standards and Technology is testing, developing, and making available to researchers a wide variety of measurement and monitoring tools and techniques to aid in the development of technologies to mitigate climate change.

Basic Science

Basic scientific research is a fundamental element of CCTP. Tackling the dual challenges of addressing climate change and meeting growing world energy demand is likely to require discoveries and innovations that can shape the future in often unexpected ways. The CCTP framework aims to strengthen the basic research enterprise through strategic research that supports ongoing or projected research activities and exploratory research involving innovative concepts. President Obama has committed to doubling federal investment in the basic sciences.

DOE will continue to support the 46 Energy Frontier Research Centers (EFRCs) that are addressing current fundamental scientific roadblocks to clean energy and energy security.[84] These centers will address the full range of energy research challenges in renewable and low-carbon energy, energy efficiency, energy storage, and cross-cutting science. The EFRCs will take advantage of new capabilities in nanotechnology, light sources that are a million times brighter than the sun, supercomputers, and other advanced instrumentation.

DOE's multidisciplinary Energy Innovation Hubs will also address basic science, technology, and economic and policy issues. The hubs will support cross-disciplinary R&D focused on the barriers to transforming energy technologies into commercially deployable materials, devices, and systems. They will advance promising areas of energy science and technology from their early stages of research to the point where the risk level will be low enough for industry to deploy them into the marketplace.

Established by DOE in 2009, ARPA-E is modeled after the Defense Advanced Research Projects Agency, which was created during the Eisenhower administration in response to the Russian Sputnik program, which launched the world's first artificial satellite. The purpose of ARPA-E is to advance high-risk energy research projects that can yield revolutionary changes in how energy is produced, distributed, and used.[85] ARRA has provided $400 million for ARPA-E.

Multilateral Research and Collaboration

The United States believes that well-designed multilateral collaborations focused on achieving practical results can accelerate development and commercialization of new technologies. Thus, the United States has initiated or joined a number of multilateral technology collaborations in hydrogen, carbon sequestration, nuclear energy, and fusion that address many energy-related concerns (e.g., energy security, climate change, and environmental protection). The following initiatives are examples of U.S. multinational collaboration.

Carbon Sequestration Leadership Forum

The Carbon Sequestration Leadership Forum (CSLF) is a multilateral U.S. initiative that provides a framework for international collaboration on sequestration technologies.[86] Established at a June 2003 ministerial meeting held in Washington, D.C., CSLF consists of 23 members, including 22 national governments representing both developed and developing countries, as well as the European Commission. The CSLF's main focus is assisting the development of technologies to separate, capture, transport, and store CO_2 safely over the long term; making carbon sequestration technologies broadly available internationally; and addressing broader issues relating to carbon capture and storage, such as regulation and policy. To date, CSLF has endorsed 20 international research projects, five of which involve the United States.

[84] See http://www.science.doe.gov/bes/EFRC.html.

[85] See http://arpa-e.energy.gov/.

[86] See http://www.cslforum.org/.

Generation IV International Forum

The Generation IV International Forum (GIF) is a multilateral partnership of 10 countries and the European Commission that is fostering international cooperation in R&D for the next generation of safer, more affordable, and more proliferation-resistant nuclear energy systems.[87] This new generation of nuclear power plants could produce electricity and hydrogen with substantially less waste and without emitting any air pollutants or GHG emissions. Since its creation in July 2001, GIF has established a legal basis for collaboration through a treaty-level Framework Agreement (2005) and implementing arrangements (2007 onward). The United States supports collaboration in two of the systems: the Very-High-Temperature Reactor (VHTR) and the Sodium Fast Reactor (SFR). The primary mission of the VHTR is to provide carbon-free process heat for cogeneration and many potential industrial applications, including hydrogen production, while the SFR's primary mission is the closing of the nuclear fuel cycle.

ITER

In January 2003, President Bush announced that the United States was joining the negotiations for the construction and operation of the international fusion experiment ITER.[88] The goal of this collaborative project is to demonstrate the scientific and technological feasibility of fusion as an energy source. If successful, ITER will advance progress toward producing clean, abundant, commercially available fusion energy by the end of the century. Toward this goal, the seven ITER partners signed an agreement in November 2006 to construct the project; site preparation began in Saint-Paul-lez-Durance, France, in January 2007; and construction began in 2009.

Asia-Pacific Partnership on Clean Development and Climate

The Asia-Pacific Partnership on Clean Development and Climate (APP) is an innovative effort to accelerate the development and deployment of clean energy technologies.[89] The seven APP partner countries (Australia, Canada, China, India, Japan, Korea, and the United States) collectively account for more than half of the world's economy, population, and energy use. They produce about 65 percent of the world's coal, 62 percent of the world's cement, 52 percent of world's aluminum, and more than 60 percent of the world's steel. They have committed to collaborate and work with the private-sector partners to meet goals for energy security, national air pollution reduction, and climate change in ways that promote sustainable economic growth and poverty reduction. The APP focuses on expanding investment and trade in cleaner energy technologies, goods, and services in key market sectors. The partners have approved eight public–private-sector task forces covering aluminum, buildings and appliances, cement, coal mining, power generation and transmission, renewable energy and distributed generation, steel, and cleaner fossil energy technologies.

---

[87] See http://www.gen-4.org/.

[88] See http://www.iter.org/.

[89] See http://www.asiapacificpartnership.org/.

9

Education, Training, and Outreach

ederal agencies' climate change education, training, and outreach efforts seek to ensure that individuals and communities understand the essential principles of Earth's climate system and the impacts of climate change, and are able to evaluate and make informed and responsible decisions with regard to actions that may affect the climate. Increasing our resilience to these impacts depends not only upon our ability to understand climate science and the implications of climate change, but also upon our ability to integrate and use that knowledge effectively. Changes in our economy and infrastructure as well as changes in individual attitudes, societal norms, and government policies will be required to alter the impact of climate change on human lives. Individuals, communities, and countries will be called on to implement

effective management strategies for critical institutional and natural resources to ensure the stability of both human and natural systems as temperatures rise.

As nations and the international community seek solutions to global climate change over the coming decades, a more comprehensive, interdisciplinary approach to fostering public climate literacy—one that includes economic and social considerations—will play a vital role in knowledgeable planning, decision making, and governance. Increased efforts to integrate social sciences into federal agencies' educational and outreach programs would help to ensure informed decision making and effective systems-level responses to climate change. This integration offers a significant challenge to the nation's diverse educational systems. It is imperative that these responses to climate change

are supported by sustained and robust educational initiatives to develop a climate-literate citizenry and skilled workforce.

UPDATES SINCE THE 2006 *U.S. CLIMATE ACTION REPORT*

Climate change education, training, and outreach efforts have expanded significantly since the last *U.S. Climate Action Report* was released in 2006. Since then, federal programs to support formal educational initiatives on climate change have expanded considerably. These programs involve K–12 and ungraduate curricula and postgraduate professional development programs, as well as informal education programs conducted in museums, parks, nature centers, zoos, and aquariums across the country.

Federal Programmatic Coordination

Increased coordination across the federal agencies is critical to increasing public climate literacy in the United States. A Climate Change Education Interagency Working Group was formed in 2008 to coordinate an integrated national approach to climate change education. This coordination mechanism is an outgrowth of discussions within the U.S. Global Change Research Program (USGCRP) Communications Interagency Working Group (CIWG) and the federal science agencies currently conducting climate research, and climate change education and outreach. This Interagency Working Group consists of senior staff members of the USGCRP member agencies that have climate and climate change education programs. Their primary responsibilities are to:

- serve as a forum in the USGCRP for the development of a national climate and climate change education strategy that is inclusive of all USGCRP members,

- coordinate climate education activities and priorities across the USGCRP members, and

- make recommendations to agency management on all aspects of federal agencies' climate and climate change educational activities.

It is critical that climate scientists and climate science agencies play a more active role in the dissemination of their findings. The public and students at all levels—in both formal and informal learning settings—must have access to climate data in ways that foster climate literacy and informed decision making. The federal agencies are working with social scientists to determine the most effective ways to communicate with students and the public about how Earth's climate is changing. In an effort to extend their education and outreach programs and maximize their impact, federal agencies are addressing the following questions: How can local high-impact activities be scaled up and serve as national models? What are effective climate change literacy professional development opportunities for policy decision makers at all levels? How do we assess changes in individuals' understanding of Earth's climate system and the decisions they make about their actions? How can nationally representative assessments of public knowledge and understanding of climate change help identify common knowledge gaps, misunderstandings, sources of confusion, and key concepts the American public needs to understand about climate change?

Table 9-1 at the end of this chapter presents an extensive listing of federal agencies' online, climate-relevant education resources.

OVERVIEW OF NATIONAL EFFORTS TO ENGAGE THE UNITED STATES ON CLIMATE CHANGE

During the period of this *U.S. Climate Action Report* (2006–2010), nongovernmental organizations (NGOs) and the federal government conducted major communications campaigns to raise awareness and educate the nation and the world about global warming. On May 24, 2006, the popular film *An Inconvenient Truth* was released, featuring former U.S. Vice President Al Gore's efforts to educate citizens about global warming. Many colleges and high schools use the film to complement their science curricula.

Roughly one year later, Vice President Gore and other celebrities conducted "Live Earth"—a 24-hour music concert spanning seven continents and delivering a worldwide call to action to address climate change. During the performances, viewers were invited to support a seven-point pledge to adopt solutions to address climate change. Subsequently, "Live Earth" launched a multi-year campaign to encourage individuals, corporations, and governments to take action.

The Intergovernmental Panel on Climate Change (IPCC) Fourth Assessment on Climate Change was published in a series of reports in 2007. The first report issued by Working Group 1, titled *The Physical Science Basis*, was published on February 2, 2007. Later in 2007, the second and third IPCC reports were released and then completed by a summary *Synthesis Report* released on November 17, 2007. Collectively, the IPCC's fourth climate assessment updated, clarified, and strengthened societal understanding of the causes and effects of climate change. The reports also bolstered climate science lessons used in classrooms and in public outreach efforts. Later in 2007, former Vice President Gore and the IPCC shared the 2007 Nobel Peace Prize.

Perhaps not coincidentally, news media coverage of climate change began to change in 2007, with the number of articles on the subject increasing substantially (Figure 9-1).[1] Recent studies on the role of mass

[1] See http://www.eci.ox.ac.uk/ research/climate/mediacoverage. php.

media in communicating climate science, mitigation, and adaptation have been mixed or more positive (Boykoff and Boykoff 2007).

In March 2008, a group of 11 organizations led by the National Oceanic and Atmospheric Administration (NOAA) produced a publication titled *Climate Literacy: The Essential Principles of Climate Science*. One year later, the USGCRP issued a revised and expanded second version of this publication, with endorsement by its 13 federal agency members and an expanded list of partners that includes 24 institutions and organizations (available online at http://www.globalchange.gov/).

Numerous reports and studies by various social scientists documented increasing public awareness and concern regarding climate change and the need for individuals and governments to take action to address the problem. The 2008 U.S. presidential election debates and campaign speeches placed an unprecedented emphasis on the importance of addressing climate change. The U.S. Congress began new climate change education-focused programs at three federal agencies—the National Aeronautics and Space Administration (NASA), the U.S. National Science Foundation (NSF), and NOAA—while other agencies launched major education and outreach programs to help promote public climate literacy.

Efforts by state and local governments, universities, schools, and NGOs are essential complements to federal programs that educate industry and the public regarding climate change. State environment and energy agencies continue to provide teacher training, often in cooperation with universities and local utility companies. Local school systems are adopting climate change curricula and activities at the middle and high school levels. Universities are joining forces with NGOs to educate staff and students about the importance of energy efficiency and are instituting new, sustainable practices on campuses across the country. From wildlife conservation groups (e.g., National Wildlife Federation, National Council for Science and the Environment, National Environmental Education Foundation, and Council of Environmental Deans and Directors), to science-based organizations (e.g., American Meteorological Society, University Corporation for Atmospheric Research, and Federation of Earth Science Information Partners), to education organizations (e.g., American Association for the Advancement of Science Project 2061, Association of Science-Technology Centers, and National Science Teachers Association), a variety of NGOs conduct programs and surveys, produce brochures and kits, and write media articles to alert the public to the science underlying, impacts of, and possible solutions to climate change.

Industry also plays a role in education, training, and outreach. Several corporations have contributed to the

Figure 9-1 **2004–2009 World Newspaper Coverage of Climate Change or Global Warming**

In 2007, news media coverage of climate change increased substantially. Recent studies on the role of mass media in communicating climate science, mitigation, and adaptation have been mixed or more positive.

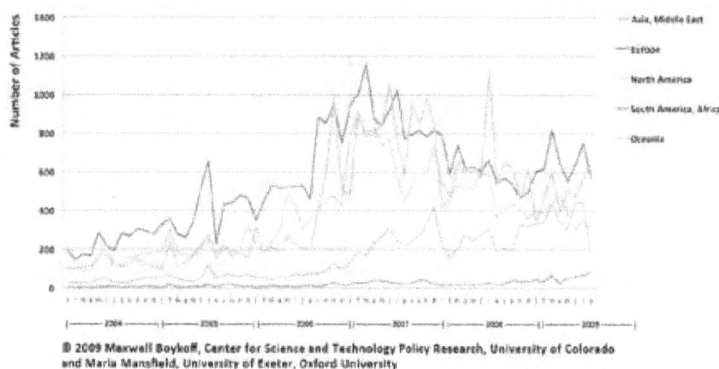

© 2009 Maxwell Boykoff, Center for Science and Technology Policy Research, University of Colorado and Maria Mansfield, University of Exeter, Oxford University

National Park Service's efforts to communicate energy efficiency messages.

Because of these efforts, segments of the American public are better informed about climate change and better equipped to act on that information. The USGCRP *2009 Global Climate Change Impacts in the United States* report states that "future climate change and its impacts depend on choices made today" (Karl et al. 2009). As such, the nation has a considerable way to go in comprehending and realizing the implications of climate change. The *Climate Literacy* publication captures this concept well in its Guiding Principle B:

> *Reducing human vulnerability to the impacts of climate change depends not only upon our ability to understand climate science, but also upon our ability to integrate that knowledge into human society. Decisions that involve Earth's climate must be made with an understanding of the complex inter-connections among the physical and biological components of the Earth system as well as the consequences of such decisions on social, economic, and cultural systems.*

FEDERAL AGENCY EDUCATION, TRAINING, AND OUTREACH PROGRAM OVERVIEWS

A number of federal agencies provide state and local governments, industry, NGOs, and the public with information about national and global climate change research and risk assessment studies, U.S. mitigation activities, and policy development. They work both independently and in partnership with other agencies, NGOs, and industry toward the common goal of increasing awareness about the potential environmental and societal challenges posed by climate change.

U.S. Global Change Research Program
USGCRP is responsible for communicating with a variety of stakeholders nationally and globally on

issues related to climate variability and climate change science and coordinating the federal agencies' climate change education programs. The CIWG leads efforts to coordinate interagency education and communications activities.

U.S. Department of Commerce

National Oceanic and Atmospheric Administration

NOAA is committed to the development of a society that is environmentally responsible and utilizes effective, science-based problem-solving skills. Improvements in societal stewardship of natural resources extend directly from effective formal and informal education systems. NOAA's climate education programs support the development of strong and comprehensive education materials about climate and oceanic and atmospheric sciences. NOAA is committed to supporting and facilitating system-wide change of the formal education system to build educators' capacity to produce climate-literate citizens. Such change requires engagement and participation across the spectrum of the education community—including policymakers, academic institutions, professional associations, teachers, and students.

Informal education plays a critical role in developing climate-literate citizens. To help equip informal education institutions with modern instructional resources and interdisciplinary methods for teaching Earth system science, NOAA pioneered Science On a Sphere® (SOS). The animated global data sets projected onto this 6-foot-diameter sphere, with live or pre-recorded interpretive narration, show members of the lay public cause-and-effect relationships at work in the global climate system. SOS® is now operating in more than 40 informal education venues across the United States and internationally, and more locations are being planned.

NOAA is engaged in the improvement of both formal and informal education systems because these venues are important to the development of literate citizens and to the long-term maintenance of their skills, knowledge, and attitudes. Partnerships and collaboration are integral to sustaining and scaling up NOAA's ability to promote public climate literacy.

U.S. Department of Energy

The U.S. Department of Energy's (DOE's) Atmospheric Radiation Measurement (ARM) Climate Research Facility Education and Outreach Program is involved in climate change educational outreach in the communities and regions hosting ARM data-gathering field sites. The program also provides educational resources to a global audience through its Web site.[2] The goal of the program is to develop basic science awareness and increase critical thinking skills focusing on environmental science and climate change for K–12 students. In addition, the program supports relationship build-

ing among teachers, students, scientists, and the community. Lesson plans, puzzles, and related materials are made available at the Web site. The site also includes the "Ask a Scientist" interface that provides the opportunity for anyone to pose questions to ARM scientists. Questions and answers are posted on the Web site.

Additionally, DOE's Office of Energy Efficiency and Renewable Energy (EERE) funds partners to develop curricula and implement standardized, high-quality training programs. These projects are aimed at creating a pipeline beginning at the K–12 level and extending through the postgraduate level to ensure the ongoing development of a workforce to invent and scale up clean energy and energy efficiency technologies and processes over the long term. Education and workforce training are critical parts of EERE's mission, which is to create an energy-literate generation of skilled workers, leaders, and innovators who will produce affordable, abundant, and clean energy, thus accelerating the transition to a low-carbon economy and ensuring U.S. global competitiveness.

U.S. Department of the Interior

National Park Service

As the steward of the world's foremost system of national parks, the fundamental mission of the National Park Service (NPS), as articulated in the 1916 NPS Organic Act is: "to conserve the scenery and the natural and historic objects and the wild life therein and to provide for the enjoyment of the same in such manner and by such means as will leave them unimpaired for future generations." Nearly 100 years later, this still holds true, as the NPS manages 392 natural and cultural sites, including 58 National Parks, 18 National Recreation Areas, and 20 National Preserves and Reserves. Together, these units cover over 84 million acres (more than 3.5 percent of the nation's total area) across 49 states. Park units are found in diverse locations, from remote areas to urban settings, and in all of the country's climatic zones. Recognizing its role as the model for national park systems around the world, NPS has increased its support of education on climate change and environmental stewardship through several innovative programs.

U.S. Geological Survey

As the nation's largest water, earth, and biological science and civilian mapping agency, the U.S. Geological Survey (USGS) collects, monitors, analyzes, and provides scientific understanding about natural resource conditions, issues, and problems. The agency's diversity of scientific expertise enables it to carry out large-scale, multidisciplinary investigations and provide impartial scientific information to resource managers, planners, and other customers. USGS climate change science efforts include the development and implementation of the framework for a comprehensive,

[2] See http://education.arm.gov.

national climate effects research and monitoring network; continuation of the rigorous scientific research that provides the data, new knowledge, inputs to modeling, and other outcomes that are required to understand, assess, adapt to, and mitigate climate change; and efforts to build partnerships and to translate scientific findings into real-life applications and decision-support tools.

U.S. Department of Transportation

The U.S. Department of Transportation (DOT) developed many programs to address the questions raised by the public, government employees, state and local agencies, and other transportation stakeholders about climate change. For example, the Federal Transit Administration (FTA) has several programs that provide information about the benefits of public transit and how to reduce the environmental impacts of transportation. FTA's Environmental Management Systems Training, in particular, offers training for public transit agencies to assess and reduce the environmental impact of their operations, including their carbon footprint. The Federal Highway Administration (FHWA) targets metropolitan planning organizations (MPOs) and local transportation agencies to provide information on their climate science and mitigation strategies. In 2008, the FHWA hosted three peer exchanges, which allowed senior staff of MPOs and state departments of transportation to come together and share information about integrating adaptation and mitigation strategies into the transportation planning process. These efforts are in addition to a DOT-wide effort to educate federal and state employees about a variety of transportation and climate change issues, which includes the Transportation and Climate Change Clearinghouse Web site. The site serves as a one-stop source of information about transportation and climate change topics, and is intended for use by the transportation community and general public. DOT is committed to engaging federal employees and all invested parties on the topic of climate change, especially as it relates to transportation.

National Aeronautics and Space Administration

NASA supports extensive education, training, and public awareness on climate change that take advantage of NASA's unique observational, research, and modeling assets. NASA outreach includes a Science Education Program, which sponsors educational activities at all levels of formal and informal education and creates inspiring new educational products based on NASA observations and technologies. NASA also produces informational materials and participates in public events in order to engage directly with young people and citizen scientists on a broad range of scientific topics, including climate change.

U.S. National Science Foundation

Consistent with its mission to support research and education across a broad range of science and engineering disciplines, NSF funds research and education in numerous areas related to global climate change, from training of undergraduate and graduate students in climate change research, to education and cognition research on how to better communicate the complex science of climate to lay audiences, to efforts to improve public understanding of climate change, to research on climate-related decision making. NSF's Directorates for Geosciences; Biological Sciences; Social, Behavioral, and Economic Sciences; Education and Human Resources; Mathematics and Physical Sciences; Computer and Information Science and Engineering; and the Office of Polar Programs participate in the USGCRP and provide access to climate-related results from principal investigators.

NSF is the principal federal agency charged with promoting science, technology, engineering, and mathematics (STEM) education. To this end, NSF supports the development of a diverse and well-prepared scientific and technical workforce, and a scientifically literate citizenry. NSF investments in climate change education occur through a variety of core programs aimed at strengthening STEM education. Among the projects currently being supported through these investments are the Communicating Climate Change (C3) project, to build the capacity of science centers and Long-Term Ecological Research centers to engage the public in climate change science; Seasons of Change, an effort to identify "dominant climate forcings, feedbacks, and component linkages driving change in the Arctic system seasonality"; Creating a Learning Community for Solutions to Climate Change, an effort to engage scientists and other experts in creating curricular resources to teach about climate change; and a National Academies Roundtable on Climate Change Education, to develop a national strategy for climate change education.

Smithsonian Institution

The Smithsonian is addressing the global challenge of climate change with special exhibitions and ongoing research. Smithsonian collections related to the evidence about, impact of, and response to climate change provide a unique and accessible resource for public education. Smithsonian scientists and curators regularly engage the museums' millions of U.S. and foreign visitors on climate change issues, from the perspectives of science, history, and art. The Smithsonian has also brought its outreach online to reach an even wider audience. In September 2009, a Smithsonian online climate change conference drew more than 3,700 participants in 82 countries and U.S. territories and in all 50 states. The three-day conference approached the subject from myriad Smithsonian disciplines, from wildlife management to paleontology to art history. Ten Smithsonian units took part, along with NSF, the GLOBE (Global Learning and Observations to Benefit the Environment) program, the Corporation for National and Community Service, and others. Such events support the Smithsonian's core mission to promote the "increase and diffusion of knowledge."

U.S. Agency for International Development

As a the foreign assistance arm of the U.S. government, the U.S. Agency for International Development (USAID) plays a leadership role in delivering climate change-related international assistance to over 40 developing and transition countries. With headquarters in Washington, D.C., USAID has field offices in many regions of the world—namely, sub-Saharan Africa, Asia, the Middle East, Latin America and the Caribbean, and Europe and Eurasia. USAID works in close partnership with private voluntary organizations, indigenous groups, universities, American businesses, international organizations, other governments, trade and professional associations, faith-based organizations, and other U.S. government agencies.

USAID's Global Climate Change Program incorporates climate change considerations into development projects, supporting on-the-ground programs to achieve climate change results and strengthen economic growth. Climate change education, training, and outreach are a cornerstone of USAID's activities, providing the foundation for sustainable actions. Capacity building for improved decision making through applied science and access to information is increasingly important. Building on clean energy, sustainable landscapes, and adaptation strategies, USAID will continue to integrate education, outreach, and training into its development mission to contribute to reducing the threat of climate change around the world.

U.S. Department of Agriculture

Agricultural Research Service

As the U.S. Department of Agriculture's (USDA's) chief intramural scientific research body, the Agricultural Research Service (ARS) is responsible for research on the impacts of agricultural practices on climate change, and the impacts of climate change on agriculture. Although ARS has no formal educational mechanism to disseminate research information to the general public, it employs a number of less formal means to communicate and make use of research advances.

All USDA scientific research publications are submitted with an Interpretive Summary that is used for timely news releases. In addition, through collaboration with university scientists, climate change research information is provided to state and county cooperative extension agencies for release to identified pro-

ducers. Also, all USDA field locations publish informative brochures and technical reports that describe their work and the impact of the research findings on stakeholders' interests.

National Institute of Food and Agriculture

The National Institute of Food and Agriculture (NIFA), formerly the USDA Cooperative State Research, Education, and Extension Service, is the primary USDA agency that supports extramural research, extension, and education by providing $1.2 billion annually in competitive and state-designated formula funds. NIFA also supports a number of research, education, and extension funding programs on climate change.

U.S. Forest Service

All U.S. Forest Service (USFS) national efforts in climate change education, training, and public awareness are based on the scientific expertise and findings of the agency's more than 500 scientists. The USFS Research and Development program conducts a wide variety of climate change research, investigating how climate change is and may be affecting terrestrial and freshwater natural resources and ecosystems. These results are made available to professional resource managers and the public through a variety of Web sites and publications. USFS also provides climate change education resources to educators and students through a variety of programs. One of these is *The Natural Inquirer*, a science education journal based on published Forest Service science, targeted for U.S. and international middle school students. A Climate Change Edition of *The Natural Inquirer*, focused on contemporary research findings regarding climate change and wildfires and the impact of a changing climate on wildlife and stream temperatures, will be published in 2010. In addition, EUGENE (Ecological Understanding as a Guideline for Evaluation of Nonformal Education)—a broadly applicable, user-friendly Web-based environmental education evaluation instrument that assesses student knowledge on limits, regulation, and adaptation related to climate change—will assist educators in the evaluation and improvement of their climate change programs and will increase accountability in climate change education.

U.S. Environmental Protection Agency

Climate change information, education, and outreach at the U.S. Environmental Protection Agency (EPA) have traditionally been coordinated through the Climate Change Division (CCD) in the Office of Air and Radiation. Besides managing EPA's Climate Change Web site, CCD has produced educational and informational materials that reach a wide range of audiences. In addition, CCD provides outreach programs that educate decision makers and the public about opportunities to reduce greenhouse gas emissions and to adapt to the impacts of climate change that humans and nature are already facing.

EPA's Environmental Education Division's grant program distributes over $3 million a year to formal and informal education programs across the country that educate learners of all ages about the reasons for and ways to solve environmental problems. For the last several years a good percentage of those funds went specifically to climate change education programs.

EPA's National Center for Environmental Research in the Office of Research and Development manages fellowship and other programs at the undergraduate, graduate, and postdoctoral levels, all of which have climate change educational components. In particular, the Science To Achieve Results (STAR) fellowship program has attracted and supported approximately 1,400 of America's new generation of environmental scientists, engineers, and policymakers since 1995.

EPA's Office of Children's Health Protection and Aging Initiatives provides information and outreach to populations particularly vulnerable to the effects of climate change.

Program Name	Description	Audiences	Learning Setting	Web Site
K–12 Students				
National Aeronautics and Space Administration (NASA)				
Global Learning and Observations to Benefit the Environment	GLOBE is a worldwide hands-on, primary and secondary school-based science and education program. GLOBE observations and measurements include atmosphere and climate, hydrology, land cover and phenology, and soils. GLOBE students, teachers, and scientists collaborate on inquiry-based investigations of the environment and the Earth system, working in close partnership with NASA and NSF Earth System Science Projects, on research topics related to the carbon cycle, watersheds, seasons, and biomes and extreme environments. Understanding Earth as an interconnected system is at the core of the GLOBE program. Partner: National Science Foundation (NSF)	K–12 Students, K–12 Teachers	Formal/Informal	http://www.globe.gov/
Signals of Spring	In Signals of Spring, middle and high school students investigate migration patterns of land and marine animals. Animal location data relayed from small satellite transmitters are overlaid onto maps of topography, vegetation, sea surface temperature, and other NASA Earth data, prompting students to pose, research, and analyze questions about the many factors affecting the migration and health of different species. Students use online journals, which are then read and commented on by Earth scientists and wildlife biologists. The program provides teacher training, which can be conducted on site or by live, interactive Webcasts.	K–12 Students, K–12 Teachers	Formal	http://www.signalsofspring.net/
3D-View (Virtual Interactive Environmental Worlds)	Project 3D-VIEW is a comprehensive curriculum engaging students in Earth system science with immersive three-dimensional (3D) views. The program combines NASA mission data with 3D technologies in grades 5 and 6 as students "explore" five units: lithosphere (land), hydrosphere (water), biosphere (life), atmosphere (air), and Earth systems. A project goal is for students to understand Earth system science topics and science-based decision making, preparing them for high school and beyond. Using simple Web interfaces and a custom viewer, students explore 3D stereo views to learn traditional science content. Project 3D-VIEW provides teacher training and is aimed at increasing student achievement in middle school science by using 3D technology to help students to truly understand abstract concepts.	K–12 Students, K–12 Teachers	Formal	http://www.3dview.org/
Students' Cloud Observations On-Line	S'COOL is a component of NASA's CERES (Clouds and the Earth's Radiant Energy System) instrument. CERES instrument measures the amount of energy reflected and emitted by the Earth system and focuses on understanding how clouds affect these energy transfers. Participating students make basic weather observations and record the types and features of clouds in the sky at the time the satellite passes over their location. The data are then submitted to NASA (by Web, e-mail, fax, or mail) for entry into an online database. Students can access their results as well as those from other participating schools using the S'COOL Web site (which is available in seven languages). Satellite observations for matching times are also posted so that students can compare their observations to those of the satellite, and scientists can evaluate CERES' performance. Participants receive instructional materials and information necessary for reporting results.	K–12 Students, K–12 Teachers	Formal/Informal	http://science-edu.larc.nasa.gov/SCOOL/index.php
Earth Climate Course: What Determines a Planet's Climate?	The Earth Climate Course is a set of student activities and teacher's guides designed to connect NASA Earth science research with the teaching and learning of core science and mathematics concepts and skills, while addressing national education standards. The four modules cover: (1) Temperature Variations and Habitability, (2) Modeling Hot and Cold Planets, (3) Using Mathematical Models to Investigate Planetary Habitability, and (4) How Do Atmospheres Affect Planetary Atmospheres. Scientific inquiry and the tools used to do research play a major role in the lessons. Presented with a science problem, students seek answers and consensus by experimenting with physical and computer models, collecting and analyzing their own measurements, and conducting comparisons with real-world data from satellites and ground-based observations.	K–12 Students, K–12 Teachers, Undergraduate Students	Formal	http://icp.giss.nasa.gov/education/modules/eccm/

Program Name	Description	Audiences	Learning Setting	Web Site
K–12 Students (Continued)				
U.S. Environmental Protection Agency (EPA)				
EPA Climate Change Kids Site	This popular environmental education Web site provides a wealth of resources for students and educators. Graphically engaging and interactive, it includes information about science, what we can do to make a difference, resources for educators and administrators, games (e.g., multiple-choice tests, hangman, word searches, crosswords), and more. One feature is a greenhouse gas calculator for classrooms, which instructs students about steps they can take to reduce their carbon "footprint"—measured in pounds of carbon dioxide annually—and what those reductions can mean for the environment. This site will be undergoing a revision to update it and improve its accessibility in 2010.	K–12 Students	Formal (K–12)	http://www.epa.gov/climatechange/kids
Climate Change Emission Calculator Kit (Climate CHECK)	High school students can investigate the link between everyday actions at their high school, greenhouse gas emissions, and climate change. Using Climate CHECK, students can learn about climate change, estimate their school's greenhouse gas emissions, and conceptualize ways to mitigate their school's climate impact. Students gain detailed understandings of climate change drivers, impacts, and science; produce an emission inventory and action plan; and can submit the results of their emission inventory to their school district.	K–12 Students	Formal (K–12)	http://www.epa.gov/climatechange/wycd
Global Warming Wheel Card Classroom Activity Kit	A hand-held wheel card and other resources in this kit created by EPA help middle school students estimate their classroom's and their household's greenhouse gas emissions. The activities in the kit encourage critical thinking skills and new ideas about ways to reduce the students' personal, family, school, and community contributions to climate change.	K–12 Students	Formal (K–12)	http://www.epa.gov/climatechange/emissions/wheel_card.htm
Climate for Action Campaign	Children suffer disproportionately from the health effects of some environmental hazards, and climate change could increase some of those hazards. EPA's Climate for Action Campaign educates young people about climate change and its effects on children's health, and encourages them to take actions to reduce greenhouse gas emissions. These "climate ambassadors" will in turn educate others and mobilize their communities to "create a new climate for action."	K–12 Students/ Public	Formal/Informal	http://www.epa.gov/climateforaction/
U.S. National Science Foundation				
Discovery Research K–12	The program seeks to enable significant advances in pre-K–12 student and teacher learning of the science, technology, engineering, and mathematics (STEM) disciplines through development, study, and implementation of resources, models, and technologies for use by students, teachers, and policymakers.	K–12 Students	Formal (K–12)	http://www.nsf.gov/funding/pgm_summ.jsp?pims_id=500047
K–12 Teachers				
U.S. Department of Commerce/National Oceanic and Atmospheric Administration (NOAA)				
American Meteorological Society Education Program	This AMS program promotes the teaching of atmospheric, oceanographic, and hydrologic sciences through pre-college teacher training and instructional resource material development. It also promotes instructional innovation at the introductory college course level. All programs promote activity directed toward greater human resource diversity in the sciences AMS represents. To date, over 100,000 teachers have received training and instructional resources, which have benefited millions of students. Partners: NSF, NASA	K–12 Teachers, Introductory College Teachers	Formal (K–12)	http://www.ametsoc.org/amsedu/
Climate Literacy ("Climate Smart") through National Marine Sanctuaries	NOAA's National Marine Sanctuaries works with partnering organizations to offer bilingual (Spanish and English) education to audiences in California on environmental threats, including climate change.	K–12 Teachers, K–12 Students, Public	Formal (K–12)	http://sanctuaries.noaa.gov/education/merito/welcome.html
Climate Program Office CommEd: Climate Literacy	This program takes an audience-focused approach to promoting climate science literacy among the public. It communicates the challenges, processes, and results of NOAA-supported climate science through stories and data visualizations on the Web and in popular media, and provides information to a range of audiences to enhance society's ability to plan for and respond to climate variability and change. Partner: U.S. Global Change Research Program	K–12 Teachers	Formal (K–12)	http://climate.noaa.gov/education/

Program Name	Description	Audiences	Learning Setting	Web Site
K–12 Teachers (Continued)				
Multicultural Education for Resource Issues Threatening Oceans	As part of the Climate Smart Sanctuaries initiative, NOAA's National Marine Sanctuaries are developing a climate literacy program called MERITO, which is based on the Climate Literacy Essential Principles.	K-12 Teachers, K-12 Students, Public	Formal/Informal	http://sanctuaries.noaa. gov/education
NSTA Learning Center	The National Science Teachers Association (NSTA) collaborates with NOAA, NASA, and NSF on the NSTA Learning Center to provide a variety of climate-focused online learning experiences to fit the learning style and content needs of 3 million science teachers. Accessible 24 hours a day, The Learning Center is NSTA's e-professional development portal for teachers of science. Partners: NSF, NASA	K–12 Teachers	Formal/Informal	http://learningcenter. nsta.org/
National Aeronautics and Space Administration (NASA)				
MY NASA DATA	MY NASA DATA works to make NASA Earth science data accessible to the K–12 and citizen scientist communities. The principal activity of the project is to create "micro sets" from large scientific data sets, and to wrap these with tools, lesson plans, and supporting documentation for teachers' classroom use. Climate change-related lesson plans are available for middle and high school.	K–12 Teachers, K–12 Students, Citizen Scientists	Formal/Informal	http://mynasadata.larc. nasa.gov/ClimChg_ lessons.html
Global Climate Change Education	The GCCE project extends the results of NASA's Earth Science Program to the education community by sponsoring unique and stimulating opportunities for global climate and Earth system science education. GCCE is designed to improve the quality of the nation's STEM (science, technology, engineering, and mathematics) education and enhance students' and teachers' literacy about global climate and Earth system change from elementary grades to lifelong learners. Partners: Related programs at NSF and NOAA	K–12 Teachers, Undergraduate Students, Formal (K–12 and Undergraduate)	Formal/Informal	http://www.nasa.gov/ offices/education/ programs/descriptions/ Global_Climate_ Change_Education_ Project.html
Earth System Science Education Alliance	ESSEA is an NSF-supported program implemented by the Institute for Global Environmental Strategies to improve the quality of geoscience instruction for pre-service and in-service K–12 teachers. The program is based on a series of online courses for teachers that are offered by more than 40 participating universities. The inquiry-based and place-based course modules provide teachers with the content knowledge and tools they need to incorporate Earth system science into their curricula. ESSEA modules are also available online as teacher resources. ESSEA was initially developed through NASA support, and many of the course modules use NASA data and content. Some examples of ESSEA course modules include black carbon, Brazilian deforestation, coral reefs, Hurricane Katrina, stratospheric ozone, and sea ice. Additional funding from NASA and NOAA is being used to expand the course modules, including climate change-related resources. Partners: NSF, NOAA	K–12 Teachers, Pre-service Teachers	Formal	http://esseacourses. strategies.org/
U.S. Department of Agriculture (USDA)/U.S. Forest Service (USFS)				
Forest Service Climate Change Educator Resources	USFS has several interrelated programs to help forests, grasslands, and humans mitigate and adapt to the effects of global climate change. This Web site contains a variety of resources for researchers, managers, educators, and the public on climate change issues and science, and provides links to free climate change education resources.	K–12 Teachers/K–12 Students/Public	Formal/Informal	http://www.fs.fed.us/ climatechange/
Green Schools	In partnership with the American Forest Foundation–Project Learning Tree, USFS selected five schools in Washington, D.C., to pilot the GreenSchools Initiative. The program provides training and funding for diverse and underserved pre-K–12 public schools. Students and teachers investigate environmental issues at their schools and engage with their community in ongoing service-learning projects that create green and healthy learning environments. Partner: American Forest Foundation – Project Learning Tree	Pre-K–12 Teachers/ Pre-K–12 Students/Public	Formal/Informal	http://www.plt.org/cms/ pages/21_23_242. html

Program Name	Description	Audiences	Learning Setting	Web Site
K–12 Teachers (Continued)				
Washington, D.C., Green Summer Jobs Program	USFS is joining Anacostia Urban Tree House partners to train the on-the-ground supervisors of the Washington, D.C., Mayor's Green Summer Job Corps program, which introduces young District residents to green-collar career paths. The program uses a combination of substantive work projects and traditional educational sessions to increase job readiness, connect youths to the environment within their communities, and improve the District's environment overall. Broadly, this program complements the District's efforts in combating climate change, restoring its waterways, and increasing its green infrastructure by engaging youths and preparing them to be part of a future skilled labor force. Partner: DC Mayor's Green Summer Job Corps program	K–12 Teachers/K–12 Students/Public	Formal/Informal	http://green. dc.gov/green/cwp/ view,a,1233,q,461478. asp
U.S. National Science Foundation				
Innovative Technology Experiences for Students and Teachers	ITEST addresses the growing U.S. demand for professionals and information technology workers and seeks solutions to help ensure the breadth and depth of the STEM workforce. ITEST supports research studies to address questions about how to find solutions.	K-12 Students, K-12 Teachers	Formal (K-12)	http://www.nsf.gov/ funding/pgm_summ. jsp?pims_id=5467
Graduate Students				
U.S. Department of Commerce/National Oceanic and Atmospheric Administration (NOAA)				
Climate and Society Master's Program	This program enables understanding of climate science, decision processes, and social needs to deliver management strategies that incorporate climate. The International Research Institute has developed the core courses of the program for Climate and Society in collaboration with renowned Columbia University faculty in climate, engineering, policy, public health, economics, political science, statistics, psychology, sociology, and anthropology.	Graduate Students	Formal (Grad)	http://www.columbia. edu/cu/climatesociety/
U.S. Environmental Protection Agency				
Science To Achieve Results (STAR) Fellowship Program	The STAR fellowship program exists to help the U.S. meet its current and projected human resource needs in the environmental science, engineering, and policy fields by encouraging promising students to obtain advanced degrees and pursue careers in these areas. The program has provided a steady stream of well-trained environmental specialists to meet such challenges as climate change, and has also supported environmental research in engineering and in the physical, biological, health, and social sciences. A new topic area in the program focuses exclusively on global change.	Graduate Students	Formal (Grad)	http://www.epa.gov/ ncer/fellow/
U.S. National Science Foundation (NSF)				
Integrative Graduate Education and Research Traineeship Program	This program has been developed to meet the challenges of educating U.S. Ph.D. scientists and engineers who will pursue careers in research and education, with the interdisciplinary backgrounds, deep knowledge in chosen disciplines, and technical, professional, and personal skills to become, in their own careers, leaders and creative agents for change.	Graduate Students, Professionals	Formal (Grad)	http://www.nsf.gov/ funding/pgm_summ. jsp?pims_
Undergraduate and Graduate Students				
U.S. Department of Agriculture (USDA)/Cooperative State Research, Education, and Extension Service (CSREES) and National Institute of Food and Agriculture (NIFA)				
Higher Education Challenge Grants	The CSREES/NIFA grant program addresses national priorities in the development of higher education programs and curricula.	Undergraduate and Graduate Students	Formal (Undergrad/ Grad)	http://www.csrees. usda.gov/
1890 Capacity-Building Grants Program	The CSREES/NIFA grant program addresses capacity building in teaching, research, and extension in national priorities. The program supports the development of courses, curricula, and faculty development in these priorities.	Undergraduate and Graduate Minority Students, Faculty, Scientists, and Extension Agents	Formal (Undergrad/ Grad) and Informal Education	http://www.csrees. usda.gov

Program Name	Description	Audiences	Learning Setting	Web Site
Undergraduate and Graduate Students (Continued)				
U.S. National Science Foundation (NSF)				
Significant Opportunities in Atmospheric Research in Science	The SOARS undergraduate-to-graduate bridge program is designed to broaden participation in the atmospheric and related sciences. The program is equal parts research internship, learning community, and mentoring. Partner: University Corporation for Atmospheric Research	Undergraduate Students	Formal (Undergrad)	http://www.soars.ucar.edu/
Ecosystem Studies Program	This program investigates whole-system ecological processes and relationships across a diversity of spatial and temporal (including paleo) scales to advance understanding of: (1) material and energy fluxes and transformations within and among ecosystems; (2) the relationships between structure, including complexity, and functioning of ecosystems; (3) ecosystem dynamics and trajectories of ecosystem development through time; and (4) linkages among ecosystems at different spatial and temporal scales.	Undergraduate Students, Graduate Students, Professionals	Formal (Undergrad/Grad)	http://www.nsf.gov/funding/pgm_summ.jsp?pims_
Population and Community Ecology Program	This program supports fundamental studies in the broadly defined areas of population and community ecology. Topics include the population dynamics of individual species, demography, and fundamental ecological interactions affecting populations, communities, and their environments. Themes include, but are not limited to: population regulation; food-web structure and trophic dynamics; competition, predation, mutualism. and parasitism; mechanisms of coexistence and the maintenance of species diversity; community assembly; paleoecology; landscape ecology; conservation and restoration biology; behavioral ecology; and macroecology. The program particularly encourages studies that can be applied to a wide range of habitats and taxa across multiple spatial and temporal scales.	Undergraduate Students, Graduate Students, Professionals	Formal (Undergrad/Grad)	http://www.nsf.gov/funding/pgm_summ.jsp?pims_
Emerging Topics in Biogeochemical Cycles	ETBC proposals should be interdisciplinary and address biogeochemical processes and dynamics within and/or across one or more of the following systems: terrestrial, aquatic, and atmospheric. The program encourages proposals that focus on nonlinear dynamics and/or on interactions and thresholds in climate, ecological, and/or hydrological systems. The goals of this effort are to increase understanding of how biological systems respond to changing physical and chemical conditions and influence the physical and chemical characteristics of soils and sediments, air, or water.	Undergraduate Students, Graduate Students, Professionals	Formal (Undergrad/Grad)	http://www.nsf.gov/pubs/2009/nsf09030/nsf09030.jsp
Multi-Scale Modeling	The MSM program seeks to support projects that focus on the development and/or integration of environmental models that link local, regional, and global scales. Proposals are encouraged that have the potential to dramatically improve understanding of how small- and large-scale processes lead to non-linearities and activation thresholds, as well as to improve predictive capabilities. Projects could address, but are not limited to, such topics as the carbon cycle, climate, population dynamics, food webs, biodiversity, biogeochemical cycles, and hydrological processes.	Undergraduate Students, Graduate Students, Professionals	Formal (Undergrad/Grad)	http://www.nsf.gov/pubs/2009/nsf09032/nsf09032.jsp
Emerging Frontiers in Research and Innovation	The EFRI Office is launching a new funding opportunity for interdisciplinary teams of researchers to embark on rapidly advancing frontiers of fundamental engineering research. For this solicitation, the EFRI Office will consider proposals that aim to investigate emerging frontiers in (1) Renewable Energy Storage and (2) Science in Energy and Environmental Design: Engineering Sustainable Buildings. This solicitation will be coordinated with other NSF Directorates, as well as with the U.S. Department of Energy (DOE), and EPA. Partners: DOE, EPA	Undergraduate Students, Graduate Students, Professionals	Formal (Undergrad/Grad)	http://www.nsf.gov/funding/pgm_summ.jsp?pims_
Energy for Sustainability	This program supports fundamental research and education in energy production, conversion, and storage, and is focused on environmentally friendly and renewable energy sources.	Undergraduate Students, Graduate Students, Professionals	Formal (Undergrad/Grad)	http://www.nsf.gov/funding/pgm_summ.jsp?pims_
Environmental Engineering	The goal of this program is to encourage transformative research that applies scientific principles to minimize solid, liquid, and gaseous discharges into land, inland and coastal waters, and air that result from human activity, and to evaluate the adverse impacts of these discharges on human health and environmental quality.	Undergraduate Students, Graduate Students, Professionals	Formal (Undergrad/Grad)	http://www.nsf.gov/funding/pgm_summ.jsp?pims_

Program Name	Description	Audiences	Learning Setting	Web Site
Undergraduate and Graduate Students (Continued)				
Environmental Implications of Emerging Technologies	This program seeks fundamental and basic research to establish and understand outcomes as a result of the implementation of new technologies, such as nanotechnology and biotechnology. The program also supports research on the development and refinement of sensors and sensor network technologies that can be used to measure a wide variety of physical, chemical, and biological properties of interest in characterizing, monitoring, and understanding environmental impacts.	Undergraduate Students, Graduate Students, Professionals	Formal (Undergrad/ Grad)	http://www.nsf.gov/ funding/pgm_summ. jsp?pims_
Environmental Sustainability	This program supports engineering research with the goal of promoting sustainable, engineered systems that support human well-being and that are also compatible with sustaining natural (environmental) systems that provide ecological services vital for human survival. The long-term viability of natural capital is critical for many areas of human endeavor. This program typically considers long time horizons and may incorporate contributions from the social sciences and ethics.	Undergraduate Students, Graduate Students, Professionals	Formal (Undergrad/ Grad)	http://www.nsf.gov/ funding/pgm_summ. jsp?pims_
Decision, Risk and Management Sciences	This program supports scientific research directed at increasing the understanding and effectiveness of decision making by individuals, groups, organizations, and society. Disciplinary and interdisciplinary research, doctoral dissertation research, and workshops are funded in the areas of judgment and decision making; decision analysis and decision aids; risk analysis, perception, and communication; societal and public policy decision making; and management science and organizational design.	Undergraduate Students, Graduate Students, Professionals	Formal (Undergrad/ Grad)	http://www.nsf.gov/ funding/pgm_summ. jsp?pims_
Climate and Large-Scale Dynamics	The goals of this program are to: (1) advance knowledge about the processes that force and regulate the atmosphere's synoptic and planetary circulation, weather, and climate; and (2) sustain the pool of human resources required for excellence in synoptic and global atmospheric dynamics and climate research.	Undergraduate Students, Graduate Students, Professionals	Formal (Undergrad/ Grad)	http://www.nsf.gov/ funding/pgm_summ. jsp?pims_
Climate Process and Modeling Teams	This concept aims to speed development of global coupled climate models and reduce uncertainties in climate models by bringing together theoreticians, field observationalists, process modelers, and the large modeling centers to concentrate on the scientific problems facing climate models today.	Undergraduate Students, Graduate Students, Professionals	Formal (Undergrad/ Grad)	http://www.nsf.gov/ funding/pgm_summ. jsp?pims_
Paleo Perspectives on Climate Change	The goal of research funded under the interdisciplinary P2C2 solicitation is to utilize key geological, chemical, and biological records of climate system variability to provide insights into the mechanisms and rate of change that characterized Earth's past climate variability, the sensitivity of Earth's climate system to changes in forcing, and the response of key components of the Earth system to these changes.	Undergraduate Students, Graduate Students, Professionals	Formal (Undergrad/ Grad)	http://www.nsf.gov/ funding/pgm_summ. jsp?pims_
Paleoclimate	NSF supports research on the natural evolution of Earth's climate, with the goal of providing a baseline for present variability and future trends through improved understanding of the physical, chemical, and biological processes that influence climate over the long term.	Undergraduate Students, Graduate Students, Professionals	Formal (Undergrad/ Grad)	http://www.nsf.gov/ funding/pgm_summ. jsp?pims_
Global Ocean Ecosystems Dynamics Program: Pan-Regional Synthesis	As the culmination of a series of solicitations for U.S. GLOBEC, this solicitation seeks a broader understanding of climate impacts on marine ecosystems that builds upon findings from the three regional U.S. GLOBEC studies: the Northwest Atlantic, the Northeast Pacific, and the Southern Ocean.	Undergraduate Students, Graduate Students, Professionals	Formal (Undergrad/ Grad)	http://www.nsf.gov/ funding/pgm_summ. jsp?pims_
Collaboration in Mathematical Geosciences	CMG aims to enable collaborative research at the intersection of mathematical sciences and geosciences, and to encourage cross-disciplinary education. Projects should fall within one of three broad themes: (1) conducting mathematical and statistical modeling of complex geosystems, (2) understanding and quantifying uncertainty in geosystems, or (3) analyzing large/complex geoscience data sets.	Undergraduate Students, Graduate Students, Professionals	Formal (Undergrad/ Grad)	http://www.nsf.gov/ funding/pgm_summ. jsp?pims_

Program Name	Description	Audiences	Learning Setting	Web Site
Undergraduate and Graduate Students (Continued)				
Environment, Society, and the Economy	The Intergovernmental Group on Earth Observations and the Directorate for Social, Behavioral, and Economic Sciences will consider proposals that describe new research efforts relating to the integrated study of environment, society, and economics. Interdisciplinary teams of researchers are strongly encouraged. Projects are expected to involve researchers in the geosciences and social and behavioral sciences, but may also include other disciplines.	Undergraduate Students, Graduate Students, Professionals	Formal (Undergrad/ Grad)	http://www.nsf.gov/ pubs/2009/nsf09031/ nsf09031.jsp
CHE-DMR-DMS Solar Energy Initiative	The SOLAR initiative supports interdisciplinary efforts by groups of researchers from the NSF Divisions of Chemistry (CHE), Materials Research (DMR), and Mathematical Sciences (DMS) to address the scientific challenges of highly efficient harvesting, conversion, and storage of solar energy.	Undergraduate Students, Graduate Students, Professionals	Formal (Undergrad/ Grad)	http://www.nsf.gov/ funding/pgm_summ. jsp?pims_id
Antarctic Earth Sciences Program	Beneath its thick ice sheets, Antarctica is a dynamic and diverse continent with mountains, volcanoes, deserts, meteorites, dinosaur fossils, and some of Earth's most ancient crust. The Antarctic Earth Sciences Program supports research to interpret this rich history and the processes that shape Antarctica today.	Undergraduate Students, Graduate Students, Professionals	Formal (Undergrad/ Grad)	http://www.nsf.gov/ funding/pgm_summ. jsp?pims_
Antarctic Glaciology Program	This program is concerned with the study of the history and dynamics of all naturally occurring forms of snow and ice, including floating ice shelves, glaciers, and continental and marine ice sheets. Program emphases include paleoenvironments from ice cores, ice dynamics, numerical modeling, glacial geology, and remote sensing of ice sheets.	Undergraduate Students, Graduate Students, Professionals	Formal (Undergrad/ Grad)	http://www.nsf.gov/ funding/pgm_summ. jsp?pims_
Antarctic Ocean and Atmospheric Sciences	Antarctic oceanic and tropospheric studies focus on the structure and processes of the ocean-atmosphere environment and their relationships with the global ocean, the atmosphere, and the marine biosphere. As part of the global heat engine, the Antarctic has a major role in the world's transfer of energy. Its ocean-atmosphere system is known to be both an indicator and a component of climate change.	Undergraduate Students, Graduate Students, Professionals	Formal (Undergrad/ Grad)	http://www.nsf.gov/ funding/pgm_summ. jsp?pims_
Antarctic Organisms and Ecosystems	The goal of this program is to improve understanding of Antarctic organisms and their interactions within the biosphere and geosphere. The program supports projects directed at all levels of biological organization, from molecular, cellular, and organismal, to communities and ecosystems, up to regional and global scales. Investigators are encouraged to develop and apply theory and innovative technologies to understand how organisms adapt to and live in high-latitude environments and how populations and ecosystems may respond to global change.	Undergraduate Students, Graduate Students, Professionals	Formal (Undergrad/ Grad)	http://www.nsf.gov/ funding/pgm_summ. jsp?pims_
Arctic Natural Sciences Program	Areas of special program interest include marine and terrestrial ecosystems, Arctic atmospheric and oceanic dynamics and climatology, Arctic geological and glaciological processes, and their connectivity to lower latitudes.	Undergraduate Students, Graduate Students, Professionals	Formal (Undergrad/ Grad)	http://www.nsf.gov/ funding/pgm_summ. jsp?pims_
Arctic Observing Network	Compared with much of the rest of the planet, the Arctic is a data-sparse region where large, rapid, and system-wide environmental change is occurring. The Arctic Observing Network aims to enhance the environmental observing infrastructure required for the scientific investigation of Arctic environmental change and its global connections.	Undergraduate Students, Graduate Students, Professionals	Formal (Undergrad/ Grad)	http://www.nsf.gov/ funding/pgm_summ. jsp?pims_
Arctic System Science Program	The Arctic system behaves in ways that we do not understand fully, and has demonstrated the capacity for rapid and unpredictable change with global ramifications. Because the Arctic is pivotal to the dynamics of our planet, it is critical that we understand this complex and interactive system.	Undergraduate Students, Graduate Students, Professionals	Formal (Undergrad/ Grad)	http://www.nsf.gov/ funding/pgm_summ. jsp?pims_
Interdisciplinary Training for Undergraduates in Biological and Mathematical Sciences	The goal of this activity is to enhance undergraduate education and training at the intersection of the biological and mathematical sciences and to better prepare undergraduate biology or mathematics students to pursue graduate study and careers in fields that integrate the mathematical and biological sciences.	Undergraduate Students, Professionals	Formal (Grad)	http://www.nsf.gov/ funding/pgm_summ. jsp?pims_

Program Name	Description	Audiences	Learning Setting	Web Site
Informal Educators, Public				
U.S. National Science Foundation (NSF)				
Informal Science Education	ISE invests in projects that promote lifelong learning of STEM in a wide variety of informal settings. Funding is provided for projects that advance understanding of informal STEM learning, develop and implement innovative strategies and resources for informal STEM education, and build the national professional capacity for research, development, and practice in the field.	Informal Educators, Public	Informal	http://www.nsf.gov/funding/pgm_summ.jsp?pims_
National Aeronautics and Space Administration (NASA)				
Earth to Sky: Climate Change Professional Development for Informal Educators	Earth to Sky is an ongoing and expanding partnership between NASA, the National Park Service (NPS), and the U.S. Fish and Wildlife Service (USFWS), providing professional development for informal educators to access and use relevant NASA science, data, and educational products in their work. Partners: NPS, USFWS	Informal Educators	Training	http://www.earthtosky.org/
U.S. Department of Commerce/National Oceanic and Atmospheric Administration (NOAA)				
Environmental Literacy Grants Program	The NOAA Office of Education supports applications for projects designed to build capacity within NOAA's Science On a Sphere (SOS)® Users Collaborative Network (Network) to enhance the educational use of spherical display systems as public exhibits, environmental literacy projects in support of K–12 education, and Free-Choice Learning projects that will create new, or capitalize on existing, networks of institutions, agencies, and/or organizations to provide common messages about key concepts in Earth system science.	Formal/Informal Educators	Formal/Informal	http://www.oesd.noaa.gov/funding_opps.html
Ocean Education Grants for AZA-Accredited Aquariums	The NOAA Office of Education issued a request for applications to support education projects in aquariums accredited by the Association of Zoos and Aquariums that are designed to engage the public in activities that increase ocean and/or climate literacy and the adoption of a stewardship ethic.	Informal Educators	Informal	http://www.oesd.noaa.gov/funding_opps.html
Science On a Sphere	SOS® is a spherical display system approximately 6 feet in diameter that shows "movies" of animated Earth system dynamics. NOAA supports the use of SOS® in public exhibits as part of a focused effort to increase environmental literacy. The institutions that currently have NOAA's SOS®, as well as other partners who are creating content and educational programming for these systems, have formed a collaborative network. Partners: NASA, DOE	Informal Educators	Informal	http://www.sos.noaa.gov/; http://www.oesd.noaa.gov/network/
International Action on Global Warming	IGLO is a project of the Association of Science-Technology Centers, an international organization of science centers and museums dedicated to furthering the public understanding of science. IGLO is designed to raise worldwide public awareness about global warming and the particular ways that the polar regions profoundly influence Earth's climate, environments, ecosystems, and human society. IGLO's activities present the best of current research to an international audience and explain how global warming affects their daily lives. Partners: NSF, NASA	Informal Educators	Informal	http://astc.org/iglo/
Formal/Informal Educators				
U.S. Global Change Research Program (USGCRP)				
Climate Literacy: The Essential Principles of Climate Sciences—A Guide for Individuals and Communities	This guide presents important information for individuals and communities to understand Earth's climate, the impacts of climate change, and approaches for adapting to and mitigating change. Principles in the guide can serve as discussion starters or launching points for scientific inquiry. The guide can also serve educators who teach climate science as part of their science curricula, and is available to help individuals of all ages understand how climate influences them—and how they influence climate. The guide was compiled by an interagency group led by NOAA.	Formal/Informal Educators	Formal/Informal	http://globalchange.gov/resources/educators/climate-literacy

Program Name	Description	Audiences	Learning Setting	Web Site
Formal/Informal Educators (Continued)				
Climate Change, Wildlife and Wildlands Toolkit for Formal and Informal Educators	This new toolkit is an updated and expanded version of the award-winning, popular Climate Change, Wildlife and Wildlands Toolkit for Teachers and Interpreters, first published in 2001. It profiles climate stewards in all 11 eco-regions. For example, students can participate in programs that grow native plants for wetland and dune restoration projects, or help protect terrapins endangered by sea level rise, or train citizen scientists to monitor bird migrations and spring "budbursts." The new kit is designed for classroom teachers and informal educators in parks, refuges, forest lands, nature centers, zoos, aquariums, science centers, etc., who work with the middle school grade level. EPA, in partnership with six other federal agencies (NPS, USFWS, NOAA, NASA, USDA/USFS, Bureau of Land Management), developed the kit to aid educators in teaching how climate change is affecting our nation's wildlife and public lands, and how everyone can become climate stewards.	Formal/Informal Educators	Formal/Informal	http://globalchange.gov/resources/educators/toolkit/
U.S. Department of Commerce/National Oceanic and Atmospheric Administration (NOAA)				
Office of Education: Environmental Literacy Grants	This program supports applications for projects designed to build capacity within NOAA's Science On a Sphere (SOS)® Users Collaborative Network to enhance the educational use of spherical display systems as public exhibits, environmental literacy projects in support of K–12 education, and Free-Choice Learning projects that create new, or capitalize on existing, networks of institutions, agencies, and/or organizations to provide common messages about key concepts in Earth system science.	Formal/Informal Educators	Formal/Informal	http://www.oesd.noaa.gov/funding_opps.html
NOAA's International Polar Year Program	Throughout the 2007–2009 period of the International Polar Year (IPY), NOAA and several other federal agencies were involved in numerous educational programs and activities to help educators and students understand the polar regions, as well as continuing stewardship of Earth. NOAA's Climate Program Office and Office of Education coordinated the NOAA-wide IPY education and outreach activities and other agencies' activities that focused primarily on teacher professional development, as well as developing materials for use in classrooms, science centers, and museums. Additionally, "The Arctic: A Friend Acting Strangely" exhibition was developed by the Smithsonian Institution's National Museum of Natural History with a grant from NOAA's Arctic Research Office. The exhibition puts a human face on warming in the Arctic by exploring how changes have been observed and documented by scientists and polar residents alike. Partners: Smithsonian Institution, NASA, NSF, USGS	Formal/Informal Educators	Formal/Informal	http://ipy.noaa.gov/ and http://forces.si.edu/arctic/
U.S. Environmental Protection Agency (EPA)				
Environmental Education Grant Program	This program distributes over $3 million annually to formal and informal education organizations across the nation to provide environmental education programs to learners of all ages. Many of these grants have been awarded to climate change education programs over the last several years, including public school districts, privately run nature centers, public and private colleges and universities, and community organizations.	Formal/Informal Educators	Formal/Informal	http://www.epa.gov/enviroed/
General Audiences (Undergraduate Students, Graduate Students, K–12 Students, K–12 Teachers, Informal Educators, Professionals, Public)				
U.S. Department of Agriculture (USDA)/Cooperative State Research, Education, and Extension Service (CSREES) and National Institute of Food and Agriculture (NIFA)				
Higher Education Challenge Grants	This grant program addresses national priorities in the development of higher education programs and curricula.	Undergraduate and Graduate Students	Formal (Undergrad/Grad)	http://www.csrees.usda.gov/
U.S. National Science Foundation (NSF)				
National STEM Education Distributed Learning	NSDL aims to establish a national network of learning environments and resources for science, technology, engineering, and mathematics (STEM) education at all levels.	Undergraduate Students, Graduate Students, K–12 Students, K–12 Teachers, Professionals, Public	Formal/Informal	http://www.nsf.gov/funding/pgm_summ.jsp?pims_

Program Name	Description	Audiences	Learning Setting	Web Site
General Audiences (Undergraduate Students, Graduate Students, K–12 Students, K–12 Teachers, Informal Educators, Professionals, Public) (Continued)				
Dynamics of Coupled Natural and Human Systems	CNH promotes quantitative, interdisciplinary analyses of relevant human and natural system processes and complex interactions among human and natural systems at diverse scales.	Undergraduate Students, Graduate Students, K–12 Teachers	Formal (K–12, Undergrad, Grad)	http://www.nsf.gov/ funding/pgm_summ. jsp?pims_
Geoscience Education	GeoEd aims to improve the quality of geoscience education at all educational levels; increase the number and competency of Earth and space science teachers at K–12 levels; demonstrate the relevance of the geosciences by identifying and promoting traditional and nontraditional career opportunities in the field; increase the number of students enrolling in geoscience courses and degree programs at all educational levels; increase the number of students drawn from groups underrepresented in STEM fields in geoscience courses and degree programs; and increase the public's understanding of geoscience-related issues.	Undergraduate Students, Graduate Students, K–12 Teachers, Professionals	Formal (K–12, Undergrad, Grad)	http://www.nsf.gov/ funding/pgm_summ. jsp?pims_
Centers for Ocean Science Education Excellence	With 12 Centers and a Central Coordinating Office located throughout the United States, each COSEE Center is a consortium of one or more ocean science research institutions, informal science education organizations, and formal education entities. COSEE is funded primarily by NSF, with support from NOAA. In addition to the work at the Centers, the program engages in network-level activities, such as scientist-educator partnerships, ocean literacy, and promoting ocean careers. Partner: NOAA	Graduate Students, K–12 Teachers and Students, Informal Educators, Professionals	Formal (K–12, Undergrad, Grad), Informal	http://www.cosee.net
Arctic Research and Education Program	Arctic research spans the major STEM fields and is often multi- or interdisciplinary. Research in the Arctic has clear applications for education and outreach at many levels. The region itself is an interesting hook for teaching about life, physical, and social sciences, and concepts such as ocean and atmosphere circulation, climate, Earth system science, animal migrations, and life in extreme environments.	Undergraduate Students, Graduate Students, K–12 Teachers, Professionals	Formal (K–12 Undergrad, Grad)	http://www.nsf.gov/ funding/pgm_summ. jsp?pims_
Organization of Projects on Environmental Research in the Arctic	This program seeks proposals for activities to foster and sustain collaboration among projects funded by NSF that contribute to the U.S. Arctic environmental change research effort.	Undergraduate Students, Graduate Students, K–12 Teachers, Professionals	Formal (K–12 Undergrad, Grad)	http://www.nsf.gov/ funding/pgm_summ. jsp?pims_
Public				
U.S. Department of Commerce/National Oceanic and Atmospheric Administration (NOAA)				
National Snow and Ice Data Center	The Center manages about 60 NOAA data sets, and publishes several new data sets each year, with an emphasis on *in situ* data, digitizing old and sometimes forgotten but valuable analog data, and data sets from operational communities, such as the U.S. Navy. The Center also helps develop educational pages, created Google Earth™ files that enable the public to overlay data-based images on a virtual globe, and houses many photographic prints of glaciers, taken from both the air and the ground. Partners: NSF, NASA	Public	Informal	http://nsidc.org/data/ virtual_globes/
Sea Grant	NOAA's Sea Grant is a nationwide network of 32 university-based programs that work with coastal communities. The National Sea Grant College Program engages this network of the nation's top universities in conducting scientific research, education, training, and extension projects designed to foster science-based decisions about the use and conservation of natural resources and to increase coastal resiliency. The Sea Grant network is engaged in a multifaceted and diverse series of programs to address climate change in coastal and Great Lakes regions.	Public	Formal/Informal	http://www.seagrant. noaa.gov/

Program Name	Description	Audiences	Learning Setting	Web Site
Public (Continued)				
U.S. Environmental Protection Agency (EPA)				
Climate Change and Health Effects on Older Adults	This Web page is the top site the public sees on Google when searching for information about older adults and climate change. A fact sheet entitled "It's Too Darn Hot: Planning for Excessive Heat Events" was developed and has been widely disseminated throughout aging and public health networks. Now in its second printing, this fact sheet was translated into 15 languages. "Beat the Heat" posters highlighting key messages about steps to take during extreme heat are available in English and Spanish and have been shared in senior centers around the country.	Public	Informal	http://www.epa.gov/ aging/resources/ climatechange/index. htm; http://www.epa. gov/aging/resources/ factsheets/itdhpfehe/ index.htm
Climate Change	Managed by EPA's Climate Change Division, this Web site is rated number one for "climate change" hits in the Google search engine. It offers comprehensive information accessible to U.S. and international visitors—communities, individuals, businesses, states and localities, and governments. The site features information about climate change science, greenhouse gas emissions and inventories, health and environmental effects, U.S. climate policy, mitigation and adaptation opportunities, educational resources, and EPA's varied activities on the issue. In response to public inquiries, a "Frequent Questions" database was added in 2009; other updates and improvements are planned for 2010. Many informational fact sheets and tools can be downloaded for printing.	Public	Formal/Informal	http://www.epa.gov/ climatechange
Individual Emissions Calculator	This calculator is one of the most popular features of the Climate Change Web site. Individuals can use it to get a "ballpark" estimate of personal or family greenhouse gas emissions, and also explore the impact of taking various actions to reduce emissions. Newspapers and other media outlets frequently feature this useful tool to help consumers learn about their own habits' effects on energy consumption and climate change.	Public	Formal/Informal	http://www.epa. gov/climatechange/ emissions/ind_ calculator.html
Climate "Back to Basics" Informational Materials	Among the resources available on the climate change site and in print form is a series of What You Can Do fact sheets and Web pages that provide over 25 easy steps readers can take to not only reduce greenhouse gas emissions, but also increase energy efficiency and save resources. A related science education resource for adults and students is the brochure "Frequently Asked Questions About Global Warming and Climate Change: Back to Basics." The brochure addresses key questions about this issue by restating in easy-to-understand language the most current climate science from widely accepted, peer-reviewed scientific literature.	Public	Formal/Informal	http://www.epa. gov/climatechange/ wycd; www.epa. gov/climatechange/ downloads/Climate_ Basics.pdf
Climate Friendly Parks	EPA worked closely with the National Park Service to establish this program to provide national parks with management tools and resources to address greenhouse gas emissions and climate change effects in the parks. In some cases, local communities, nonprofit organizations, and others have taken on projects and programs to enhance this joint partnership over the last few years. This program provides awareness-raising and educational activities through staff and community workshops, instruction in the use of greenhouse gas inventory tools, and pledges for park visitors to reduce their carbon footprint. Partner: National Park Service	Public/ Professionals	Informal	http://www.nps.gov/ climatefriendlyparks
Climate Change Indicators in the United States	This path-breaking report presents 24 indicators, each describing trends in some way related to the state, causes, and effects of climate change. The indicators focus primarily on the United States, but in some cases present global trends to provide context or a basis for comparison. Indicators are divided into five chapters: Greenhouse Gases, Weather and Climate, Oceans, Snow and Ice, and Human Society and Ecosystems. Data sources come from EPA, other government agencies, academia, and nongovernmental organizations.	Public/ Professionals	Formal/Informal	Under development

Program Name	Description	Audiences	Learning Setting	Web Site
Public (Continued)				
Climate Ready Estuaries	This program works with EPA's National Estuary Program and other coastal managers to assess climate change vulnerabilities, develop and implement adaptation strategies, engage and educate stakeholders, and share the lessons learned with other coastal managers. The program's Web site offers information on climate change impacts on estuaries in various U.S. regions, access to tools and resources to monitor changes, and information to help managers develop adaptation plans for estuaries and coastal communities. By fostering information sharing among estuary managers and communities around America's coasts, this program promotes adaptation to a changing climate.	Public/ Professionals	Training	www.epa.gov/cre
U.S. Department of Agriculture (USDA)/U.S. Forest Service (USFS)				
Forest Service Research Web Site	This site provides online access to USFS climate change research.	Public	Formal/Informal	http://www.fs.fed.us/ research/climate/usfs-cc-research.shtml
Climate Change in the Southern Region	This Web site provides information on upcoming climate change seminars, climate-related reading materials, regional and agency climate initiatives, and tips for reducing one's carbon footprint. Additionally, leaders from various resource areas participate in region-wide climate change seminars, discussing such topics as region-specific information, adaptation, carbon, and planning.	Public	Formal/Informal	http://fsweb.r8.fs.fed. us/climate/index.php
Treesearch	This online search engine provides access to almost 30,000 USFS publications, including over 4,500 climate change-related publications for the general public and land managers.	Public	Formal/Informal	http://www.treesearch. fs.fed.us
i-Tree	i-Tree is a state-of-the-art, peer-reviewed software suite that provides urban forestry analysis and benefits assessment tools. The i-Tree tools help professionals in communities of all sizes to strengthen their urban forest management and advocacy efforts by quantifying the structure of community trees and the environmental services that trees provide, including those that mitigate the effects of climate change.	Public/ Professionals	Formal/Informal	http://www.itreetools. org/
Task Force on Traditional Forest Knowledge	This USFS Research and Development and International Union of Forest Research Organizations task force provides information on traditional forest knowledge and practices related to climate change.	Public/ Professionals	Formal/Informal	http://www.iufro.org/ science/task-forces/ traditional-forest-knowledge/
National Report on Sustainable Forests—2010	This draft report provides access to a public review draft in preparation for its 2010 release. It is designed for the general public and managers.	Public/ Professionals	Formal/Informal	http://www.fs.fed.us/ research/sustain/
U.S. Department of Energy (DOE)				
Atmospheric Radiation Measurement Climate Research Facility	The ARM program provides online materials to develop basic science awareness related to climate change and supports community outreach in ARM program site regions.	Public	Formal/Informal	http://education.arm. gov
U.S. Department of Transportation (DOT)				
Hydrogen Working Group	In 2009, the Hydrogen Working Group distributed $900,000 to five projects with a focus on outreach.	Public, Government	Informal	http://hydrogen.dot. gov/
University Transportation Centers	This program awards grants to universities across the United States to advance the state of the art in transportation research and develop the next generation of transportation professionals.	Public	University	http://utc.dot.gov/
Transportation and Climate Change Clearinghouse	This Web site aims to serve as a one-stop source of information for the transportation community on transportation and climate change issues. Intended for use by all levels of government, private industry, and nonprofit organizations, the site provides a forum to share information, learn about new research, and understand practices and approaches being used to address the linkages between transportation and climate change. The Clearinghouse is funded jointly through the National Cooperative Highway Research Program and DOT's Center for Climate Change and Environmental Forecasting.	Public	Online	http://www.climate. dot.gov

Program Name	Description	Audiences	Learning Setting	Web Site
Public (Continued)				
National Aeronautics and Space Administration (NASA)				
Global Climate Change: NASA's Eyes on the Earth	This Web resource for educators, citizen scientists, and the public includes the planet's vital signs, feature stories, visualizations, and links to NASA missions involved in investigating climate change.	Public, K–12 Students, K–12 Teachers, Informal Educators	Informal	http://climate.nasa.gov/
Earth Observatory	The award-winning Earth Observatory's Web site's mission is to share with the public the images, stories, and discoveries about climate and the environment that emerge from NASA and Earth science research, including satellite missions, in-the-field research, and climate models.	Public, K–12 Students, K–12 Teachers, Informal Educators	Informal	http://earthobservatory.nasa.gov/
Professionals				
U.S. Department of Commerce/National Oceanic and Atmospheric Administration (NOAA)				
Coastal Training Program	This program provides up-to-date scientific information and skill-building opportunities to individuals who are responsible for making decisions that affect coastal resources. Through this program, National Estuarine Research Reserves can ensure that coastal decision makers have the knowledge and tools they need to address critical resource management issues of concern to local communities.	Professionals	Training	http://www8.nos.noaa.gov/publicnerrs/training.aspx
Coastal Resource Managers Training and Capacity Building	NOAA's Coastal Services Center works with other federal agencies to impart information, services, and technology to the nation's coastal resource managers. This community includes state coastal zone management and natural resource management offices, research reserves, sanctuaries, and Sea Grant offices. Each of these organizations has the difficult task of helping coastal communities balance the often competing demands for coastal resources.	Professionals	Training	http://oceanservice.noaa.gov/topics/coasts/training/
Responding to Climate Change: A Workshop for Coral Reef Managers	Resources from a global series of workshops are distributed to coral reef managers to support their learning of how to predict where coral bleaching will occur, measure coral reef resilience, and assess the socioeconomic impacts of climate damage. The aim is to help managers develop response strategies for coping with climate change. NOAA, the Great Barrier Reef Marine Park Authority, and The Nature Conservancy host the workshops. These partners joined with the International Union for the Conservation of Nature on *A Reef Manager's Guide to Coral Bleaching*, the book that inspired these workshops. Partners: Great Barrier Reef Marine Park Authority, The Nature Conservancy	Professionals	Training	http://coralreefwatch.noaa.gov/satellite/education/workshop/index.html
Training Program in Climate Services	NOAA's National Weather Service initiated a training program in climate services in 2001 to increase the knowledge base of its field staff. It included about 25 hours of online distance learning material, a 5-day virtual course on Climate Variability and Change, and a 3-day residence course on Operational Climate Services. Due to the continuing interest in global and regional climate variability and change as well as their local impacts on socioeconomic development, this training program is expanding.	Professionals, Graduate Students, Educators	Training	http://www.nws.noaa.gov/om/csd/pds/DistanceLearning.shtml
U.S. Department of Agriculture (USDA)/U.S. Forest Service (USFS)				
Climate Change Resource Center: Information and Tools for Land Managers	The Center is a joint project of USFS's Pacific Northwest Research Station and Rocky Mountain Research Station. This Web-based resource summarizes climate change research for resource managers, provides implications for management based on the scientific findings, and contains video presentations from scientists describing their findings.	Professionals	Formal/Informal	http://www.fs.fed.us/ccrc/
Western Wildland Environmental Threat Assessment Center	The Center's mission is to generate and integrate knowledge and information to provide credible prediction, early detection, and quantitative assessment of environmental threats in the western United States. The Center provides regional online access to the general public and land managers.	Professionals	Formal/Informal	http://www.fs.fed.us/wwetac/threats/climate_change.html
Eastern Forest Environmental Threat Assessment Center	The Center is an interdisciplinary resource that develops new technology and tools to anticipate and respond to emerging eastern forest threats, including climate change. Center researchers work with other scientists nationally as well as with a variety of federal, state, and local government agencies, universities, and nongovernmental partners to address these threats. The Center provides regional online access to the general public and land managers.	Professionals	Formal/Informal	http://www.forestthreats.org/climate-change

Program Name	Description	Audiences	Learning Setting	Web Site
Professionals (Continued)				
U.S. Department of the Interior (DOI)/National Park Service (NPS)				
Climate Leadership In Parks (CLIP) Tool	This Microsoft Excel-based calculator is designed for parks to assess their greenhouse gas emissions. It focuses on operational activities—electricity use, transportation, waste and wastewater treatment, and other greenhouse gas-emitting activities—inside parks. While this tool has a method for calculating forest carbon flux, it is not the most up-to-date tool, nor is it specific enough to adequately represent park forest carbon storage/emissions. For parks that want to include forest carbon in their reporting, NPS recommends that they use the latest forest models to calculate the flux, and then enter the numbers into the CLIP tool.	Professionals	Training	http://www.nps.gov/climatefriendlyparks/index.html
U.S. Department of the Interior (DOI)				
Regional Climate Change Response Centers	The eight DOI regional centers—serving Alaska, the Northeast, the Southeast, the Southwest, the Midwest, the West, the Northwest, and the Pacific regions—synthesize existing climate change impact data and management strategies, help resource managers put them into action on the ground, and engage the public through education initiatives.	Professionals	Training	http://www.doi.gov/news/09_News_Releases/091409.html
U.S. Department of Transportation (DOT)				
Climate Change Forums	DOT's Center for Climate Change and Environmental Forecasting offers this ongoing series to raise awareness of the many and varied perspectives of American industry, government, and nonprofit organizations.	Professionals (government employees)	Classroom/Briefing Style	
U.S. Department of Transportation/Federal Aviation Administration (FAA)				
Partnership for AiR Transportation Noise and Emissions Reduction	PARTNER is a leading aviation cooperative research organization and an FAA/NASA/Transport Canada-sponsored Center of Excellence. It fosters break-through technological, operational, policy, and workforce advances for the betterment of mobility, economy, national security, and the environment. The organization comprises 9 universities and 51 advisory board members. PARTNER has funded outreach and educational activities, and the research of more than 200 master's and Ph.D. students, many in climate research.	Professionals (aviation stakeholders, including airlines, airports, manufacturers, the public, and government organizations)	Formal/Informal	http://www.partner.aero
U.S. Department of Transportation/Federal Highway Administration (FHWA)				
Peer Exchanges on Transportation and Climate Change Mitigation and Adaptation Issues	Conducted in 2008, these workshops allowed senior staff from a variety of metropolitan planning organizations (MPOs) and state departments of transportation (DOTs) from across the country to come together to share information, experiences, and challenges regarding how both climate change mitigation and adaptation issues can be integrated into the transportation planning process. FHWA (with support from the American Association of State Highway Transportation Officials (AASHTO) conducted a peer exchange on adaptation of transportation infrastructure to climate change impacts in December 2008. Partner: AASHTO	Professionals (state DOTs, local transportation agencies, MPOs)	Formal/Informal	http://www.fhwa.dot.gov/hep/climate/resources.htm or http://www.fhwa.dot.gov/planning/statewide/pwsacci.htm
Highways & Climate Change	DOT's Office of Planning, Environment and Realty's Web site provides information on FHWA research, publications, and resources related to climate change science, policies, and actions. Visitors will also find some current state and local practices in adapting to climate change and reducing greenhouse gas emissions, and the *Transportation and Climate Change Newsletter*.	Professionals (state DOTs, local transportation agencies, MPOs, public)	Formal/Informal	http://www.fhwa.dot.gov/hep/climate/index.htm; http://www.fhwa.dot.gov/hep/climatechange/newsletter/index.htm

Program Name	Description	Audiences	Learning Setting	Web Site
Professionals (Continued)				
U.S. Department of Transportation/Federal Transit Administration (FTA)				
Education and Outreach Conferences	FTA organizes, sponsors, and participates in numerous conferences as part of its outreach efforts, including conferences and sessions geared toward education on environmental and climate change issues. In 2009, FTA sponsored and participated in climate change panels at the annual Transportation Research Board conference, the Rail-Volution conference, the American Public Transportation Association sustainability workshop, and the New Partners for Smart Growth conference.	Professionals (transit agencies, state and local governments, academics)	Conferences	http://www.fta.dot.gov/news/news_events_415.html
Transit and Environmental Sustainability	This Web site provides the public and transit agencies with information on the environmental benefits of transit (including reducing greenhouse gas emissions), FTA activities to support sustainability, and information on what transit agencies are doing across the country to reduce the environmental impacts of transportation.	Professionals (transit agencies, state and local governments, public)	Web Site	http://www.fta.dot.gov/planning/planning_environment_8510.html
National Transit Institute	Established under the Intermodal Surface Transportation Efficiency Act of 1991, the NTI provides training, education, and clearinghouse services in support of U.S. public transportation and quality of life. NTI courses on transportation planning, environmental review, transit-oriented development, and transportation and land use are particularly relevant to climate change issues.	Professionals (transit agency staff, public transportation, transit industry private companies)	Classroom and Online Courses	http://www.ntionline.com/
Environmental Management Systems Training and Assistance	FTA sponsors training for public transit agencies to continually assess and reduce the environmental impact of their operations. Training and technical assistance include workshops, on-site technical support visits, electronic software, and consultation. During the 18-month training period, each agency develops an environmental management system suited to its needs.	Professionals (transit agencies)	Workshops, On-Site Technical Support Visits, Electronic Software, Consultation	http://www.fta.dot.gov/planning/environment/planning_environment_227.html

Appendices

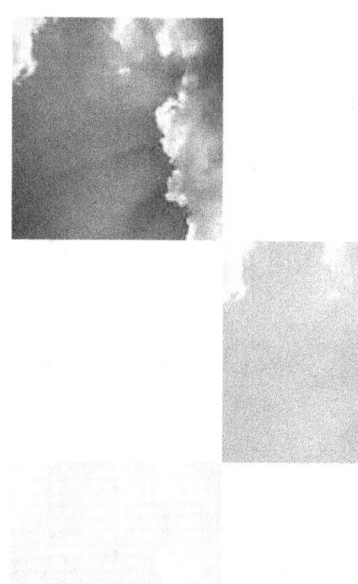

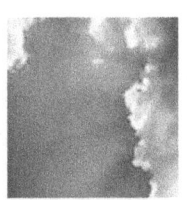

Inventory 2007
Submission 2009 v1.1
UNITED STATES OF AMERICA

TABLE 10 EMISSION TRENDS
CO₂
(Part 1 of 2)

GREENHOUSE GAS SOURCE AND SINK CATEGORIES	Base year (1990)	1991	1992	1993	1994	1995	1996	1997	1998	1999
	(Gg)	(Gg)	(Gg)	(Gg)	(Gg)	(Gg)	(Gg)	(Gg)	(Gg)	(Gg)
1. Energy	**4,870,953.27**	**4,836,985.34**	**4,936,119.34**	**5,063,791.85**	**5,149,752.90**	**5,201,233.13**	**5,375,999.90**	**5,464,358.29**	**5,497,051.13**	**5,577,645.24**
A. Fuel Combustion (Sectoral Approach)	4,836,844.15	4,803,782.01	4,903,588.99	5,029,995.51	5,115,919.64	5,167,081.83	5,344,184.55	5,432,694.24	5,467,404.39	5,547,011.49
1. Energy Industries	1,820,634.08	1,818,266.55	1,829,956.81	1,912,854.25	1,938,171.63	1,954,574.11	2,028,073.81	2,097,640.10	2,185,937.48	2,196,703.57
2. Manufacturing Industries and Construction	834,203.71	820,587.80	859,330.47	845,258.56	854,409.80	862,557.10	896,469.91	900,700.99	853,821.95	828,918.19
3. Transport	1,436,605.15	1,396,569.98	1,446,753.58	1,492,574.68	1,536,849.54	1,567,436.09	1,603,248.65	1,627,139.89	1,650,444.90	1,720,181.80
4. Other Sectors	552,259.71	566,412.85	571,144.07	583,555.67	577,249.83	578,843.59	619,202.36	596,848.13	549,032.05	569,344.17
5. Other	193,141.50	201,944.84	196,404.07	195,752.36	209,238.84	203,670.93	197,189.84	210,365.15	228,168.01	231,863.76
B. Fugitive Emissions from Fuels	34,109.12	33,203.32	32,530.35	33,796.33	33,833.26	34,151.30	31,815.34	31,664.05	29,646.74	30,633.76
1. Solid Fuels	IE,NE,NO	IE,NE,NO	IE,NE,NO	IE,NE,NO	IE,NE,NO	IE,NE,NO	IE,NE,NO	IE,NE,NO	IE,NE,NO	IE,NE,NO
2. Oil and Natural Gas	34,109.12	33,203.32	32,530.35	33,796.33	33,833.26	34,151.30	31,815.34	31,664.05	29,646.74	30,633.76
2. Industrial Processes	**197,622.92**	**185,599.17**	**188,123.22**	**185,342.00**	**191,408.79**	**198,584.35**	**198,200.97**	**201,177.86**	**197,639.68**	**194,664.61**
A. Mineral Products	54,079.29	52,453.88	53,152.90	54,951.90	57,411.73	61,127.15	62,617.93	63,699.61	65,031.29	66,054.32
B. Chemical Industry	23,566.36	23,369.71	24,588.15	24,770.76	25,900.88	25,335.69	25,304.46	25,875.88	27,055.43	25,609.06
C. Metal Production	119,977.27	109,775.58	110,382.17	105,619.33	108,096.18	112,121.50	110,278.58	111,602.36	105,552.95	103,001.23
D. Other Production	NE	NE	NE	NE	NE	NE	NE	NE	NE	NE
E. Production of Halocarbons and SF₆										
F. Consumption of Halocarbons and SF₆	NA,NO	NA,NO	NA,NO	NA,NO	NA,NO	NA,NO	NA,NO	NA,NO	NA,NO	NA,NO
G. Other	NA,NE	NA,NE	NA,NE	NA,NE	NA,NE	NA,NE	NA,NE	NA,NE	NA,NE	NA,NE
3. Solvent and Other Product Use										
4. Agriculture										
A. Enteric Fermentation										
B. Manure Management										
C. Rice Cultivation										
D. Agricultural Soils										
E. Prescribed Burning of Savannas										
F. Field Burning of Agricultural Residues										
G. Other										
5. Land Use, Land-Use Change and Forestry[(2)]	**-833,312.75**	**-851,158.63**	**-837,851.17**	**-779,963.93**	**-904,804.98**	**-842,885.19**	**-761,102.39**	**-836,046.07**	**-732,305.85**	**-686,851.10**
A. Forest Land	-529,273.49	-571,489.63	-564,608.35	-524,816.68	-563,902.11	-568,151.87	-543,329.69	-566,674.83	-500,461.50	-435,441.31
B. Cropland	-20,175.45	-22,118.72	-33,574.35	-35,057.79	-36,745.27	-12,947.30	-21,148.24	-19,889.29	-16,014.07	5,877.31
C. Grassland	-69,001.74	-49,703.91	-29,828.28	-13,972.28	-96,315.76	-58,906.42	-113.60	-45,106.17	-12,728.90	-47,943.69
D. Wetlands	1,033.48	962.28	919.55	980.58	937.87	1,017.94	872.25	1,037.48	1,084.22	1,154.91
E. Settlements	-60,648.74	-62,825.20	-65,001.67	-67,178.13	-69,354.59	-71,531.05	-73,707.52	-75,883.98	-78,060.44	-80,236.90
F. Other Land	NE	NE	NE	NE	NE	NE	NE	NE	NE	NE
G. Other	-155,246.82	-145,983.44	-145,758.07	-139,919.64	-139,425.12	-132,366.49	-123,675.60	-129,529.28	-126,125.16	-130,261.42
6. Waste	**IE,NA,NE**	**IE,NA,NE**	**IE,NA,NE**	**IE,NA,NE**	**IE,NA,NE**	**IE,NA,NE**	**IE,NA,NE**	**IE,NA,NE**	**IE,NA,NE**	**IE,NA,NE**
A. Solid Waste Disposal on Land	NA,NE	NA,NE	NA,NE	NA,NE	NA,NE	NA,NE	NA,NE	NA,NE	NA,NE	NA,NE
B. Waste-water Handling										
C. Waste Incineration	IE	IE	IE	IE	IE	IE	IE	IE	IE	IE
D. Other	NA	NA	NA	NA	NA	NA	NA	NA	NA	NA
7. Other (as specified in Summary 1.A)	NA	NA	NA	NA	NA	NA	NA	NA	NA	NA
Total CO₂ emissions including net CO₂ from LULUCF	**4,235,263.44**	**4,171,425.88**	**4,286,391.39**	**4,469,169.91**	**4,436,356.71**	**4,556,932.29**	**4,813,098.48**	**4,829,490.08**	**4,962,384.96**	**5,085,458.75**
Total CO₂ emissions excluding net CO₂ from LULUCF	**5,068,576.19**	**5,022,584.51**	**5,124,242.56**	**5,249,133.85**	**5,341,161.69**	**5,399,817.48**	**5,574,200.87**	**5,665,536.15**	**5,694,690.81**	**5,772,309.85**
Memo Items:										
International Bunkers	**114,329.60**	**123,567.53**	**115,588.68**	**104,003.73**	**106,489.22**	**101,620.41**	**108,308.43**	**111,092.04**	**120,332.95**	**102,459.82**
Aviation	46,377.60	49,489.37	52,132.13	51,205.39	56,297.74	51,195.57	57,693.42	56,587.38	61,848.03	55,561.92
Marine	67,952.00	74,078.16	63,456.54	52,798.34	50,191.48	50,424.84	50,615.00	54,504.66	58,484.91	46,897.90
Multilateral Operations	NE	NE	NE	NE	NE	NE	NE	NE	NE	NE
CO₂ Emissions from Biomass	219,340.79	220,082.20	230,515.03	225,686.48	232,174.48	236,774.53	241,236.81	235,458.98	218,102.50	221,354.19

Note: All footnotes for this table are given at the end of the table on sheet 5.

TABLE 10 EMISSION TRENDS
CO_2
(Part 2 of 2)

GREENHOUSE GAS SOURCE AND SINK CATEGORIES	2000	2001	2002	2003	2004	2005	2006	2007	Change from base to latest reported year
	(Gg)	(Gg)	(Gg)	(Gg)	(Gg)	(Gg)	(Gg)	(Gg)	%
1. Energy	5,753,192.29	5,676,392.94	5,725,316.43	5,785,075.45	5,865,681.40	5,910,830.29	5,830,206.44	5,919,451.52	21.53
A. Fuel Combustion (Sectoral Approach)	5,723,473.25	5,647,271.04	5,695,363.30	5,756,310.03	5,837,253.92	5,881,080.19	5,800,379.21	5,890,484.85	21.78
1. Energy Industries	2,300,662.09	2,263,809.72	2,273,168.45	2,302,469.44	2,333,871.91	2,400,534.15	2,347,137.23	2,417,976.57	32.81
2. Manufacturing Industries and Construction	844,553.58	839,115.92	842,053.96	844,277.95	844,607.28	828,008.27	844,505.28	845,415.82	1.34
3. Transport	1,772,568.94	1,748,893.27	1,786,899.10	1,782,181.60	1,830,672.19	1,855,548.29	1,858,382.71	1,864,111.44	29.76
4. Other Sectors	597,284.45	582,938.69	581,272.90	615,244.74	595,566.44	579,796.52	527,901.66	554,976.35	0.49
5. Other	208,404.20	212,513.44	211,968.88	212,136.30	232,536.11	217,192.97	222,452.33	208,004.68	7.70
B. Fugitive Emissions from Fuels	29,719.04	29,121.90	29,953.13	28,765.42	28,427.48	29,750.10	29,827.23	28,966.67	-15.08
1. Solid Fuels	IE,NE,NO	IE,NE,NO	IE,NE,NO	IE,NE,NO	IE,NE,NO	IE,NE,NO	IE,NE,NO	IE,NE,NO	0.00
2. Oil and Natural Gas	29,719.04	29,121.90	29,953.13	28,765.42	28,427.48	29,750.10	29,827.23	28,966.67	-15.08
2. Industrial Processes	193,216.55	174,662.49	173,359.36	168,930.68	173,691.29	171,074.83	175,897.22	174,938.59	-11.48
A. Mineral Products	64,514.74	63,868.84	65,321.71	65,056.15	70,248.55	71,285.60	73,859.38	69,442.20	28.41
B. Chemical Industry	24,210.16	20,080.99	21,391.06	19,886.44	21,016.52	20,334.61	19,832.55	21,526.11	-8.66
C. Metal Production	104,491.64	90,712.66	86,646.59	83,988.09	82,426.22	79,454.62	82,205.28	83,970.28	-30.01
D. Other Production	NE	NE	NE	NE	NE	NE	NE	NE	0.00
E. Production of Halocarbons and SF_6									
F. Consumption of Halocarbons and SF_6									
G. Other	NA,NO	NA,NO	NA,NO	NA,NO	NA,NO	NA,NO	NA,NO	NA,NO	0.00
3. Solvent and Other Product Use	NA,NE	NA,NE	NA,NE	NA,NE	NA,NE	NA,NE	NA,NE	NA,NE	0.00
4. Agriculture									
A. Enteric Fermentation									
B. Manure Management									
C. Rice Cultivation									
D. Agricultural Soils									
E. Prescribed Burning of Savannas									
F. Field Burning of Agricultural Residues									
G. Other									
5. Land Use, Land-Use Change and Forestry[2]	-708,737.38	-736,975.85	-1,019,122.24	-1,256,125.55	-1,285,822.59	-1,113,812.87	-1,041,773.03	-1,053,548.12	26.43
A. Forest Land	-399,675.24	-489,904.29	-754,103.55	-1,028,558.70	-1,044,794.02	-871,739.16	-791,694.96	-809,631.52	52.97
B. Cropland	-20,272.80	-845.54	1,873.13	-3,472.07	-4,619.69	-4,465.72	-5,247.42	-5,698.80	-71.75
C. Grassland	-83,352.93	-57,734.31	-71,396.80	-31,199.75	-31,240.83	-31,290.80	-31,326.57	-31,362.65	-54.55
D. Wetlands	1,227.28	1,140.27	1,000.95	983.07	1,194.82	1,078.91	878.94	1,010.50	-2.22
E. Settlements	-82,413.36	-84,589.83	-86,766.29	-88,942.75	-91,119.21	-93,295.68	-95,472.14	-97,648.60	61.01
F. Other Land	NE	NE	NE	NE	NE	NE	NE	NE	0.00
G. Other	-124,250.32	-105,042.16	-109,729.67	-104,935.34	-115,243.66	-114,100.42	-118,910.87	-110,217.05	-29.01
6. Waste	IE,NA,NE	IE,NA,NE	IE,NA,NE	IE,NA,NE	IE,NA,NE	IE,NA,NE	IE,NA,NE	IE,NA,NE	0.00
A. Solid Waste Disposal on Land	NA,NE	NA,NE	NA,NE	NA,NE	NA,NE	NA,NE	NA,NE	NA,NE	0.00
B. Waste-water Handling									
C. Waste Incineration	IE	IE	IE	IE	IE	IE	IE	IE	0.00
D. Other	NA	NA	NA	NA	NA	NA	NA	NA	0.00
7. Other (as specified in Summary 1.A)	NA	NA	NA	NA	NA	NA	NA	NA	0.00
Total CO_2 emissions including net CO_2 from LULUCF	5,237,671.46	5,114,079.58	4,879,553.55	4,697,880.58	4,753,550.10	4,968,092.25	4,964,330.63	5,040,841.99	19.02
Total CO_2 emissions excluding net CO_2 from LULUCF	5,946,408.84	5,851,055.43	5,898,675.79	5,954,006.13	6,039,372.69	6,081,905.12	6,006,103.66	6,094,390.11	20.24
Memo Items:									
International Bunkers	98,965.90	97,046.95	104,628.27	100,392.03	114,869.20	111,487.07	110,519.62	108,755.77	-4.88
Aviation	57,693.74	58,180.80	63,215.60	55,615.00	60,784.92	56,423.97	54,563.91	52,739.81	13.72
Marine	41,272.16	38,866.16	41,412.67	44,777.03	54,084.29	55,063.10	55,955.71	56,015.96	-17.57
Multilateral Operations	NE	NE	NE	NE	NE	NE	NE	NE	0.00
CO_2 Emissions from Biomass	227,276.44	203,163.58	204,350.53	209,537.43	224,825.39	231,481.40	240,385.63	247,829.09	12.99

Note: All footnotes for this table are given at the end of the table on sheet 5.

TABLE 10 EMISSION TRENDS
CH₄
(Part 1 of 2)

GREENHOUSE GAS SOURCE AND SINK CATEGORIES	Base year (1990)	1991	1992	1993	1994	1995	1996	1997	1998	1999
	(Gg)	(Gg)	(Gg)	(Gg)	(Gg)	(Gg)	(Gg)	(Gg)	(Gg)	(Gg)
1. Energy	12,651.50	12,582.00	12,488.81	11,963.67	12,076.10	11,969.96	12,064.61	11,871.58	11,677.96	11,257.74
A. Fuel Combustion (Sectoral Approach)	577.52	578.58	590.88	565.39	553.71	546.77	551.23	510.59	475.04	470.87
1. Energy Industries	26.99	26.69	26.83	28.25	28.61	27.45	28.54	29.57	31.54	31.57
2. Manufacturing Industries and Construction	82.52	81.01	83.52	83.56	87.00	88.63	90.19	91.98	86.30	86.04
3. Transport	213.73	207.45	207.47	204.27	200.47	194.20	185.55	176.68	167.56	157.24
4. Other Sectors	250.56	259.58	269.33	245.59	233.85	233.00	243.62	208.93	186.18	192.57
5. Other	3.71	3.86	3.74	3.72	3.77	3.48	3.33	3.43	3.46	3.46
B. Fugitive Emissions from Fuels	12,073.98	12,003.42	11,897.93	11,398.28	11,522.39	11,423.20	11,513.38	11,360.99	11,202.92	10,786.87
1. Solid Fuels	4,290.85	4,155.70	4,077.32	3,551.44	3,631.23	3,585.27	3,583.39	3,521.73	3,508.75	3,329.88
2. Oil and Natural Gas	7,783.13	7,847.72	7,820.61	7,846.84	7,891.16	7,837.92	7,929.99	7,839.26	7,694.17	7,457.00
2. Industrial Processes	88.46	85.23	89.35	93.82	97.71	100.30	101.31	104.49	104.76	103.41
A. Mineral Products	NA	NA	NA	NA	NA	NA	NA	NA	NA	NA
B. Chemical Industry	42.22	44.19	45.30	49.02	51.81	53.02	55.37	58.40	59.96	60.38
C. Metal Production	46.24	41.04	44.05	44.79	45.90	47.28	45.93	46.09	44.80	43.04
D. Other Production										
E. Production of Halocarbons and SF₆										
F. Consumption of Halocarbons and SF₆										
G. Other	NA,NO	NA,NO	NA,NO	NA,NO	NA,NO	NA,NO	NA,NO	NA,NO	NA,NO	NA,NO
3. Solvent and Other Product Use	NA,NO	NA,NO	NA,NO	NA,NO	NA,NO	NA,NO	NA,NO	NA,NO	NA,NO	NA,NO
4. Agriculture	8,161.33	8,230.38	8,419.81	8,420.24	8,713.50	8,873.04	8,668.59	8,643.63	8,751.29	8,715.01
A. Enteric Fermentation	6,342.30	6,357.38	6,557.05	6,565.84	6,687.95	6,837.01	6,723.17	6,594.89	6,528.64	6,498.54
B. Manure Management	1,446.92	1,509.02	1,451.07	1,491.07	1,595.61	1,641.54	1,577.75	1,655.91	1,808.70	1,784.94
C. Rice Cultivation	339.21	333.19	374.79	334.24	391.13	362.90	331.75	356.24	376.26	394.87
D. Agricultural Soils	NA,NE	NA,NE	NA	NA	NA	NA,NE	NA	NA	NA	NA
E. Prescribed Burning of Savannas	NA	NA	NA	NA	NA	NA	NA	NA	NA	NA
F. Field Burning of Agricultural Residues	32.89	30.80	36.22	29.09	38.82	31.59	35.92	36.60	37.69	36.66
G. Other	NA	NA	NA	NA	NA	NA	NA	NA	NA	NA
5. Land Use, Land-Use Change and Forestry	218.36	190.15	284.79	175.92	525.89	292.85	832.85	162.34	217.74	772.75
A. Forest Land	218.36	190.15	284.79	175.92	525.89	292.85	832.85	162.34	217.74	772.75
B. Cropland	NE	NE	NE	NE	NE	NE	NE	NE	NE	NE
C. Grassland	NE	NE	NE	NE	NE	NE	NE	NE	NE	NE
D. Wetlands	NE	NE	NE	NE	NE	NE	NE	NE	NE	NE
E. Settlements	NE	NE	NE	NE	NE	NE	NE	NE	NE	NE
F. Other Land	NE	NE	NE	NE	NE	NE	NE	NE	NE	NE
G. Other	NA,NO	NA,NO	NA,NO	NA,NO	NA,NO	NA,NO	NA,NO	NA,NO	NA,NO	NA,NO
6. Waste	8,240.20	8,328.94	8,406.04	8,408.19	8,379.05	8,088.92	7,937.43	7,619.79	7,306.02	7,243.56
A. Solid Waste Disposal on Land	7,105.30	7,173.42	7,227.68	7,221.48	7,171.62	6,870.91	6,711.18	6,373.85	6,056.73	5,984.13
B. Waste-water Handling	1,119.66	1,138.10	1,158.76	1,161.67	1,176.65	1,183.28	1,186.66	1,202.14	1,201.60	1,206.02
C. Waste Incineration	NE	NE	NE	NE	NE	NE	NE	NE	NE	NE
D. Other	15.24	17.42	19.60	25.04	30.77	34.73	39.59	43.80	47.68	53.42
7. Other (as specified in Summary 1.A)	NA	NA	NA	NA	NA	NA	NA	NA	NA	NA
Total CH₄ emissions including CH₄ from LULUCF	29,359.85	29,416.72	29,688.80	29,061.84	29,792.25	29,325.06	29,604.79	28,401.83	28,057.76	28,092.48
Total CH₄ emissions excluding CH₄ from LULUCF	29,141.49	29,226.57	29,404.01	28,885.92	29,266.36	29,032.21	28,771.94	28,239.49	27,840.02	27,319.73
Memo Items:										
International Bunkers	8.04	8.71	7.73	6.68	6.55	6.43	6.61	6.95	7.48	6.19
Aviation	1.51	1.60	1.63	1.60	1.73	1.58	1.75	1.71	1.86	1.68
Marine	6.53	7.11	6.10	5.08	4.82	4.85	4.86	5.24	5.63	4.51
Multilateral Operations	NE	NE	NE	NE	NE	NE	NE	NE	NE	NE
CO₂ Emissions from Biomass										

Note: All footnotes for this table are given at the end of the table on sheet 5.

TABLE 10 EMISSION TRENDS
CH₄
(Part 2 of 2)

GREENHOUSE GAS SOURCE AND SINK CATEGORIES	2000	2001	2002	2003	2004	2005	2006	2007	Change from base to latest reported year
	(Gg)	(Gg)	(Gg)	(Gg)	(Gg)	(Gg)	(Gg)	(Gg)	%
1. Energy	11,380.90	11,252.09	11,001.27	10,872.90	10,468.41	9,831.58	9,795.39	9,795.51	-22.57
A. Fuel Combustion (Sectoral Approach)	478.23	452.15	435.90	437.43	436.83	439.33	415.01	423.81	-26.62
1. Energy Industries	32.58	32.31	32.49	33.61	34.02	35.49	33.70	34.79	28.88
2. Manufacturing Industries and Construction	87.42	83.20	81.02	80.74	84.89	84.12	86.93	84.54	-2.44
3. Transport	150.25	144.54	127.74	117.45	111.72	105.24	98.95	92.67	-56.64
4. Other Sectors	204.56	187.92	190.83	201.35	201.69	210.15	191.14	207.77	-17.08
5. Other	3.42	4.18	3.81	4.28	4.50	4.33	4.28	4.05	9.08
B. Fugitive Emissions from Fuels	10,902.67	10,799.94	10,565.37	10,435.47	10,031.58	9,392.24	9,380.39	9,371.69	-22.38
1. Solid Fuels	3,231.16	3,195.35	2,998.34	2,990.45	3,046.74	2,984.45	3,043.19	3,016.85	-29.69
2. Oil and Natural Gas	7,671.50	7,604.60	7,567.03	7,445.02	6,984.84	6,407.79	6,337.20	6,354.85	-18.35
2. Industrial Processes	103.73	93.84	93.61	92.34	98.45	86.16	83.25	82.46	-6.78
A. Mineral Products	NA	NA	NA	NA	NA	NA	NA	NA	0.00
B. Chemical Industry	59.34	54.86	56.39	54.78	59.44	51.70	48.21	48.83	15.66
C. Metal Production	44.39	38.98	37.22	37.56	39.01	34.46	35.03	33.63	-27.27
D. Other Production									
E. Production of Halocarbons and SF₆									
F. Consumption of Halocarbons and SF₆									
G. Other	NA,NO	NA,NO	NA,NO	NA,NO	NA,NO	NA,NO	NA,NO	NA,NO	0.00
3. Solvent and Other Product Use									
4. Agriculture	8,596.63	8,634.38	8,665.36	8,692.89	8,663.34	8,832.54	8,894.49	9,047.06	10.85
A. Enteric Fermentation	6,398.46	6,362.80	6,381.62	6,410.32	6,368.96	6,474.14	6,579.88	6,617.96	-4.35
B. Manure Management	1,803.59	1,870.96	1,924.78	1,916.11	1,892.31	1,991.37	1,993.34	2,093.30	44.67
C. Rice Cultivation	356.84	363.78	325.20	328.37	360.22	326.10	281.97	293.32	-13.53
D. Agricultural Soils	NA,NE	NA,NE	NA,NE	NA,NE	NA,NE	NA,NE	NA,NE	NA,NE	0.00
E. Prescribed Burning of Savannas	NA	NA	NA	NA	NA	NA	NA	NA	0.00
F. Field Burning of Agricultural Residues	37.73	36.84	33.77	38.09	41.84	40.93	39.30	42.47	29.14
G. Other	NA	NA	NA	NA	NA	NA	NA	NA	0.00
5. Land Use, Land-Use Change and Forestry	982.60	573.34	863.00	553.19	313.62	676.21	1,488.53	1,381.15	532.51
A. Forest Land	982.60	573.34	863.00	553.19	313.62	676.21	1,488.53	1,381.15	532.51
B. Cropland	NE	NE	NE	NE	NE	NE	NE	NE	0.00
C. Grassland	NE	NE	NE	NE	NE	NE	NE	NE	0.00
D. Wetlands	NE	NE	NE	NE	NE	NE	NE	NE	0.00
E. Settlements	NE	NE	NE	NE	NE	NE	NE	NE	0.00
F. Other Land	NE	NE	NE	NE	NE	NE	NE	NE	0.00
G. Other	NA,NO	NA,NO	NA,NO	NA,NO	NA,NO	NA,NO	NA,NO	NA,NO	0.00
6. Waste	7,083.96	6,926.00	7,038.35	7,346.49	7,252.75	7,321.56	7,451.46	7,566.06	-8.18
A. Solid Waste Disposal on Land	5,824.65	5,688.59	5,802.62	6,109.60	6,009.43	6,088.08	6,211.43	6,327.42	-10.95
B. Waste-water Handling	1,199.62	1,177.35	1,174.98	1,167.65	1,169.04	1,158.91	1,164.62	1,159.85	3.59
C. Waste Incineration	NE	NE	NE	NE	NE	NE	NE	NE	0.00
D. Other	59.69	60.06	60.75	69.24	74.28	74.57	75.41	78.78	416.90
7. Other (as specified in Summary 1.A)	NA	NA	NA	NA	NA	NA	NA	NA	0.00
Total CH₄ emissions including CH₄ from LULUCF	28,147.82	27,479.65	27,661.59	27,557.81	26,796.56	26,748.04	27,713.12	27,872.24	-5.07
Total CH₄ emissions excluding CH₄ from LULUCF	27,165.21	26,906.31	26,798.59	27,004.62	26,482.94	26,071.83	26,224.59	26,491.09	-9.09
Memo Items:									
International Bunkers	5.70	5.49	5.87	5.99	7.03	6.99	7.00	6.96	-13.47
Aviation	1.74	1.76	1.89	1.68	1.83	1.70	1.63	1.58	4.42
Marine	3.96	3.73	3.98	4.31	5.21	5.29	5.38	5.38	-17.62
Multilateral Operations	NE	NE	NE	NE	NE	NE	NE	NE	0.00
CO₂ Emissions from Biomass									

Note: All footnotes for this table are given at the end of the table on sheet 5.

TABLE 10 EMISSION TRENDS

N_2O

(Part 1 of 2)

GREENHOUSE GAS SOURCE AND SINK CATEGORIES	Base year (1990)	1991	1992	1993	1994	1995	1996	1997	1998	1999
	(Gg)	(Gg)	(Gg)	(Gg)	(Gg)	(Gg)	(Gg)	(Gg)	(Gg)	(Gg)
1. Energy	**183.77**	**189.97**	**200.39**	**207.87**	**213.77**	**217.62**	**221.87**	**224.47**	**223.63**	**220.09**
A. Fuel Combustion (Sectoral Approach)	183.77	189.97	200.39	207.87	213.77	217.62	221.87	224.47	223.63	220.09
1. Energy Industries	27.56	27.22	27.67	28.83	29.01	29.08	30.57	31.28	31.91	31.92
2. Manufacturing Industries and Construction	13.47	13.25	13.68	13.64	14.18	14.40	14.69	14.96	14.15	14.15
3. Transport	136.38	143.01	152.66	159.41	164.83	168.49	170.76	172.95	172.67	169.02
4. Other Sectors	4.69	4.83	4.93	4.59	4.43	4.40	4.64	4.11	3.69	3.84
5. Other	1.67	1.65	1.45	1.41	1.32	1.24	1.20	1.18	1.20	1.16
B. Fugitive Emissions from Fuels	IE,NA,NE	IE,NA,NE	IE,NA,NE	IE,NA,NE	IE,NA,NE	IE,NA,NE	IE,NA,NE	IE,NA,NE	IE,NA,NE	IE,NA,NE
1. Solid Fuels	IE,NE	IE,NE	IE,NE	IE,NE	IE,NE	IE,NE	IE,NE	IE,NE	IE,NE	IE,NE
2. Oil and Natural Gas	IE,NA,NE	IE,NA,NE	IE,NA,NE	IE,NA,NE	IE,NA,NE	IE,NA,NE	IE,NA,NE	IE,NA,NE	IE,NA,NE	IE,NA,NE
2. Industrial Processes	**113.84**	**112.65**	**108.58**	**112.57**	**119.72**	**127.76**	**130.20**	**110.34**	**95.18**	**90.83**
A. Mineral Products	NA	NA	NA	NA	NA	NA	NA	NA	NA	NA
B. Chemical Industry	113.84	112.65	108.58	112.57	119.72	127.76	130.20	110.34	95.18	90.83
C. Metal Production	NA	NA	NA	NA	NA	NA	NA	NA	NA	NA
D. Other Production										
E. Production of Halocarbons and SF_6										
F. Consumption of Halocarbons and SF_6										
G. Other	NA,NO	NA,NO	NA,NO	NA,NO	NA,NO	NA,NO	NA,NO	NA,NO	NA,NO	NA,NO
3. Solvent and Other Product Use	**14.21**	**13.81**	**13.02**	**14.80**	**14.80**	**14.80**	**14.80**	**15.74**	**15.74**	**15.74**
4. Agriculture	**686.34**	**704.66**	**671.32**	**724.08**	**698.35**	**695.55**	**750.35**	**709.30**	**749.63**	**676.89**
A. Enteric Fermentation										
B. Manure Management	38.92	40.69	39.71	41.15	41.28	41.62	40.90	42.35	43.20	43.53
C. Rice Cultivation										
D. Agricultural Soils	646.22	662.80	630.27	681.83	655.59	652.70	708.08	665.50	704.95	631.91
E. Prescribed Burning of Savannas	NA	NA	NA	NA	NA	NA	NA	NA	NA	NA
F. Field Burning of Agricultural Residues	1.20	1.17	1.33	1.10	1.47	1.23	1.37	1.45	1.48	1.44
G. Other	NA	NA	NA	NA	NA	NA	NA	NA	NA	NA
5. Land Use, Land-Use Change and Forestry	**4.89**	**4.77**	**5.73**	**5.66**	**8.34**	**6.43**	**10.14**	**5.59**	**5.39**	**9.43**
A. Forest Land	1.65	1.45	2.19	1.50	3.94	2.46	6.39	1.97	2.44	6.59
B. Cropland	IE,NE	IE,NE	IE,NE	IE,NE	IE,NE	IE,NE	IE,NE	IE,NE	IE,NE	IE,NE
C. Grassland	IE	IE	IE	IE	IE	IE	IE	IE	IE	IE
D. Wetlands	0.02	0.02	0.02	0.01	0.01	0.02	0.01	0.02	0.01	0.02
E. Settlements	3.22	3.31	3.53	4.14	4.39	3.96	3.74	3.61	2.93	2.83
F. Other Land	NE	NE	NE	NE	NE	NE	NE	NE	NE	NE
G. Other	IE,NA,NO	IE,NA,NO	IE,NA,NO	IE,NA,NO	IE,NA,NO	IE,NA,NO	IE,NA,NO	IE,NA,NO	IE,NA,NO	IE,NA,NO
6. Waste	**13.04**	**13.48**	**14.05**	**14.62**	**15.32**	**15.62**	**16.25**	**16.59**	**17.19**	**18.09**
A. Solid Waste Disposal on Land	11.90	12.18	12.58	12.74	13.01	13.02	13.28	13.30	13.62	14.08
B. Waste-water Handling	IE	IE	IE	IE	IE	IE	IE	IE	IE	IE
C. Waste Incineration	1.14	1.31	1.47	1.88	2.31	2.60	2.97	3.28	3.58	4.01
D. Other	NA	NA	NA	NA	NA	NA	NA	NA	NA	NA
7. Other (as specified in Summary 1.A)	**NA**	**NA**	**NA**	**NA**	**NA**	**NA**	**NA**	**NA**	**NA**	**NA**
Total N_2O emissions including N_2O from LULUCF	**1,016.08**	**1,039.35**	**1,013.10**	**1,079.60**	**1,070.29**	**1,077.78**	**1,143.61**	**1,082.03**	**1,106.77**	**1,031.06**
Total N_2O emissions excluding N_2O from LULUCF	**1,011.20**	**1,034.57**	**1,007.37**	**1,073.94**	**1,061.95**	**1,071.35**	**1,133.46**	**1,076.44**	**1,101.38**	**1,021.64**
Memo Items:										
International Bunkers	3.40	3.64	3.42	3.13	3.21	3.05	3.25	3.29	3.56	3.07
Aviation	1.74	1.84	1.87	1.84	1.99	1.82	2.01	1.96	2.13	1.93
Marine	1.66	1.81	1.55	1.29	1.22	1.23	1.23	1.33	1.43	1.14
Multilateral Operations	**NE**	**NE**	**NE**	**NE**	**NE**	**NE**	**NE**	**NE**	**NE**	**NE**
CO_2 Emissions from Biomass										

Note: All footnotes for this table are given at the end of the table on sheet 5.

TABLE 10 EMISSION TRENDS
N₂O
(Part 2 of 2)

GREENHOUSE GAS SOURCE AND SINK CATEGORIES	2000	2001	2002	2003	2004	2005	2006	2007	Change from base to latest reported year
	(Gg)	(Gg)	(Gg)	(Gg)	(Gg)	(Gg)	(Gg)	(Gg)	%
1. Energy	**218.53**	**208.43**	**194.98**	**184.95**	**177.09**	**167.32**	**156.37**	**145.76**	**-20.68**
A. Fuel Combustion (Sectoral Approach)	218.53	208.43	194.98	184.95	177.09	167.32	156.37	145.76	-20.68
1. Energy Industries	33.40	32.55	32.73	33.50	33.72	34.49	33.74	34.38	24.77
2. Manufacturing Industries and Construction	14.41	14.14	13.85	13.93	14.61	14.58	15.07	14.71	9.27
3. Transport	165.51	156.62	143.40	132.15	123.34	112.88	102.68	91.58	-32.85
4. Other Sectors	4.07	3.86	3.81	4.07	4.08	4.13	3.72	3.95	-15.81
5. Other	1.14	1.26	1.19	1.31	1.35	1.23	1.15	1.14	-32.06
B. Fugitive Emissions from Fuels	IE,NA,NE	IE,NA,NE	IE,NA,NE	IE,NA,NE	IE,NA,NE	IE,NA,NE	IE,NA,NE	IE,NA,NE	0.00
1. Solid Fuels	IE,NE	IE,NE	IE,NE	IE,NE	IE,NE	IE,NE	IE,NE	IE,NE	0.00
2. Oil and Natural Gas	IE,NA,NE	IE,NA,NE	IE,NA,NE	IE,NA,NE	IE,NA,NE	IE,NA,NE	IE,NA,NE	IE,NA,NE	0.00
2. Industrial Processes	**90.78**	**73.85**	**81.74**	**78.85**	**77.02**	**79.21**	**77.96**	**89.16**	**-21.68**
A. Mineral Products	NA	NA	NA	NA	NA	NA	NA	NA	NA
B. Chemical Industry	90.78	73.85	81.74	78.85	77.02	79.21	77.96	89.16	-21.68
C. Metal Production	NA	NA	NA	NA	NA	NA	NA	NA	0.00
D. Other Production									
E. Production of Halocarbons and SF₆									
F. Consumption of Halocarbons and SF₆									
G. Other	NA,NO	NA,NO	NA,NO	NA,NO	NA,NO	NA,NO	NA,NO	NA,NO	0.00
3. Solvent and Other Product Use	**15.74**	**15.74**	**14.15**	**14.15**	**14.15**	**14.15**	**14.15**	**14.15**	**-0.38**
4. Agriculture	**706.18**	**758.59**	**717.00**	**700.43**	**728.15**	**726.91**	**721.01**	**719.60**	**4.85**
A. Enteric Fermentation									
B. Manure Management	45.13	46.19	45.91	44.74	45.33	45.85	46.99	47.40	21.80
C. Rice Cultivation									
D. Agricultural Soils	659.56	710.91	669.69	654.26	681.16	679.42	672.40	670.65	3.78
E. Prescribed Burning of Savannas	NA	NA	NA	NA	NA	NA	NA	NA	0.00
F. Field Burning of Agricultural Residues	1.49	1.50	1.39	1.44	1.66	1.63	1.62	1.56	29.50
G. Other	NA	NA	NA	NA	NA	NA	NA	NA	0.00
5. Land Use, Land-Use Change and Forestry	**11.75**	**9.67**	**11.79**	**9.81**	**8.31**	**10.55**	**16.21**	**15.71**	**221.43**
A. Forest Land	7.83	5.05	7.24	5.03	3.16	5.65	11.24	10.50	535.55
B. Cropland	IE,NE	IE,NE	IE,NE	IE,NE	IE,NE	IE,NE	IE,NE	IE,NE	0.00
C. Grassland	IE	IE	IE	IE	IE	IE	IE	IE	0.00
D. Wetlands	0.02	0.02	0.02	0.02	0.02	0.02	0.01	0.02	-2.37
E. Settlements	3.90	4.60	4.54	4.76	5.13	4.88	4.96	5.19	61.29
F. Other Land	NE	NE	NE	NE	NE	NE	NE	NE	0.00
G. Other	IE,NA,NO	IE,NA,NO	IE,NA,NO	IE,NA,NO	IE,NA,NO	IE,NA,NO	IE,NA,NO	IE,NA,NO	0.00
6. Waste	**18.85**	**19.28**	**19.09**	**19.87**	**20.55**	**20.93**	**21.14**	**21.61**	**65.72**
A. Solid Waste Disposal on Land									
B. Waste-water Handling	14.37	14.78	14.53	14.68	14.98	15.33	15.49	15.70	31.98
C. Waste Incineration	IE	IE	IE	IE	IE	IE	IE	IE	0.00
D. Other	4.48	4.50	4.56	5.19	5.57	5.59	5.66	5.91	416.90
7. Other (as specified in Summary 1.A)	**NA**	**NA**	**NA**	**NA**	**NA**	**NA**	**NA**	**NA**	**0.00**
Total N₂O emissions including N₂O from LULUCF	**1,061.83**	**1,085.56**	**1,038.75**	**1,008.07**	**1,025.28**	**1,019.07**	**1,006.85**	**1,005.99**	**-0.99**
Total N₂O emissions excluding N₂O from LULUCF	**1,050.08**	**1,075.89**	**1,026.96**	**998.26**	**1,016.97**	**1,008.51**	**990.64**	**990.29**	**-2.07**
Memo Items:									
International Bunkers	**3.00**	**2.97**	**3.18**	**3.02**	**3.42**	**3.29**	**3.24**	**3.18**	**-6.33**
Aviation	2.00	2.02	2.17	1.93	2.10	1.95	1.87	1.82	4.42
Marine	1.01	0.95	1.01	1.09	1.32	1.34	1.37	1.37	-17.62
Multilateral Operations	**NE**	**NE**	**NE**	**NE**	**NE**	**NE**	**NE**	**NE**	**0.00**
CO₂ Emissions from Biomass									

Note: All footnotes for this table are given at the end of the table on sheet 5.

TABLE 10 EMISSION TRENDS
HFCs, PFCs and SF_6
(Part 1 of 2)

GREENHOUSE GAS SOURCE AND SINK CATEGORIES	Base year (1990)	1991	1992	1993	1994	1995	1996	1997	1998	1999
	(Gg)	(Gg)	(Gg)	(Gg)	(Gg)	(Gg)	(Gg)	(Gg)	(Gg)	(Gg)
Emissions of HFCs[3] - (Gg CO_2 equivalent)	**36,924.10**	**33,540.69**	**38,282.65**	**38,755.86**	**44,050.24**	**61,803.33**	**71,168.28**	**81,040.87**	**96,880.42**	**94,853.65**
HFC-23	3.13	2.81	3.12	2.85	2.72	2.84	2.69	2.60	3.41	2.64
HFC-32	C,IE,NA,NE,NO	C,IE,NA,NE,NO	C,IE,NA,NE,NO	C,IE,NA,NE,NO	C,IE,NA,NE,NO	C,IE,NA,NE,NO	0.00	0.01	0.01	0.02
HFC-41	IE,NA,NO	IE,NA,NO	IE,NA,NO	IE,NA,NO	IE,NA,NO	IE,NA,NO	IE,NA,NO	IE,NA,NO	IE,NA,NO	IE,NA,NO
HFC-43-10mee	C,IE,NA,NO	C,IE,NA,NO	C,IE,NA,NO	C,IE,NA,NO	C,IE,NA,NO	C,IE,NA,NO	C,IE,NA,NO	C,IE,NA,NO	C,IE,NA,NO	C,IE,NA,NO
HFC-125	IE,NA,NO	IE,NA,NO	IE,NA,NO	0.01	0.04	0.29	0.58	0.91	1.21	1.53
HFC-134	C,IE,NA,NO	C,IE,NA,NO	C,IE,NA,NO	C,IE,NA,NO	C,IE,NA,NO	C,IE,NA,NO	C,IE,NA,NO	C,IE,NA,NO	C,IE,NA,NO	C,IE,NA,NO
HFC-134a	IE,NA,NO	IE,NA,NO	0.83	3.57	8.58	19.54	26.20	32.88	36.05	40.00
HFC-152a	C,IE,NA,NO	C,IE,NA,NO	C,IE,NA,NO	C,IE,NA,NO	C,IE,NA,NO	C,IE,NA,NO	C,IE,NA,NO	C,IE,NA,NO	C,IE,NA,NO	C,IE,NA,NO
HFC-143	C,IE,NA,NO	C,IE,NA,NO	C,IE,NA,NO	C,IE,NA,NO	C,IE,NA,NO	C,IE,NA,NO	C,IE,NA,NO	C,IE,NA,NO	C,IE,NA,NO	C,IE,NA,NO
HFC-143a	IE,NA,NO	IE,NA,NO	IE,NA,NO	0.02	0.05	0.13	0.25	0.40	0.58	0.80
HFC-227ea	C,IE,NA,NO	C,IE,NA,NO	C,IE,NA,NO	C,IE,NA,NO	C,IE,NA,NO	C,IE,NA,NO	C,IE,NA,NO	C,IE,NA,NO	C,IE,NA,NO	C,IE,NA,NO
HFC-236fa	IE,NA,NO	IE,NA,NO	IE,NA,NO	0.01	0.02	0.04	0.04	0.05	0.06	0.08
HFC-245ca	C,IE,NA,NO	C,IE,NA,NO	C,IE,NA,NO	C,IE,NA,NO	C,IE,NA,NO	C,IE,NA,NO	C,IE,NA,NO	C,IE,NA,NO	C,IE,NA,NO	C,IE,NA,NO
Unspecified mix of listed HFCs[4] - (Gg CO_2 equivalent)	331.04	640.05	648.53	656.99	658.63	1,590.70	2,780.98	3,452.19	4,076.77	4,203.48
Emissions of PFCs[3] - (Gg CO_2 equivalent)	**20,759.93**	**17,774.74**	**16,539.87**	**16,507.74**	**15,167.42**	**15,587.02**	**16,600.19**	**15,222.69**	**14,029.04**	**13,961.47**
CF_4	2.55	2.16	1.99	1.96	1.75	1.75	1.86	1.68	1.47	1.46
C_2F_6	0.45	0.40	0.39	0.41	0.41	0.45	0.49	0.47	0.49	0.49
C_3F_8	0.00	0.00	0.00	0.00	0.00	0.00	0.00	0.00	0.00	0.00
C_4F_{10}	C,IE,NA,NE,NO	C,IE,NA,NE,NO	C,IE,NA,NE,NO	C,IE,NA,NE,NO	C,IE,NA,NE,NO	C,IE,NA,NE,NO	C,IE,NA,NE,NO	C,IE,NA,NE,NO	C,IE,NA,NE,NO	C,IE,NA,NE,NO
$c\text{-}C_4F_8$	C,IE,NA,NE,NO	C,IE,NA,NE,NO	C,IE,NA,NE,NO	C,IE,NA,NE,NO	C,IE,NA,NE,NO	C,IE,NA,NE,NO	C,IE,NA,NE,NO	C,IE,NA,NE,NO	C,IE,NA,NE,NO	C,IE,NA,NE,NO
C_5F_{12}	C,IE,NA,NE,NO	C,IE,NA,NE,NO	C,IE,NA,NE,NO	C,IE,NA,NE,NO	C,IE,NA,NE,NO	C,IE,NA,NE,NO	C,IE,NA,NE,NO	C,IE,NA,NE,NO	C,IE,NA,NE,NO	C,IE,NA,NE,NO
C_6F_{14}	C,IE,NA,NE,NO	C,IE,NA,NE,NO	C,IE,NA,NE,NO	C,IE,NA,NE,NO	C,IE,NA,NE,NO	C,IE,NA,NE,NO	C,IE,NA,NE,NO	C,IE,NA,NE,NO	C,IE,NA,NE,NO	C,IE,NA,NE,NO
Unspecified mix of listed PFCs[4] - (Gg CO_2 equivalent)	NA,NE,NO	NA,NE,NO	NA,NE,NO	NA,NE,NO	NA,NE,NO	NA,NE,NO	NA,NE,NO	NA,NE,NO	NA,NE,NO	NA,NE,NO
Emissions of SF6[3] - (Gg CO_2 equivalent)	**32,787.70**	**31,399.66**	**31,592.00**	**31,043.81**	**29,535.10**	**28,080.98**	**27,314.39**	**25,551.89**	**22,535.93**	**22,891.96**
SF_6	1.37	1.31	1.32	1.30	1.24	1.17	1.14	1.07	0.94	0.96

Note: All footnotes for this table are given at the end of the table on sheet 5.

TABLE 10 EMISSION TRENDS
HFCs, PFCs and SF$_6$
(Part 2 of 2)

GREENHOUSE GAS SOURCE AND SINK CATEGORIES	2000	2001	2002	2003	2004	2005	2006	2007	Change from base to latest reported year
	(Gg)	(Gg)	(Gg)	(Gg)	(Gg)	(Gg)	(Gg)	(Gg)	%
Emissions of HFCs(a) - (Gg CO$_2$ equivalent)	100,098.22	96,905.86	104,307.81	101,361.21	112,376.34	116,073.33	119,072.79	125,531.42	239.97
HFC-23	2.47	1.70	1.82	1.07	1.49	1.37	1.21	1.48	-52.78
HFC-32	0.04	0.09	0.17	0.27	0.40	0.56	0.91	1.33	100.00
HFC-41	IE,NA,NO	IE,NA,NO	IE,NA,NO	IE,NA,NO	IE,NA,NO	IE,NA,NO	IE,NA,NO	IE,NA,NO	0.00
HFC-43-10mee	C,IE,NA,NO	C,IE,NA,NO	C,IE,NA,NO	C,IE,NA,NO	C,IE,NA,NO	C,IE,NA,NO	C,IE,NA,NO	C,IE,NA,NO	0.00
HFC-125	1.87	2.15	2.44	2.80	3.22	3.67	4.39	5.25	100.00
HFC-134	C,IE,NA,NO	C,IE,NA,NO	C,IE,NA,NO	C,IE,NA,NO	C,IE,NA,NO	C,IE,NA,NO	C,IE,NA,NO	C,IE,NA,NO	0.00
HFC-134a	44.01	46.89	49.33	51.27	53.28	54.23	54.36	52.78	100.00
HFC-152a	C,IE,NA,NO	C,IE,NA,NO	C,IE,NA,NO	C,IE,NA,NO	C,IE,NA,NO	C,IE,NA,NO	C,IE,NA,NO	C,IE,NA,NO	0.00
HFC-143	C,IE,NA,NO	C,IE,NA,NO	C,IE,NA,NO	C,IE,NA,NO	C,IE,NA,NO	C,IE,NA,NO	C,IE,NA,NO	C,IE,NA,NO	0.00
HFC-143a	1.09	1.41	1.78	2.19	2.65	3.20	3.78	4.40	100.00
HFC-227ea	C,IE,NA,NO	C,IE,NA,NO	C,IE,NA,NO	C,IE,NA,NO	C,IE,NA,NO	C,IE,NA,NO	C,IE,NA,NO	C,IE,NA,NO	0.00
HFC-236fa	0.09	0.09	0.10	0.11	0.12	0.12	0.13	0.14	100.00
HFC-245ca	C,IE,NA,NO	C,IE,NA,NO	C,IE,NA,NO	C,IE,NA,NO	C,IE,NA,NO	C,IE,NA,NO	C,IE,NA,NO	C,IE,NA,NO	0.00
Unspecified mix of listed HFCs(a) - (Gg CO$_2$ equivalent)	4,045.51	3,965.97	4,517.98	5,178.10	5,559.65	5,917.87	6,194.60	6,482.85	1,858.34
Emissions of PFCs(b) - (Gg CO$_2$ equivalent)	13,479.45	6,979.60	8,711.06	7,077.89	6,125.08	6,194.63	6,030.44	7,479.87	-63.97
CF$_4$	1.48	0.67	0.88	0.67	0.55	0.56	0.51	0.69	-73.03
C$_2$F$_6$	0.41	0.27	0.31	0.28	0.27	0.26	0.28	0.32	-30.10
C$_3$F$_8$	0.02	0.01	0.01	0.01	0.01	0.00	0.01	0.01	1,417.03
C$_4$F$_{10}$	C,IE,NA,NE,NO	C,IE,NA,NE,NO	C,IE,NA,NE,NO	C,IE,NA,NE,NO	C,NA,NE,NO	C,NA,NE,NO	C,NA,NE,NO	C,NA,NE,NO	0.00
c-C$_4$F$_8$	C,IE,NA,NE,NO	C,IE,NA,NE,NO	0.01	0.01	0.01	0.01	0.01	0.01	100.00
C$_5$F$_{12}$	C,IE,NA,NE,NO	C,IE,NA,NE,NO	C,IE,NA,NE,NO	C,IE,NA,NE,NO	C,NA,NE,NO	C,NA,NE,NO	C,NA,NE,NO	C,NA,NE,NO	0.00
C$_6$F$_{14}$	C,IE,NA,NE,NO	C,IE,NA,NE,NO	C,IE,NA,NE,NO	C,IE,NA,NE,NO	C,NA,NE,NO	C,NA,NE,NO	C,NA,NE,NO	C,NA,NE,NO	0.00
Unspecified mix of listed PFCs(a) - (Gg CO$_2$ equivalent)	NA,NE,NO	NA,NE,NO	NA,NE,NO	NA,NE,NO	NA,NE,NO	NA,NE,NO	NA,NE,NO	NA,NE,NO	0.00
Emissions of SF$_6$(b) - (Gg CO$_2$ equivalent)	19,199.54	18,761.05	18,122.79	18,207.16	17,756.95	17,893.06	17,026.87	16,458.76	-49.80
SF$_6$	0.80	0.78	0.76	0.76	0.74	0.75	0.71	0.69	-49.80

Note: All footnotes for this table are given at the end of the table on sheet 5.

TABLE 10 EMISSION TRENDS
SUMMARY
(Part 1 of 2)

GREENHOUSE GAS EMISSIONS	Base year (1990)	1991	1992	1993	1994	1995	1996	1997	1998	1999
	CO₂ equivalent (Gg)	CO₂ equivalent (Gg)	CO₂ equivalent (Gg)	CO₂ equivalent (Gg)	CO₂ equivalent (Gg)	CO₂ equivalent (Gg)	CO₂ equivalent (Gg)	CO₂ equivalent (Gg)	CO₂ equivalent (Gg)	CO₂ equivalent (Gg)
CO₂ emissions including net CO₂ from LULUCF	4,235,263.44	4,171,425.88	4,286,391.39	4,469,169.91	4,436,356.71	4,556,932.29	4,813,098.48	4,829,490.08	4,962,384.96	5,085,458.75
CO₂ emissions excluding net CO₂ from LULUCF	5,068,576.19	5,022,584.51	5,124,242.56	5,249,133.85	5,341,161.69	5,399,817.48	5,574,200.87	5,665,536.15	5,694,690.81	5,772,309.85
CH₄ emissions including CH₄ from LULUCF	616,556.93	617,751.11	623,464.81	610,298.60	625,637.28	615,826.21	621,700.54	596,438.37	589,213.01	589,942.12
CH₄ emissions excluding CH₄ from LULUCF	611,971.33	613,757.88	617,484.23	606,604.31	614,593.49	609,676.45	604,210.74	593,029.25	584,640.47	573,714.28
N₂O emissions including N₂O from LULUCF	314,985.77	322,197.59	314,060.21	334,677.36	331,790.72	334,113.02	354,518.38	335,429.09	343,099.44	319,630.09
N₂O emissions excluding N₂O from LULUCF	313,470.93	320,717.41	312,283.69	332,922.90	329,204.20	332,119.16	351,373.55	333,695.17	341,428.62	316,707.45
HFCs	36,924.10	33,540.69	38,282.65	38,755.86	44,050.24	61,803.33	71,168.28	81,040.87	96,880.42	94,853.65
PFCs	20,759.93	17,774.74	16,539.87	16,507.74	15,167.42	15,587.02	16,600.19	15,222.69	14,029.04	13,961.47
SF₆	32,787.70	31,399.66	31,592.00	31,043.81	29,535.10	28,080.98	27,314.39	25,551.89	22,535.93	22,891.96
Total (including LULUCF)	**5,257,277.87**	**5,194,089.67**	**5,310,330.93**	**5,500,453.29**	**5,482,537.48**	**5,612,342.86**	**5,904,400.25**	**5,883,172.99**	**6,028,142.80**	**6,126,738.05**
Total (excluding LULUCF)	**6,084,490.17**	**6,039,774.89**	**6,140,425.01**	**6,274,968.47**	**6,373,712.14**	**6,447,084.43**	**6,644,868.01**	**6,714,076.03**	**6,754,205.30**	**6,794,438.66**

GREENHOUSE GAS SOURCE AND SINK CATEGORIES	Base year (1990)	1991	1992	1993	1994	1995	1996	1997	1998	1999
	CO₂ equivalent (Gg)	CO₂ equivalent (Gg)	CO₂ equivalent (Gg)	CO₂ equivalent (Gg)	CO₂ equivalent (Gg)	CO₂ equivalent (Gg)	CO₂ equivalent (Gg)	CO₂ equivalent (Gg)	CO₂ equivalent (Gg)	CO₂ equivalent (Gg)
1. Energy	5,193,603.06	5,160,098.15	5,260,504.40	5,379,469.73	5,469,619.64	5,520,063.22	5,698,135.18	5,783,247.60	5,811,613.44	5,882,285.73
2. Industrial Processes	325,243.41	305,025.99	310,075.03	308,517.81	319,325.85	345,768.84	355,773.21	359,393.67	362,791.93	356,700.76
3. Solvent and Other Product Use	4,404.02	4,281.69	4,037.02	4,587.52	4,587.52	4,587.52	4,587.52	4,879.50	4,879.50	4,879.50
4. Agriculture	384,152.60	391,282.02	384,925.21	401,290.41	399,470.61	401,954.09	414,647.53	401,397.92	416,163.86	392,850.32
5. Land Use, Land-Use Change and Forestry[5]	-827,212.31	-845,685.22	-830,094.08	-774,515.18	-891,174.66	-834,741.57	-740,467.76	-830,903.04	-726,062.50	-667,700.62
6. Waste	177,087.09	179,087.04	180,883.36	181,103.00	180,708.52	174,710.76	171,724.56	165,157.34	158,756.57	157,722.35
7. Other	NA	NA	NA	NA	NA	NA	NA	NA	NA	NA
Total (including LULUCF)[5]	**5,257,277.87**	**5,194,089.67**	**5,310,330.93**	**5,500,453.29**	**5,482,537.48**	**5,612,342.86**	**5,904,400.25**	**5,883,172.99**	**6,028,142.80**	**6,126,738.05**

(1) The column "Base year" should be filled in only by those Parties with economies in transition that use a base year different from 1990 in accordance with the relevant decisions of the COP. For these Parties, this different base year is used to calculate the percentage change in the final column of this table.

(2) Fill in net emissions/removals as reported in table Summary 1.A. For the purposes of reporting, the signs for removals are always negative (-) and for emissions positive (+).

(3) Enter actual emissions estimates. If only potential emissions estimates are available, these should be reported in this table and an indication for this be provided in the documentation box. Only in these rows are the emissions expressed as CO2 equivalent emissions.

(4) In accordance with the UNFCCC reporting guidelines, HFC and PFC emissions should be reported for each relevant chemical. However, if it is not possible to report values for each chemical (i.e. mixtures, confidential data, lack of disaggregation), this row could be used for reporting aggregate figures for HFCs and PFCs, respectively. Note that the unit used for this row is Gg of CO2 equivalent and that appropriate notation keys should be entered in the cells for the individual chemicals.

(5) Includes net CO2, CH4, and N2O from LULUCF.

TABLE 10 EMISSION TRENDS SUMMARY
(Part 2 of 2)

GREENHOUSE GAS EMISSIONS	2000 CO₂ equivalent (Gg)	2001 CO₂ equivalent (Gg)	2002 CO₂ equivalent (Gg)	2003 CO₂ equivalent (Gg)	2004 CO₂ equivalent (Gg)	2005 CO₂ equivalent (Gg)	2006 CO₂ equivalent (Gg)	2007 CO₂ equivalent (Gg)	Change from base to latest reported year (%)
CO₂ emissions including net CO₂ from LULUCF	5,237,671.46	5,114,079.58	4,879,553.55	4,697,880.58	4,753,550.10	4,968,092.25	4,964,330.63	5,040,841.99	19.02
CO₂ emissions excluding net CO₂ from LULUCF	5,946,408.84	5,851,055.43	5,898,675.79	5,954,006.13	6,039,372.69	6,081,905.12	6,006,103.66	6,094,390.11	20.24
CH₄ emissions including CH₄ from LULUCF	591,104.15	577,072.62	580,893.47	578,713.95	562,727.77	561,708.94	581,975.56	585,317.05	-5.07
CH₄ emissions excluding CH₄ from LULUCF	570,469.46	565,032.48	562,770.39	567,097.01	556,141.75	547,508.47	550,716.44	556,312.80	-9.09
N₂O emissions including N₂O from LULUCF	329,168.37	336,524.87	322,013.18	312,501.78	317,835.74	315,910.27	312,124.05	311,857.84	-0.99
N₂O emissions excluding N₂O from LULUCF	325,524.81	333,525.98	318,357.20	309,459.55	315,259.52	312,638.86	307,099.01	306,988.74	-2.07
HFCs	100,098.22	96,905.86	104,307.81	101,361.21	112,376.34	116,073.33	119,072.79	125,531.42	239.97
PFCs	13,479.45	6,979.60	8,711.06	7,077.89	6,125.08	6,194.63	6,030.44	7,479.87	-63.97
SF₆	19,199.54	18,761.05	18,122.79	18,207.16	17,756.95	17,893.06	17,026.87	16,458.76	-49.80
Total (including LULUCF)	**6,290,721.21**	**6,150,323.57**	**5,913,601.85**	**5,715,742.55**	**5,770,371.99**	**5,985,872.47**	**6,000,560.33**	**6,087,486.93**	**15.79**
Total (excluding LULUCF)	**6,975,180.33**	**6,872,260.40**	**6,910,945.04**	**6,957,208.94**	**7,047,032.33**	**7,082,213.46**	**7,006,049.20**	**7,107,161.69**	**16.81**

GREENHOUSE GAS SOURCE AND SINK CATEGORIES	2000 CO₂ equivalent (Gg)	2001 CO₂ equivalent (Gg)	2002 CO₂ equivalent (Gg)	2003 CO₂ equivalent (Gg)	2004 CO₂ equivalent (Gg)	2005 CO₂ equivalent (Gg)	2006 CO₂ equivalent (Gg)	2007 CO₂ equivalent (Gg)	Change from base to latest reported year (%)
1. Energy	6,059,934.22	5,977,299.97	6,016,786.57	6,070,742.36	6,140,416.84	6,169,162.03	6,084,385.29	6,170,343.23	18.81
2. Industrial Processes	356,315.05	322,171.53	331,806.83	321,960.17	335,893.15	337,598.92	343,943.73	353,779.52	8.77
3. Solvent and Other Product Use	4,879.50	4,879.50	4,387.15	4,387.15	4,387.15	4,387.15	4,387.15	4,387.15	-0.38
4. Agriculture	399,444.64	416,485.79	404,242.71	399,683.72	407,657.36	410,825.55	410,297.98	413,064.72	7.53
5. Land Use, Land-Use Change and Forestry [5]	-684,459.12	-721,936.83	-997,343.18	-1,241,466.39	-1,276,660.34	-1,096,340.99	-1,005,488.87	-1,019,674.76	23.27
6. Waste	154,606.92	151,423.60	153,721.78	160,435.54	158,677.84	160,239.81	163,035.05	165,587.07	-6.49
7. Other	NA	NA	NA	NA	NA	NA	NA	NA	0.00
Total (including LULUCF) [5]	**6,290,721.21**	**6,150,323.57**	**5,913,601.85**	**5,715,742.55**	**5,770,371.99**	**5,985,872.47**	**6,000,560.33**	**6,087,486.93**	**15.79**

[1] The column "Base year" should be filled in only by those Parties with economies in transition that use a base year different from 1990 in accordance with the relevant decisions of the COP. For these Parties, this different base year is used to calculate the percentage change in the final column of this table.

[2] Fill in net emissions/removals as reported in table Summary 1.A. For the purposes of reporting, the signs for removals are always negative (-) and for emissions positive (+).

[3] Enter actual emissions estimates. If only potential emissions estimates are available, these should be reported in this table and an indication for this be provided in the documentation box. Only in these rows are the emissions expressed as CO₂ equivalent emissions.

[4] In accordance with the UNFCCC reporting guidelines, HFC and PFC emissions should be reported for each relevant chemical. However, if it is not possible to report values for each chemical (i.e. mixtures, confidential data, lack of disaggregation), this row could be used for reporting aggregate figures for HFCs and PFCs, respectively. Note that the unit used for this row is Gg of CO₂ equivalent and that appropriate notation keys should be entered in the cells for the individual chemicals.

[5] Includes net CO₂, CH₄ and N₂O from LULUCF.

GREENHOUSE GAS SOURCE AND SINK CATEGORIES	CO₂ [1]	CH₄	N₂O	HFCs [2]	PFCs [2]	SF₆ [2]	Total
	CO₂ equivalent (Gg)						
Total (Net Emissions) [1]	5,040,841.99	585,317.05	311,857.84	125,531.42	7,479.87	16,458.76	6,087,486.93
1. Energy	5,919,451.52	205,705.62	45,186.09				6,170,343.23
A. Fuel Combustion (Sectoral Approach)	5,890,484.85	8,900.08	45,186.09				5,944,571.01
1. Energy Industries	2,417,976.57	730.57	10,657.92				2,429,365.07
2. Manufacturing Industries and Construction	845,415.82	1,775.24	4,561.61				851,752.68
3. Transport	1,864,111.44	1,946.01	28,389.41				1,894,446.85
4. Other Sectors	554,976.35	4,363.22	1,224.81				560,564.37
5. Other	208,004.68	85.03	352.34				208,442.05
B. Fugitive Emissions from Fuels	28,966.67	196,805.54	IE,NA,NE				225,772.22
1. Solid Fuels	IE,NE,NO	63,353.76	IE,NE				63,353.76
2. Oil and Natural Gas	28,966.67	133,451.79	IE,NA,NE				162,418.46
2. Industrial Processes	174,938.59	1,731.76	27,639.12	125,531.42	7,479.87	16,458.76	353,779.52
A. Mineral Products	69,442.20	NA	NA				69,442.20
B. Chemical Industry	21,526.11	1,025.51	27,639.12	NA	NA	NA	50,190.74
C. Metal Production	83,970.28	706.25	NA	NA	3,836.03	2,973.25	91,485.81
D. Other Production	NE						NE
E. Production of Halocarbons and SF₆				16,999.76	NA,NE	NA,NE	16,999.76
F. Consumption of Halocarbons and SF₆ [2]				108,531.66	3,643.84	13,485.51	125,661.00
G. Other	NA,NO	NA,NO	NA,NO	NA	NA	NA	NA,NO
3. Solvent and Other Product Use	NA,NE		4,387.15				4,387.15
4. Agriculture		189,988.21	223,076.51				413,064.72
A. Enteric Fermentation		138,977.19					138,977.19
B. Manure Management		43,959.32	14,693.81				58,653.14
C. Rice Cultivation		6,159.75					6,159.75
D. Agricultural Soils [3]		NA,NE	207,900.52				207,900.52
E. Prescribed Burning of Savannas		NA	NA				NA
F. Field Burning of Agricultural Residues		891.95	482.18				1,374.13
G. Other		NA	NA				NA
5. Land Use, Land-Use Change and Forestry [1]	-1,053,548.12	29,004.25	4,869.11				-1,019,674.76
A. Forest Land	-809,631.52	29,004.25	3,255.32				-777,371.95
B. Cropland	-5,698.80	NE	IE,NE				-5,698.80
C. Grassland	-31,362.65	NE	IE				-31,362.65
D. Wetlands	1,010.50	NE	5.09				1,015.59
E. Settlements	-97,648.60	NE	1,608.69				-96,039.91
F. Other Land	NE	NE	NE				NE
G. Other	-110,217.05	NA,NO	IE,NA,NO				-110,217.05
6. Waste	IE,NA,NE	158,887.20	6,699.87				165,587.07
A. Solid Waste Disposal on Land	NA,NE	132,875.92					132,875.92
B. Waste-water Handling		24,356.87	4,868.21				29,225.08
C. Waste Incineration	IE	NE	IE,NE				IE,NE
D. Other	NA	1,654.41	1,831.66				3,486.07
7. Other (as specified in Summary 1.A)	NA	NA	NA	NA	NA	NA	NA

Memo Items: [4]							
International Bunkers	108,755.77	146.09	986.08				109,887.94
Aviation	52,739.81	33.16	562.70				53,335.67
Marine	56,015.96	112.93	423.39				56,552.27
Multilateral Operations	NE	NE	NE				NE
CO₂ Emissions from Biomass	247,829.09						247,829.09

Total CO₂ Equivalent Emissions without Land Use, Land-Use Change and Forestry	7,107,161.69
Total CO₂ Equivalent Emissions with Land Use, Land-Use Change and Forestry	6,087,486.93

[1] For CO₂ from Land Use, Land-use Change and Forestry the net emissions/removals are to be reported. For the purposes of reporting, the signs for removals are always negative (-) and for emissions positive (+).

[2] Actual emissions should be included in the national totals. If no actual emissions were reported, potential emissions should be included.

[3] Parties which previously reported CO₂ from soils in the Agriculture sector should note this in the NIR.

[4] See footnote 8 to table Summary 1.A.

Backlund et al. 2008—Backlund, P., A. Janetos, D. Schimel, J. Hatfield, K. Boote, P. Fay, L. Hahn, C. Izaurralde, B.A. Kimball, T. Mader, J. Morgan, D. Ort, W. Polley, A. Thomson, D. Wolfe, M.G. Ryan, S.R. Archer, R. Birdsey, C. Dahm, L. Heath, J. Hicke, D. Hollinger, T. Huxman, G. Okin, R. Oren, J. Randerson, W. Schlesinger, D. Lettenmaier, D. Major, L. Poff, S. Running, L. Hansen, D. Inouye, B.P. Kelly, L. Meyerson, B. Peterson, and R. Shaw. *The Effects of Climate Change on Agriculture, Land Resources, Water Resources, and Biodiversity in the United States*. Final Report—Synthesis and Assessment Product 4.3. A Report by the U.S. Climate Change Science Program and the Subcommittee on Global Change Research. Washington, DC: U.S. Department of Agriculture, 362 pp. <http://www.climatescience.gov/Library/sap/sap4-3/final-report/default.htm>

Baron et al. 2008—Baron, J.S., L.A. Joyce, P. Kareiva, B.D. Keller, M.A. Palmer, C.H. Peterson, and J.M. Scott. *Preliminary Review of Adaptation Options for Climate-Sensitive Ecosystems and Resources*. Final Report—Synthesis and Assessment Product 4.4. A Report by the U.S. Climate Change Science Program and the Subcommittee on Global Change Research. Ed. S.H. Julius and J.M. West. Washington, DC: U.S. Environmental Protection Agency, 873 pp. <http://www.climatescience.gov/Library/sap/sap4-4/final-report/>

Bell et al. 2007— Bell, M.L., R. Goldberg, C. Hogrefe, P.L. Kinney, K. Knowlton, B. Lynn, J. Rosenthal, C. Rosenzweig, and J. Patz . "Climate Change, Ambient Ozone, and Health in 50 U.S. Cities." *Climate Change* 82(1-2): 61-76. DOI 10.1007/s10584-0069166-7. <http://www.springerlink.com/content/7380101v7q674581/>

Beller-Simms et al. 2008—Beller-Simms, N., H. Ingram, D. Feldman, N. Mantua, K.L. Jacobs, and A.M. Waple, eds. *Decision-Support Experiments and Evaluations Using Seasonal-to-Interannual Forecasts and Observational Data: A Focus on Water Resources*. Final Report—Synthesis and Assessment Product 5.3. A Report by the U.S. Climate Change Science Program and the Subcommittee on Global Change Research. Asheville, NC: National Oceanic and Atmospheric Administration, National Climatic Data Center, 192 pp. <http://www.climate-science.gov/Library/sap/sap5-3/final-report/>

Boykoff and Boykoff 2007— Boykoff, M.T., and J.M. Boykoff. "Climate Change and Journalistic Norms: A Case-Study of U.S. Mass-Media Coverage." *Geoforum*. <www.eci.ox.ac.uk/publications/downloads/boykoff07-geoforum.pdf>

Brekke et al. 2009—Brekke, L.D., J.E. Kiang, J.R. Olsen, R.S. Pulwarty, D.A. Raff, D.P. Turnipseed, R.S. Webb, and K.D. White. *Climate Change and Water Resources Management: A Federal Perspective*. U.S. Geological Survey Circular 1331. Washington, DC: U.S. Department of the Interior, U.S. Geological Survey, 65 pp. <http://pubs.usgs.gov/circ/1331/>

CCTP 2005—U.S. Climate Change Technology Program. *Vision and Framework for Strategy and Planning*. DOE/PI-0003. Washington, DC: U.S. Department of Energy. <http://www.climatetechnology.gov/vision2005/>

CCTP 2006—U.S. Climate Change Technology Program. *Strategic Plan*. Washington, DC: U.S. Department of Energy. <http://www.climatetechnology.gov/stratplan/final/index.htm>

CENR 2008—Committee on Environment and Natural Resources. *Scientific Assessment of the Effects of Global Change on the United States*. Washington, DC: National Science and Technology Council. <http://www.climatescience.gov/Library/scientific-assessment/>

CEPA 2006—California Environmental Protection Agency. *Report to the Governor and Legislature*. State of California, Climate Action Team. <http://www.climatechange.ca.gov/climate_action_team/reports/index.html>

Chang et al. 2009—Chang, P.S., J. Zorana, J.M. Sienkiewicz, R. Knabb, M. Brennan, D.G. Long, and M. Freeber. "Operational Use and Impact of Satellite Remotely Sensed Ocean Surface Vector Winds in the Marine Warning and Forecasting Environment." *Oceanography* 22(2). <manati.orbit.nesdis.noaa.gov/SVW_nextgen/22.2_chang.pdf>

Creyts et al. 2007—Creyts, J., A. Derkach, S. Nyquist, K.Ostrowski, and J. Stephenson. *Reducing Greenhouse Gas Emissions: How Much at What Cost?* McKinsey & Company. December 2007. <http://www.mckinsey.com/clientservice/ccsi/greenhousegas.asp>

CTIC 2004—Conservation Technology Information Center. *National Crop Residue Management Survey 2004*. "Conservation Tillage and Other Tillage Types in the United States: 1990–2004." <http://www.conservation-information.org/index.asp?site=1&action=crm_results>

Dillaha 2007—Dillaha, Theo. "Global Assessment of Best Practices in Payments for Ecosystem Services Programs." Paper presented at the annual meeting of the Soil and Water Conservation Society, Tampa, Florida, July 21, 2007. <http://www.allacademic.com/meta/p168284_index.html>

Ebi et al. 2008—Ebi, K.L., F.G. Sussman, and T.J. Wilbanks. *Analyses of the Effects of Global Change on Human Health and Welfare and Human Systems.* Final Report—Synthesis and Assessment Product 4.6. A Report by the U.S. Climate Change Science Program and the Subcommittee on Global Change Research. Ed. J.L. Gamble. Washington, DC: U.S. Environmental Protection Agency, 204 pp. <http://www.climatescience.gov/Library/sap/sap4-6/final-report/>

EOP/CEA 2010—Executive Office of the President, Council of Economic Advisers. *The Economic Impact of the American Recovery and Reinvestment Act of 2009.* Second Quarterly Report. January 13, 2010. <www.whitehouse.gov/administration/.../cea/Economic-Impact/>

Fagre et al. 2009—Fagre, D.B., C.W. Charles, C.D. Allen, C. Birkeland, F.S. Chapin III, P.M. Groffman, G.R. Guntenspergen, A.K. Knapp, A.D. McGuire, P.J. Mulholland, D.P.C. Peters, D.D. Roby, and G. Sugihara. *Thresholds of Climate Change in Ecosystems.* Final Report—Synthesis and Assessment Product 4.2. A Report by the U.S. Climate Change Science Program and the Subcommittee on Global Change Research. Washington, DC: U.S. Department of the Interior, U.S. Geological Survey. <http://www.climatescience.gov/Library/sap/sap4-2/final-report/>

FR 2010—*Federal Register* 75(88): 25324; May 7, 2010.

Franco and Sanstad 2006—Franco, G., and A.H. Sanstad. *Climate Change and Electricity Demand in California.* Final white paper. California Climate Change Center. CEC-500-2005-201-SF. <http://www.energy.ca.gov/.../CEC-500.../CEC-500-2005-201-SF.PDF>

GEO 2005—Group on Earth Observations. *The Global Earth Observation System of Systems (GEOSS) 10-Year Implementation Plan.* <http://earthobservations.org/ag_partorg.shtml>

Haynes et al. 2007—Haynes, R.W., D.M. Adams, R.J. Alig, P.J. Ince, R. John, and X. Zhou. *The 2005 RPA Timber Assessment Update.* Gen. Tech. Rep. PNW-GTR-699. U.S. Department of Agriculture, Forest Service, Pacific Northwest Research Station. 212 pp. <http://www.fs.fed.us/pnw/publications/gtr699/>

Holdren 2009—Holdren, John P. *Climate Services: Solutions from Commerce to Communities: Hearing before the United States Senate Committee on Science, Transportation, and Commerce.* 111th Cong., 1st Sess. <http://commerce.senate.gov/public/index.cfm?FuseAction=Hearings.Testimony&Hearing_ID=f878f04f-68c7-4ee8-812f-059eb7207478&Witness_ID=ab57b04d-ecc7-4b74-8029-f39cd6450819>

Houghton et al. 2001—Houghton, J.T., Y. Ding, D.J. Griggs, M. Noguer, P.J. van der Linden, X. Dai, K. Maskell, and C.A. Johnson, eds. *Climate Change 2001: The Scientific Basis. Contribution of Working Group I to the Third Assessment Report of the Intergovernmental Panel on Climate Change.* Cambridge, UK, and New York, NY: Cambridge University Press, 881 pp. <http://www.grida.no/publications/other/ipcc_tar/?src=/climate/ipcc_tar/>

IPCC 1996—Intergovernmental Panel on Climate Change. *IPCC Second Assessment—Climate Change 1995. A Report of the Intergovernmental Panel on Climate Change.* Geneva, Switzerland: United Nations Environment Programme, World Meteorological Organization, IPCC. <http://www.ipcc.ch/pdf/climate-changes-1995/ipcc-2nd-assessment/2nd-assessment-en.pdf>

IPCC 2000—Intergovernmental Panel on Climate Change. *Good Practice Guidance and Uncertainty Management in National Greenhouse Gas Inventories.* Ed. J. Penman et al. <http://www.ipcc-nggip.iges.or.jp/public/gp/english/>

IPCC 2001—Intergovernmental Panel on Climate Change. *Climate Change 2001: The Scientific Basis. Contribution of Working Group I to the Third Assessment Report of the Intergovernmental Panel on Climate Change.* Ed. J.T. Houghton, Y. Ding, D.J. Griggs, M. Noguer, P.J. van der Linden, X. Dai, K. Maskell, and C.A. Johnson. Cambridge, United Kingdom, and New York, NY: Cambridge University Press. <http://www.grida.no/publications/other/ipcc_tar/>

IPCC 2003—Intergovernmental Panel on Climate Change. *Good Practice Guidance for Land Use, Land-Use Change and Forestry.* Ed. J. Penman, M. Gytarsky, T. Hiraishi, T. Krug, D. Kruger, R. Pipatti, L. Buendia, K. Miwa, T. Ngara, K. Tanabe, and F. Wagner. IPCC National Greenhouse Gas Inventories Programme. Hayama, Japan: Institute for Global Environmental Strategies. <http://www.ipcc-nggip.iges.or.jp/public/gpglulucf/gpglulucf.html>

IPCC 2006—Intergovernmental Panel on Climate Change. *2006 IPCC Guidelines for National Greenhouse Gas Inventories.* Ed. S. Eggleston, L. Buendia, K. Miwa, T. Ngara, and K. Tanabe. Prepared by the IPCC Task Force on National Greenhouse Gas Inventories. Hayama, Japan: Institute for Global Environmental Strategies. <http://www.ipcc-nggip.iges.or.jp/public/2006gl/index.html>

IPCC 2007—Intergovernmental Panel on Climate Change. *Climate Change 2007: The Physical Science Basis. Contribution of Working Group I to the Fourth Assessment Report of the Intergovernmental Panel on Climate Change.* Ed. S. Solomon, D. Qin, M. Manning, Z. Chen, M. Marquis, K.B. Averyt, M. Tignor, and H.L. Miller. Cambridge, UK: Cambridge University Press. <http://www.ipcc.ch/publications_and_data/publications_ipcc_fourth_assessment_report_synthesis_report.htm>

IPCC/UNEP/OECD/IEA 1997—Intergovernmental Panel on Climate Change, United Nations Environment Programme, Organisation for Economic Co-operation and Development, International Energy Agency. *Revised 1996 IPCC Guidelines for National Greenhouse Gas Inventories.* Ed. J.T. Houghton, L.G. Meira Filho, B. Lim, K. Treanton, I. Mamaty, Y. Bonduki, D.J. Griggs, and B.A. Callender. Paris: OECD. <http://www.ipcc-nggip.iges.or.jp/public/gl/invs1.html>

IRG and CRC-URI 2009—International Resources Group and The Coastal Resources Center at the University of Rhode Island. *Adapting to Coastal Climate Change: A Guidebook for Development Planners.* Washington, DC: U.S. Agency for International Development. <http://www.usaid.gov/our_work/environment/climate/pub_outreach/index.html>

IWGEO and NSTC/CENR 2005—Interagency Working Group on Earth Observations and National Science and Technology Council, Committee on Environment and Natural Resources. *Strategic Plan for the U.S. Integrated Earth Observation System.* Washington, DC. <usgeo.gov/docs/EOCStrategic_Plan.pdf>

Jacob and Winner 2009—Jacob, D.J., and D.A. Winner. "Effect of Climate Change on Air Quality." *Atmospheric Environment* 43: 51-63. <http://acmg.seas.harvard.edu/gcap/>

Karl et al. 2006—Karl, T.R., S.J. Hassol, C.D. Miller, and W.L. Murray, eds. *Temperature Trends in the Lower Atmosphere: Steps for Understanding and Reconciling Differences.* Synthesis and Assessment Product 1.1. Report by the U.S. Climate Change Science Program and the Subcommittee on Global Change Research. Washington, DC. <http://www.climatescience.gov/Library/sap/sap1-1/finalreport/>

Karl et al. 2009—Karl, T.R., S.J. Hassol, C.D. Miller, and W.L. Murray, eds. *Global Climate Change Impacts in the United States: A State of Knowledge Report from the U.S. Global Change Research Program.* Cambridge, UK, and New York, NY: Cambridge University Press. <http://www.globalchange.gov/publications/reports/scientific-assessments/us-impacts/download-the-report>

Leibensperger et al. 2008—Leibensperger, E.M., L.J. Mickley, and D.J. Jacob. "Sensitivity of U.S. Air Quality to Mid-latitude Cyclone Frequency and Complications of 1980-2006 Climate Change." *Atmospheric Chemistry and Physics* 8: 7075-7086. <www.atmos-chem-phys-discuss.net/8/.../acpd-8-12253-2008-print.pdf>

Lubowski et al. 2006a—Lubowski, Ruben, Marlow Vesterby, Shawn Bucholtz, Alba Baez, and Michael J. Roberts. *Major Uses of Land in the United States, 2002.* Economic Information Bulletin No. (EIB-14). May. <http://www.ers.usda.gov/publications/eib14/>

Lubowski et al. 2006b—Lubowski, Ruben, Andrew Plantinga, and Robert Stavins. "Land-Use Change and Carbon Sinks: Econometric Estimation of the Carbon Sequestration Supply Function." *Journal of Environmental Economics and Management* 51(2): 135-152. <ksghome.harvard.edu/.../Carbon_Sequestration_Costs_w_Lubowski_Plantinga.pdf>

McCarthy et al. 2001—McCarthy, J.J., O.F. Canziani, N.A. Leary, D.J. Dokken, and K.S. White, eds. *Climate Change 2001: Impacts, Adaptation, and Vulnerability. Contribution of Working Group II to the Third Assessment Report of the Intergovernmental Panel on Climate Change.* Cambridge, UK, and New York, NY: Cambridge University Press, 1,032 pp. <http://www.grida.no/publications/other/ipcc_tar/?src=/climate/ipcc_tar/>

Mills 2005—Mills, E. "Insurance in a Climate of Change." *Science* 309(5737): 1040-1044. LBNL-57943. <http://www.sciencemag.org/cgi/content/full/309/5737/1040>

NAS/NRC 2001—National Academies, National Research Council. *Climate Change Science: An Analysis of Some Key Questions.* Washington, DC: The National Academies Press. <http://www.nap.edu/openbook.php?record_id=10139>

NAS/NRC 2007—National Academies, National Research Council, Committee on Earth Science and Applications from Space. *Earth Science and Applications from Space: National Imperatives for the Next Decade and Beyond.* Washington, DC: The National Academies Press. <http://books.nap.edu/catalog.php?record_id=11820 - orgs>

NAS/NRC 2009—National Academies, National Research Council. *Restructuring Federal Climate Research to Meet the Challenges of Climate Change.* Washington, DC: The National Academies Press, p. 3. <http://www.nap.edu/catalog/12595.html>

NAST 2000—National Assessment Synthesis Team. *Climate Change Impacts on the United States: The Potential Consequences of Climate Variability and Change.* Report for the U.S. Global Change Research Program. Cambridge, UK: Cambridge University Press, 620 pp. <http://www.usgcrp.gov/usgcrp/Library/nationalassessment/00Intro.pdf>

OMB 2009—Office of Management and Budget. *Budget of the United States Government: Fiscal Year 2010.* <http://www.gpoaccess.gov/USbudget/fy10/index.html>

ORNL 2009—Oak Ridge National Laboratory, Energy and Transportation Science Division. *Results from the U.S. DOE 2007 Save Energy Now Assessment Initiative: DOE's Partnership with U.S. Industry to Reduce Energy Consumption, Energy Costs, and Carbon Dioxide Emission.* Detailed Assessment Opportunity Data Report. ORNL/TM-2009/074. Oak Ridge, TN. April 2009. <http://apps1.eere.energy.gov/industry/saveenergynow/partners/results.cfm>

Parmesan and Yohe 2003—Parmesan, C., and G. Yohe. "A Globally Coherent Fingerprint of Climate Impacts Across Natural Systems." *Nature* 421(January 2, 2003): 37-42. <http://www.nature.com/nature/journal/v421/n6918/abs/nature01286.html>

Peart et al. 1995—Peart, R.M., R.B. Curry, C. Rosenzweig, J.W. Jones, K.J. Boote, and L.H. Allen, Jr. "Energy and Irrigation in South Eastern U.S. Agriculture Under Climate Change." *Journal of Biogeography* 22(4/5): 635-642. <http://links.jstor.org/sici?sici=0305-0270(199507%2F09)22%3A4%2F5%3C635%3AEAIISE%3E 2.0.CO%3B2-9>

Pew 2009—Pew Center on Climate Change. "Climate Change 101: State Action." <http://www.pewclimate.org/docUploads/Climate101-State-Jan09.pdf>

Prymak 2009—Prymak, Bill. "Energy Assessments: What Are the Benefits to Small and Medium Facilities?" Golden, CO: U.S. Department of Energy, Office of Energy Efficiency and Renewable Energy, February 19, 2009. <http://www1.eere.energy.gov/industry/pdfs/webcast_2009-0219_small_medium_assessment_benefits.pdf>

Rose et al. 2001—Rose, J.B., P.R. Epstein. E.K. Lipp, B.H. Sherman, S.M. Bernard, and J.A. Patz. "Climate Variability and Change in the United States: Potential Impacts on Water- and Foodborne Diseases Caused by Microbiologic Agents." *Environmental Health Perspectives* 109(suppl 2): 211–221. <http://www.ncbi.nlm.nih.gov/pmc/articles/PMC1240668/>

Rosenzweig et al. 2007—Rosenzweig, C., D.C. Major, K. Demong, R. Horton, C. Stanton, and M. Stults. 2007. "Managing Climate Change Risks in New York City's Water System: Assessment and Adaptation Planning." *Mitigation and Adaptation Strategies for Global Change* 12(8): 1391-1409. <http://www.springerlink.com/content/u1k8435q45q47615/>

Rubin 2009—Rubin, E.S. "A Performance Standards Approach to Reducing CO_2 Emissions from Electric Power Plants." Coal Initiative Series. Pew Center on Global Climate Change. <http://www.pewclimate.org/white-papers/coal-initiative/performance-standards-electric>

Savonis et al. 2008—Savonis, M.J., V.R. Burkett, and J.R. Potter, eds. *Impacts of Climate Variability and Change on Transportation Systems and Infrastructure—Gulf Coast Study.* Final Report of Synthesis and Assessment Product 4.7. A Report by the U.S. Climate Change Science Program and the Subcommittee on Global Change Research. Washington, DC: U.S. Department of Transportation, 445 pp. <http://www.climatescience.gov/Library/sap/sap4-7/final-report/>

Scheffer et al. 2001—Scheffer, M., S. Carpenter, J.A. Foley, C. Folke, and B. Walker. "Catastrophic Shifts in Ecosystems." *Nature* 413 (October 11, 2001): 591-596. <http://www.nature.com/nature/journal/v413/n6856/abs/413591a0.html>

Skog 2008—Skog, K. "Carbon Storage in Forest Products for the United States." *Forest Products Journal* 58(6): 56-72.

Smith and Heath 2004—Smith, J.E., and L.S. Heath. "Carbon Stocks and Projections on Public Forestlands in the United States, 1952–2040." *Environmental Management* 33(4): 433-442. <http://www.treesearch.fs.fed.us/pubs/7948>

Smith et al. 2008—Smith, J.E., L.S. Heath, and M.C. Nichols. *U.S. Forest Carbon Calculation Tool: Forest-Land Carbon Stocks and Net Annual Stock Change.* Gen. Tech. Rep. NRS-13. U.S. Department of Agriculture, Forest Service, Northern Research Station. 28 pp. <http://www.treesearch.fs.fed.us/pubs/12394>

Smith et al. 2009—Smith, W. Brad, Patrick D. Miles, Charles H. Perry, and Scott A. Pugh. *Forest Resource of the United States, 2007.* Gen. Tech. Rept. WO-78. Washington, DC: U.S. Department of Agriculture, Forest Service. 336 pp. <http://www.nrs.fs.fed.us/pubs/7334>

Tester et al. 2006—Tester, J.W., et al. *The Future of Geothermal Energy: Impact of Enhanced Geothermal Systems (EGS) on the United States in the 21st Century: Interdisciplinary Panel Report.* Washington, DC: U.S. Department of Energy. <http://www1.eere.energy.gov/geothermal/future_geothermal.html>

Titus et al. 2009—Titus, J.G., K.E. Anderson, D.R. Cahoon, D.B. Gesch, S.K. Gill, B.T. Gutierrez, E.R. Thieler, and S.J. Williams. *Coastal Sensitivity to Sea-Level Rise: A Focus on the Mid-Atlantic Region.* Final Report—Synthesis and Assessment Product 4.1. A report by the U.S. Climate Change Science Program and the Subcommittee on Global Change Research. Washington, DC: U.S. Environmental Protection Agency, 320 pp. <http://www.climatescience.gov/Library/sap/sap4-1/final-report/>

Tompkins and Adger 2004—Tompkins, E.L., and W.N. Adger. "Does Adaptive Management of Natural Resources Enhance Resilience to Climate Change?" *Ecology and Society* 9(2): 10. <http://www.ecologyandsociety.org/vol9/iss2/art10/>

Turner et al. 2003—Turner, B.L.I., R.E. Kasperson, P.A. Matson, J.J. McCarthy, R.W. Corell, L. Christensen, N. Eckley, J.X. Kasperson, A. Luers, M.L. Martello, C. Polsky, A. Pulsipher, and A. Schiller. "A Framework for Vulnerability Analysis in Sustainability Science." *Proceedings of the National Academy of Sciences* 100(14): 8074-8079. <http://www.pnas.org/content/100/14/8074.abstract>

UNFCCC 2003— United Nations Framework Convention on Climate Change. *Review of the Implementation of Commitments and of Other Provisions of the Convention. National Communications: Greenhouse Gas Inventories From Parties Included in Annex I to the Convention. UNFCCC Guidelines on Reporting and Review.* FCCC/CP/2002/8. March 28 2003. <http://unfccc.int/resource/docs/cop8/08.pdf>

UNFCCC 2006—United Nations Framework Convention on Climate Change. *Updated UNFCCC Reporting Guidelines on Annual Inventories Following Incorporation of the Provisions of Decision 14/CP.11.* FCCC/SBSTA/2006/9. August 18, 2006. <http://unfccc.int/resource/docs/2006/sbsta/eng/09.pdf>

USAID 2007—U.S. Agency for International Development. *Adapting to Climate Variability and Change: A Guidance Manual for Development Planning.* Washington, DC. <http://www.usaid.gov/our_work/environment/climate/docs/reports/cc_vamanual.pdf>

USAID 2010—U.S. Agency for International Development. "USAID Global Climate Change Program." <http://www.usaid.gov/our_work/environment/climate/>

U.S. CCSP 2007—U.S. Climate Change Science Program. *Scenarios of Greenhouse Gas Emissions and Atmospheric Concentrations.* Synthesis and Assessment Product 2.1a. Washington, DC: July 2007. <http://www.gcrio.org/orders/popup_image.php?pID=187&osCsid=8im1m1iakfd9j8306sk1v5qlr0>

U.S. CCSP 2008—U.S. Climate Change Science Program, Observations Working Group. *The United States National Report on Systematic Observations for Climate for 2008: National Activities with Respect to the Global Climate Observing System (GCOS) Implementation Plan.* Prepared for submission to the United Nations Framework Convention on Climate Change. <http://www.climatescience.gov/Library/UNFCCC-report.htm>

U.S. CCSP/GCRP 2006–2009—U.S. Climate Change Science Program and Subcommittee on Global Change Research. "Information on Synthesis and Assessment Products: Summary Information." <http://www.climatescience.gov/Library/sap/sap-summary.php>

USDA 2008—U.S. Department of Agriculture, Global Change Program Office, Office of the Chief Economist. *U.S. Agriculture and Forestry Greenhouse Gas Inventory: 1990–2005.* Technical Bulletin No. 1921. Washington, DC. August 2008. <http://usda.gov/oce/global_change/AFGGInventory1990_2005.htm>

USDA/USFS 2009—U.S. Department of Agriculture, U.S. Forest Service, Forest Inventory and National Analysis Program. "Forest Inventory Data Online." <http://fiatools.fs.fed.us/fido/>

U.S. DOC/BEA 2008—U.S. Department of Commerce, Bureau of Economic Analysis. "Current-Dollar and "Real" Gross Domestic Product." <http://www.bea.gov/national/xls/gdplev.xls>

U.S. DOC/BEA 2009a—U.S. Department of Commerce, Bureau of Economic Analysis. "Table 1.1.6. Real Gross Domestic Product, Chained Dollars (Billions of Chained 2005 Dollars)." Last updated August 27, 2009. <http://www.bea.gov/national/nipaweb/TableView.asp?SelectedTable=6&FirstYear=2006&LastYear=2007&Freq=Qtr>

U.S. DOC/BEA 2009b—U.S. Department of Commerce, Bureau of Economic Analysis. "Table 1.1.1: % Change From Preceding Period in Real Gross Domestic Product (Seasonally Adjusted at Annual Rates)." Last updated August 27, 2009. <www.bea.gov/scb/pdf/2009/08%20August/0809_gdpecon.pdf>

U.S. DOC/BEA 2009c—U.S. Department of Commerce, Bureau of Economic Analysis. "Value Added by Industry: 1998–2008." <www.bea.gov/industry/gdpbyind_data.htm>

U.S. DOC/BEA 2009d—U.S. Department of Commerce, Bureau of Economic Analysis. "U.S. International Trade in Goods on a Balance of Payments Basis. Exhibits 1 & 18: U.S. International Trade in Goods and Services—Annual Revision for 2008." <www.census.gov/foreign-trade/Press-Release/2008pr/.../exh18.pdf>

U.S. DOC/Census 2008a—U.S. Department of Commerce, Bureau of the Census. "International Database (IDB)." <http://www.census.gov/ipc/www/idbnew.html>.

U.S. DOC/Census 2008b—U.S. Department of Commerce, Bureau of the Census. "U.S. National Population Projections (2008)." <http://www.census.gov/population/www/projections/files/nation/summary/np2008-t1.xls>

U.S. DOC/Census 2009a—U.S. Department of Commerce, Bureau of the Census. "New Privately Owned Housing Units Started in the United States by Purpose and Design." U.S. Census Manufacturing and Construction Division. <http://www.census.gov/const/www/newresconstindex.html>

U.S. DOC/Census 2009b—U.S. Department of Commerce, Bureau of the Census. "U.S. & World Population Clock." Accessed November 1, 2009. <http://www.census.gov/main/www/popcld.html>

U.S. DOC/NOAA 2001—U.S. Department of Commerce, National Oceanic and Atmospheric Administration. *The United States Detailed National Report on Systematic Observations for Climate: United States Global Climate Observing System Program.* Submitted to the Conference of the Parties to the United Nations Framework Convention on Climate Change. Silver Spring, MD. <http://www.ncdc.noaa.gov/oa/usgcos/documents/soc_long.pdf>

U.S. DOE/EERE 2010—U.S. Department of Energy, Office of Energy Efficiency and Renewable Energy, "Building Technologies Program. Building America." <http://www1.eere.energy.gov/buildings/building_america/>

U.S. DOE/EIA 2002—U.S. Department of Energy, Energy Information Administration. *Annual Energy Outlook 2002 With Projections to 2030.* DOE/EIA-0383(2002). Washington, DC. <http://www.eia.doe.gov/oiaf/archive/aeo02/index.html>

U.S.DOE/EIA 2003—U.S. Department of Energy, Energy Information Administration. *2003 Commercial Buildings Energy Consumption Survey—Overview of Commercial Buildings Characteristics.* <http://www.eia.doe.gov/emeu/cbecs/cbecs2003/introduction.html>

U.S. DOE/EIA 2004—U.S. Department of Energy, Energy Information Administration. "U.S. Uranium Reserves by State." June 2004. <http://www.eia.doe.gov/cneaf/nuclear/page/reserves/ures.html>

U.S. DOE/EIA 2005—U.S. Department of Energy, Energy Information Administration. "Energy Policy Act of 2005 Summary." <http://www.eia.doe.gov/oiaf/aeo/otheranalysis/aeo_2006analysispapers/epa2005_summary.html>

U.S. DOE/EIA 2006—U.S. Department of Energy, Energy Information Administration. *Annual Energy Outlook 2006 With Projections to 2030.* DOE/EIA-0383(2006). Washington, DC. <http://www.eia.doe.gov/oiaf/archive/aeo06/index.html>

U.S. DOE/EIA 2007—U.S. Department of Energy, Energy Information Administration. "Capacity Additions, Retirements, and Changes by Energy Source, 2007." <http://www.eia.doe.gov/cneaf/electricity/epm/table1_16_a.html>

U.S. DOE/EIA 2008a— U.S. Department of Energy, Energy Information Administration. *Monthly Energy Review.* "Supplemental Tables on Petroleum Product Prices." DOE/EIA-0035(2008/09). Washington, DC. <http://www.eia.doe.gov/mer/>

U.S. DOE/EIA 2008b—U.S. Department of Energy, Energy Information Administration. *International Energy Annual 2006.* <http://www.eia.doe.gov/iea/>

U.S. DOE/EIA 2008c—U.S. Department of Energy, Energy Information Administration. Form EIA-923, "Power Plant Operations Report." <http://www.eia.doe.gov/cneaf/electricity/2008forms/consolidate_923.html>

U.S. DOE/EIA 2008d—U.S. Department of Energy, Energy Information Administration. "Renewable Energy Consumption and Electricity Preliminary Statistics 2008." <http://www.eia.doe.gov/cneaf/alternate/page/renew_energy_consump/rea_prereport.html>

U.S. DOE/EIA 2009a—U.S. Department of Energy, Energy Information Administration. *An Updated Annual Energy Outlook 2009 Reference Case Reflecting Provisions of the American Recovery and Reinvestment Act and Recent Changes in the Economic Outlook.* SR-OIAF/2009-03. Washington, DC: April 2009. <http://www.eia.doe.gov/oiaf/servicerpt/stimulus/index.html>

U.S. DOE/EIA 2009b—U.S. Department of Energy, Energy Information Administration. *Annual Energy Review 2008*. DOE/EIA-0384(2008). Washington, DC: June 2009. <http://www.eia.doe.gov/emeu/aer/pdf/aer.pdf>

U.S. DOE/EIA 2009c. U.S. Department of Energy, Energy Information Administration. *International Energy Outlook 2009*. DOE/EIA-0484(2009). Washington, DC. <http://www.eia.doe.gov/oiaf/ieo/emissions.html>

U.S. DOE/EIA 2009d—U.S. Department of Energy, Energy Information Administration. "Table 1.1 Primary Energy Overview." *Monthly Energy Overview*. Washington, DC. August 2009. <www.eia.doe.gov/emeu/aer/txt/stb0101.xls>

U.S. DOE/EIA 2009e—U.S. Department of Energy, Energy Information Administration. "Coal Reserves: Current and Back Issues." February 3, 2009. <http://www.eia.doe.gov/cneaf/coal/reserves/reserves.html>

U.S. DOE/EIA 2009f—U.S. Department of Energy, Energy Information Administration. *Annual Coal Report: 2007*. DOE/EIA-0584 (2007). February 2009. <www.eia.doe.gov/cneaf/coal/page/acr/acr.pdf>

U.S. DOE/EIA 2009g—U.S. Department of Energy, Energy Information Administration. "Energy in Brief." April 23, 2009. <http://tonto.eia.doe.gov/energy_in_brief/foreign_oil_dependence.cfm>

U.S. DOE/EIA 2009h—U.S. Department of Energy, Energy Information Administration. *Annual Energy Outlook 2009*. Washington, DC. <http://www.eia.doe.gov/oiaf/aeo/leg_reg.html>

U.S. DOE/EIA 2009i—U.S. Department of Energy, Energy Information Administration. "Net Generation by Energy Source: Total (All Sectors)." August 14, 2009. <http://www.eia.doe.gov/cneaf/electricity/epm/table1_1.html>

U.S. DOE/EIA 2009j—U.S. Department of Energy, Energy Information Administration. "Electricity Net Generation From Renewable Energy by Energy Use Sector and Energy Source, 2004–2008." July 2009. <http://www.eia.doe.gov/cneaf/solar.renewables/page/trends/table11.html>

U.S. DOE/EIA 2009k—U.S. Department of Energy, Energy Information Administration. "Electricity Demand." <http://www.eia.doe.gov/oiaf/aeo/electricity.html>

U.S. DOE/EIA 2009l—U.S. Department of Energy, Energy Information Administration. "Weekly United States Spot Price FOB Weighted by Estimated Import Volume (Dollars per Barrel)." September 10, 2009. <http://tonto.eia.doe.gov/dnav/pet/hist/LeafHandler.ashx?n=PET&s=WTOTUSA&f=W>

U.S. DOE/EIA 2009m—U.S. Department of Energy, Energy Information Administration. "Oil: Crude and Petroleum Products Explained—Where Our Oil Comes From." <http://tonto.eia.doe.gov/energyexplained/index.cfm?page=oil_where>

U.S. DOE/EIA 2009n—U.S. Department of Energy, Energy Information Administration. *Emissions of Greenhouse Gases Report*. DOE/EIA-0573(2008); revised December 8, 2009. <http://www.eia.doe.gov/oiaf/1605/ggrpt/carbon.html#industrial>

U.S. DOE/FEMP 2007—U.S. Department of Energy, Federal Energy Management Program. "Energy Independence and Security Act of 2007." <http://www.eere.energy.gov/femp/regulations/eisa.html>

U.S. DOI/BOR 2005—U.S. Department of the Interior, Bureau of Reclamation. *Water 2025: Preventing Crises and Conflict in the West*. Washington, DC. <http://www.usbr.gov/uc/crsp/GetSiteInfo>

U.S. DOI/USGS 2008—U.S. Department of the Interior, United States Geological Survey. "Assessment of Moderate- and High-Temperature Geothermal Resources of the United States." USGS Fact Sheet 2008-3082. <http://pubs.usgs.gov/fs/2008/3082/>

U.S. DOL/BLS 2009—U.S. Department of Labor, Bureau of Labor Statistics. "Labor Force Statistics from the Current Population Survey." Series ID: 14000000. Series title: Seasonally Adjusted Unemployment Rate. Viewed on October 20, 2009. <http://data.bls.gov/PDQ/servlet/SurveyOutputServlet?series_id=LNS14000000>

U.S. DOS 2002—U.S. Department of State. *U.S. Climate Action Report 2002: Third National Communication of the United States of America Under the United Nations Framework Convention on Climate Change*. Washington, DC: May 2002. <http://www.gcrio.org/CAR2002/>

U.S. DOS 2007—U.S. Department of State. *U.S. Climate Action Report 2006: Fourth National Communication of the United States of America Under the United Nations Framework Convention on Climate Change*. Washington, DC: July 2007. <http://www.state.gov/g/oes/rls/rpts/car/>

U.S. DOT 2006—U.S. Department of Transportation, Center for Climate Change and Environmental Forecasting. *Strategic Plan 2006–2010*. Washington, DC. <http://climate.volpe.dot.gov/plan/splan_2006.pdf>

U.S. DOT/BTS 2008—U.S. Department of Transportation, Bureau of Transportation Statistics. *National Transportation Statistics.* Tables 1-3 1-11, 1-32, 1-37, 1-38, 1-46b, 3-1a, 4-6. Washington, DC. October 2008. <http://www.bts.gov/publications/national_transportation_statistics/>

U.S. DOT/BTS 2010—U.S. Department of Transportation, Bureau of Transportation Statistics. "National Transportation Statistics." <http://www.bts.gov/publications/national_transportation_statistics≥

U.S. DOT/BTS 2009a—U.S. Department of Transportation, Bureau of Transportation Statistics. "Airline Fuel Costs and Consumption (U.S. Carriers–Scheduled): January 2000–October 2009." <http://www.transtats.bts.gov/fuel.asp>

U.S. DOT/BTS 2009b—U.S. Department of Transportation, Bureau of Transportation Statistics. "All Carriers–All Airports." <http://www.transtats.bts.gov/Data_Elements.aspx?Data=1>

U.S. DOT/BTS 2009c—U.S. Department of Transportation, Bureau of Transportation Statistics. U.S. Air Carrier Traffic Statistics. <http://www.bts.gov/xml/air_traffic/src/datadisp.xml>

U.S. EPA and Jacoby 2008—U.S. Environmental Protection Agency and Jacoby Development Inc. "Atlantic Station: 2008 Project XL Report." <http://www.atlanticstation.com/concept_green_projectXL08.php>

U.S. EPA/OAP 2006a—U.S. Environmental Protection Agency, Office of Atmospheric Programs. *Excessive Heat Events Guidebook.* EPA 430-B-06-005. Washington DC. <http://www.epa.gov/heatisland/about/heatguidebook.html>

U.S. EPA/OAP 2006b—U.S. Environmental Protection Agency, Office of Atmospheric Programs. *Global Anthropogenic Emissions of Non-CO_2 Greenhouse Gases 1990–2020.* Washington, DC: June 2006. <http://www.epa.gov/climatechange/economics/international.html>

U.S. EPA/OAP 2008—U.S. Environmental Protection Agency, Office of Atmospheric Programs. *2008 Inventory of U.S. Greenhouse Gas Emissions and Sinks: 1990–2006.* <http://www.epa.gov/climatechange/emissions/usgginventory.html>

U.S. EPA/OAP 2009—U.S. Environmental Protection Agency, Office of Atmospheric Programs. *Inventory of U.S. Greenhouse Gas Emissions and Sinks: 1990–2007.* EPA 430 R 09-004. Washington, DC: April 2009. <http://www.epa.gov/climatechange/emissions/usinventoryreport.html>

U.S. EPA/OAQPS 2008—U.S. Environmental Protection Agency, Office of Air Quality Planning and Standards. National Emissions Inventory (NEI) Air Pollutant Emissions Trends Data. "1970–2007 Average Annual Emissions, All Criteria Pollutants in MS Excel." <http://www.epa.gov/ttn/chief/trends/index.html>

U.S. EPA/ORD 2009—U.S. Environmental Protection Agency, Office of Research and Development. *Assessment of the Impacts of Global Change on Regional U.S. Air Quality: A Synthesis of Climate Change Impacts on Ground-Level Ozone.* An Interim Report of the U.S. EPA Global Change Research Program. EPA/600/R-07/094. Washington, DC. <http://cfpub.epa.gov/ncea/CFM/recordisplay.cfm?deid=203459>

U.S. EPA/OSW 2008— U.S. Environmental Protection Agency, Office of Solid Waste. *Municipal Solid Waste in the United States: 2007 Facts and Figures.* EPA530-R-08-010. Washington, DC. November 2008. <www.epa.gov/waste/nonhaz/municipal/pubs/msw07-rpt.pdf>

U.S. EPA/OTAQ 2010—U.S. Environmental Protection Agency, Office of Transportation and Air Quality. "EPA and NHTSA Finalize Historic National Program to Reduce Greenhouse Gases and Improve Fuel Economy for Cars and Trucks." EPA-420-F-10-014. April. <http://www.epa.gov/oms/climate/regulations/420f10014.htm#2>

USGCRP and SGCR 2006—U.S. Global Change Research Program and Subcommittee on Global Change Research. *Our Changing Planet: The U.S. Climate Change Science Program for Fiscal Year 2007.* A Supplement to the President's Budget for Fiscal Year 2007. <http://www.usgcrp.gov/usgcrp/Library/ocp2007/default.htm>

Williams et al. 2008—Williams, C.F., M.J. Reed, R.H. Mariner, J. DeAngelo, and S.P. Galanis, Jr. "Assessment of Moderate-and High-Temperature Geothermal Resources of the United States." USGS Fact Sheet 2008-3082. U.S. Department of the Interior, U.S. Geological Survey. <http://pubs.usgs.gov/fs/2008/3082/>

Zervas 2001—Zervas, C. *Sea Level Variations of the United States: 1854–1999.* NOAA Technical Report NOS CO-OPS 36. Silver Spring, MD: National Oceanic and Atmospheric Administration, National Ocean Service. <http://tidesandcurrents.noaa.gov/pub.html#sltrends>

www.ingramcontent.com/pod-product-compliance
Lightning Source LLC
Chambersburg PA
CBHW080807180526
45168CB00006B/2354